Learning and Analytics in Intelligent Systems

Volume 58

The main aim of the series is to make available a publication of books in hard copy form and soft copy form on all aspects of learning, analytics and advanced intelligent systems and related technologies. The mentioned disciplines are strongly related and complement one another significantly. Thus, the series encourages cross-fertilization highlighting research and knowledge of common interest. The series allows a unified/integrated approach to themes and topics in these scientific disciplines which will result in significant cross-fertilization and research dissemination. To maximize dissemination of research results and knowledge in these disciplines, the series publishes edited books, monographs, handbooks, textbooks and conference proceedings.

Indexed by EI Compendex.

George A. Tsihrintzis · Maria Virvou ·
Vangelis Marinakis · Lakhmi C. Jain
Editors

Human-Smart City Interactions and User-Citizen Experiences

Springer

Editors
George A. Tsihrintzis
Department of Informatics
University of Piraeus
Piraeus, Greece

Maria Virvou
Department of Informatics
University of Piraeus
Piraeus, Greece

Vangelis Marinakis
School of Electrical and Computer Engineering
National Technical University of Athens
Athens, Greece

Lakhmi C. Jain
KES International
Shoreham-by-Sea, UK

ISSN 2662-3447 ISSN 2662-3455 (electronic)
Learning and Analytics in Intelligent Systems
ISBN 978-3-032-06363-2 ISBN 978-3-032-06364-9 (eBook)
https://doi.org/10.1007/978-3-032-06364-9

This Springer imprint is published by the registered company Springer Nature Switzerland AG
The registered company address is: Gewerbestrasse 11, 6330 Cham, Switzerland

人聚于乡而治, 聚于城而乱.

—顾炎武 的著作《日知录·人聚》

When the masses dwell in villages, order prevails; when the masses flock to the cities, disorder ensues

Gu Yanwu (1613–1682), Record of Daily Study

Preface

Smart cities embody a transformative approach to urban development, leveraging digital technologies, data analytics, and artificial intelligence to improve citizens' quality of life. By integrating information and communication technologies into urban systems, cities aim to streamline operations, lower costs, and create more seamless and responsive experiences for residents. These technologies process vast and often real-time datasets collected from various sources—such as citizens, sensors, city records, and digital infrastructure—allowing for the generation of actionable insights into key urban services. Such services include transportation, energy and water management, waste and recycling, urban forestry, education, law enforcement, healthcare, and infrastructure upkeep.

The insights derived from these systems empower a wide range of city stakeholders—including local and national governments, businesses, organizations, and individuals—to make data-informed decisions that enhance a city's adaptability, efficiency and livability.

Despite extensive research and development in smart city technologies, their full potential to improve urban life remains largely untapped. There are still considerable opportunities for progress in areas like predictive policing, efficient transportation, healthcare optimization, environmental sustainability, and fostering social cohesion.

This book explores the concept of "human-smart city interaction" which refers to how people engage with and are impacted by technological advancements in urban settings. It also explores human factors to be considered in enhancing "citizen experience", including citizen engagement and trust, privacy and data protection, social equity, service personalization, and the broader influence of technology on shaping cities to better align with human needs and preferences. As smart cities continue to evolve, understanding these human–technology interactions becomes essential, as they increasingly shape the lived experiences of urban residents.

With this book, we are attempting to update both specialized researchers and general readers on the most recent advances in aspects of human-smart city interactions and user-citizen experiences. The book consists of a total of 13 chapters, each of which has been authored by active and recognized researchers and reports on recent research and development findings. More specifically, the book includes an editorial note (introductory chapter) and is further structured into five parts, namely: (1) *Sustainable Urban Planning* (two chapters), (2) *Data-Driven Governance* (four chapters), (3) *Smart Mobility Systems* (one chapter), (4) *IoT-Enabled Infrastructure* (two chapters), and (5) *Digital Public Services* (three chapters).

Targeted towards academics, researchers, practitioners, and students in Engineering, Computer Science, Artificial Intelligence, Policy Making, and Governance, this work is also accessible to individuals from other disciplines interested in the cutting-edge developments in Smart City technologies. An extensive list of bibliographic references at the end of each chapter guides the readers to probe deeper into the application areas of interest to them.

Even though the area of Human-Smart City Interactions and User-Citizen Experiences is very broad and researchers are approaching it from various points of interest, we attempted to cover it in terms of both *breadth* and *depth*. We hope that we were successful, at least to some extent. Of course, several aspects and areas of application of Human-Smart City Interactions and User-Citizen Experiences may have been insufficiently addressed in this edition of the book. We hope to address them in future editions, especially as Society continuously presses for more livable cities and novel and better city services.

Piraeus, Greece — George A. Tsihrintzis
Piraeus, Greece — Maria Virvou
Athens, Greece — Vangelis Marinakis
Shoreham-by-Sea, UK — Lakhmi C. Jain

Contents

About the Editors

George A. Tsihrintzis received the Diploma of Electrical Engineering from the National Technical University of Athens, Greece (with honors), and the M.Sc. and Ph.D. degrees in Electrical and Computer Engineering from Northeastern University, Boston, MA, USA. He is currently a Professor in the University of Piraeus, Greece, and, during 2016–2020, he served as Head of its Department of Informatics. From 2008 to 2022, he served as the Director of the Graduate Program of Study in "Advanced Computing and Informatics Systems" of the department. During the period 2016–2020, he was a Member of the first Administration Council of the University of Piraeus. Currently, he is serving again as a Member of the Administration Council of the University of Piraeus for a 4-year term (2022–2026). His current research interests include Artificial and Computational Intelligence, Machine Learning, Pattern Recognition, Decision Theory, and Statistical Signal Processing and Their Applications in Internet-of-Things Technologies, Multimedia Interactive Services, User Modeling, Knowledge-Based Software Systems, Human–Computer Interaction and Information Retrieval. He has authored or co-authored over 400 research publications in these areas, which have appeared in international journals, book chapters, and conference proceedings. He has served as the principal investigator or co-investigator in several R&D projects. He has (co-)edited over 45 international volumes. He is the editor-in-chief of *Intelligent Decision Technologies Journal* (SAGE-IOS Press) and an associate editor of *Engineering Applications of Artificial Intelligence* (Elsevier). During 2012–2022, he was the Editor-in-Chief of the *International Journal of Computational Intelligence Studies* (Inderscience). He has been the Editor-in-Chief of the Engineering Section of *SpringerPlus*. He is the Founder and Editor-in-Chief of the *Learning and Analytics in Intelligent Systems* book series (Springer 2019–) and the *Artificial Intelligence-Enhanced Software and Systems Engineering* book series (Springer 2022–). He has guest (co-)edited 17 special issues of international journals. He is a Co-founder of the *Information, Intelligence, Systems and Applications (IISA)* conference series, technically sponsored by IEEE and currently in its 17th edition. He has organized and chaired over 45 international conferences. He has received best paper awards from several international conferences, and he has been a keynote speaker of several international conferences.

He is ranked among the top 2% of the most influential scientists worldwide of Scientists List at Stanford University ranking in the area of Artificial Intelligence, both career-wise and single year, for 2023 and 2024.

Maria Virvou is the Dean of the School of Informatics and Communication Technologies at the University of Piraeus, Greece, and Full Professor of the Department of Informatics. She is the Director and Founding Member of the Postgraduate Program in "Informatics" and serves as the Director of the Software Technology Research Laboratory.

In addition, she is the Co-founder and Co-editor-in-chief of the Springer book series *Learning and Analytics in Intelligent Systems* and *Artificial Intelligence-Enhanced Software and Systems Engineering*. She is also the Co-founder and Co-chair of the IEEE Intelligent Information Systems and Applications (IISA) International Conference series, which is held annually. She previously served as Editor-in-Chief of the open-access journal *SpringerPlus* and currently serves as an Associate Editor of *Knowledge-Based Systems* (Elsevier). In 2024, she was the General Co-chair of the 14th IEEE IISA Conference and the 14th AI-Enhanced Software Engineering Conference.

Professor Virvou received her Ph.D. from the University of Sussex, UK, funded by a prestigious scholarship in Artificial Intelligence from the Greek State Scholarships Foundation following competitive national examinations. She also holds M.Sc. in Computer Science from University College London (UCL), University of London, and Degree in Mathematics from the National and Kapodistrian University of Athens. Her scholarly output comprises over 400 publications, including 45 editorials and seven monographs, all indexed in Scopus. She has successfully supervised 15 Ph.D. candidates and 10 postdoctoral researchers. She has delivered numerous keynote speeches at prestigious international conferences, including the IEEE International Conference on Software Engineering Education and Training (CSEE&T 2023) in Tokyo, Japan.

Professor Virvou has received numerous honors, including an award from UNESCO for her outstanding contributions to computer science. In recognition of her impact, Springer published the edited volume Advances in Intelligent Healthcare Delivery and Management: *Research Papers in Honour of Professor Maria Virvou for Invaluable Contributions* (Springer, Germany). She holds top positions in international research rankings. According to Scopus and Microsoft Academic Search, she ranks first in User Modelling (among 150,900 publications), and first in both Intelligent Software (out of 60,779 publications) and Educational Software (out of 40,585 publications) on Scopus. She is also ranked among the top 0.02% of AI researchers worldwide, based on analysis of over 670,000 scholarly documents by Scopus. She is also recognized among the leading researchers in *User Interface, Multimedia, and Human-Computer Interaction*, according to Microsoft Academic Search (2022). Furthermore, she is consistently ranked among the top 2% of researchers worldwide with the highest scientific impact, according to the global academic ranking by Stanford University, particularly in the field of Artificial Intelligence.

Vangelis Marinakis is an Assistant Professor in the School of Electrical and Computer Engineering (ECE) of the National Technical University of Athens (NTUA). He is an Electrical and Computer Engineer of NTUA and holds a Ph.D. in the research domain of decision support systems for sustainable energy planning (ECE, NTUA). His research is focused on the design and development of methodologies and decision support systems, using leading-edge ICTs technologies (Internet of Things, Artificial Intelligence, Big Data Technologies) for intelligent energy management at different levels (Smart Homes, Smart Buildings, Smart Cities, Smart Districts), as well as increased energy efficiency gains and renewables. In the abovementioned fields, Dr. Marinakis has worked on more than 25 European (Horizon Europe, Horizon 2020, FP7, IEE, etc.) and national co-funded projects. He has more than 80 scientific publications in international scientific journals with reviewers and book chapters, as well as numerous publications/announcements in national and international conferences.

Lakhmi C. Jain Ph.D., M.E., B.E.(Hons.), Fellow (Engineers Australia) is with KES International, Shoreham-by-Sea, UK. Professor Jain's career is characterized by his solid commitment to research, education, and mentorship, as he has been a distinguished professor at numerous universities. His research works encompass Automated Information System Design; Modelling, Design, and Implementation of Knowledge-Based Intelligent Information Systems; Intelligent Agent Learning, Coordination, Collaboration, and Web-Based Agents. He has authored/co-authored/edited/co-edited more than 500 volumes. He is the Co-founder and Editor-in-Chief of the international scientific journals *International Journal of Knowledge-Based Intelligent Engineering Systems* (IOS Press, 1997–) and *Intelligent Decision Technologies* (IOS Press, 2007–). He is the Co-founder and/or Chief Editor of several scientific series for publishing houses such as Springer, Elsevier, IOS Press, among others. He is the Co-founder and Chair of the Steering Committee for four significant international conferences: the International Conference on Knowledge-Based and Intelligent Information and Engineering Systems (KES), Agents and Multi-agent Systems—Technologies and Applications (AMSTA), Intelligent Interactive Multimedia Systems and Services (IIMSS), and Intelligent Decision Technologies (IDT). He has been an Invited Speaker at numerous important international conferences in Europe, Asia, and Australia. He is the Founder of the research professional organization "Knowledge-Based and Intelligent Engineering Systems International," a global community of over 5000 researchers aimed at collaboration, knowledge exchange, and research dissemination in the fields of Artificial and Computational Intelligence and their applications, such as Complex Systems, Systems Security, Educational Technologies, Healthcare Technologies, Autonomous Systems, and Vehicles, Intelligent Agents, and more (http://www.kesinternational.org/). In September 2023, he received a Doctorate degree (Honoris Causa) from the Department of Informatics of the University of Piraeus, Greece.

Chapter 1
Introduction to Human-Smart City Interactions and User-Citizen Experiences

George A. Tsihrintzis, Maria Virvou, Vangelis Marinakis, and Lakhmi C. Jain

Abstract Smart cities offer a transformative vision for urban living by integrating digital technologies, data analytics, and artificial intelligence to improve quality of life. Through the widespread use of information and communication technologies, smart cities can optimize operations, reduce costs, and enhance citizen experiences. These systems gather and analyze real-time data from sensors, citizens, and city infrastructure to inform services such as transportation, healthcare, energy, policing, and waste management. Despite progress, many smart city technologies remain underutilized. This limited uptake is often not due to technical inadequacies, but rather to insufficient integration of technological solutions with human factors, ethical and legal considerations, system-level coordination, domain-specific requirements, and the unique challenges associated with artificial intelligence. This book explores "human-smart city interaction"—the ways individuals engage with and are shaped by smart technologies—and human factors to be considered in enhancing "citizen experience". As cities grow smarter, understanding these interactions is key to building more adaptive, efficient, and livable urban environments. Targeted towards academics, researchers, practitioners, and students in Engineering,

G. A. Tsihrintzis (✉) · M. Virvou
Department of Informatics, University of Piraeus, Piraeus, Greece
e-mail: geoatsi@unipi.gr

M. Virvou
e-mail: mvirvou@unipi.gr

V. Marinakis
Environmental Policy Unit, School of Electrical and Computer Engineering, National Technical University of Athens, Athens, Greece
e-mail: vmarinakis@epu.ntua.gr

L. C. Jain
KES International, Shoreham-by-Sea, UK

University of Piraeus, Piraeus, Greece

L. C. Jain
e-mail: jainlakhmi@gmail.com

G. A. Tsihrintzis et al. (eds.), *Human-Smart City Interactions and User-Citizen Experiences*, Learning and Analytics in Intelligent Systems 58,
https://doi.org/10.1007/978-3-032-06364-9_1

Computer Science, Artificial Intelligence, Policy Making, and Governance, this work is also accessible to individuals from other disciplines interested in the cutting-edge developments in Smart City technologies. An extensive list of bibliographic references at the end of each chapter guides the readers to probe deeper into the application areas of interest to them.

Keywords Smart cities · Urban living · Digital technologies · Data analytics · Artificial Intelligence (AI) · Information and Communication Technologies (ICT) · Quality of life · Citizen experience · City Infrastructure · Human-smart city interaction · Citizen engagement · Livable urban environments · Urban mobility · Sustainability

1.1 Introduction

Smart cities represent a transformative vision for urban development, integrating digital technologies, data analytics, and Artificial Intelligence to enhance the quality of life for citizens.

More specifically, the European Commission states that "**a smart city is a place where traditional networks and services are made more efficient with the use of digital solutions for the benefit of its inhabitants and business. A smart city goes beyond the use of digital technologies for better resource use and less emissions. It means smarter urban transport networks, upgraded water supply and waste disposal facilities and more efficient ways to light and heat buildings. It also means a more interactive and responsive city administration, safer public spaces and meeting the needs of an ageing population**" [1].

To enhance the efficiency of operations, reduce costs, and offer a more seamless interaction experience for citizens in smart cities, information and communication technologies (ICT) are becoming deeply infused in urban systems. These technologies handle diverse datasets, often in real time, gathered from various sources including citizens' data, sensors, city records, and information systems.

Urban services such as transportation, energy and water management, waste and recycling, urban forestry, education, policing, healthcare, and infrastructure maintenance can be monitored to gather useful information. This information supports decisions made by city stakeholders, including governments, businesses, organizations, and individuals. As a result, cities become more adaptive, efficient, and easier to live in. Despite years of research and development in smart city technologies, their potential to enhance many aspects of urban life remains underutilized [2].

This book focuses on the interaction between humans and smart cities. This is what we term "human-smart city interaction". This concept encompasses how individuals engage with and are affected by technological innovations in urban environments. As smart cities evolve, understanding the interactions between humans and smart systems becomes increasingly critical as these interactions affect user-citizen experiences in contemporary urban environments.

1.2 Smart Cities, Artificial Intelligence and Human-Smart City Interactions

Smart cities are evolving rapidly, driven by advancements in technology, especially artificial intelligence (AI). Over time, the concept of smart cities has expanded to include the use of digital technologies and AI for improving urban management and sustainability [3]. AI has developed over the past decades, starting with knowledge-based systems [4] and intelligent environments using multimedia [5], which enabled personalised and reasoning-based software solutions. More recently, the growth of the World Wide Web and the Internet of Things (IoT) has led to the collection of large volumes of data, known as big data, further accelerating AI development. As a result, AI has become a major driver of innovation, transforming industries by combining with other technologies such as Big Data, IoT, Blockchain, and 5G communications [6]. A significant contributor to this progress is the confluence of multiple AI techniques, ranging from machine learning and natural language processing to computer vision and data management, highlighting the complexity and depth of data collection and handling in contemporary AI systems [7].

Surveys and studies further reflect the breadth of AI applications, from intelligent decision technologies to learning analytics in adaptive systems [8, 9]. More recent advances underscore the sophistication of AI methodologies, such as adaptive boundary-aware artificial immune systems for classification [10], enterprise architecture models for managing machine learning projects [11], and neural networks aiding code review through program image analysis [12]. Additionally, deep learning's transformative impact on healthcare [13] illustrates AI's growing relevance and reliability in critical sectors. This multifaceted growth underscores AI's ongoing evolution as a field defined by continual innovation, integration, and interdisciplinary collaboration.

As artificial intelligence has matured, its influence has increasingly extended beyond data processing and automation to reform the way humans interact with intelligent systems. This evolution is particularly evident in the field of Human-AI interaction, where AI technologies are being designed to better understand, anticipate, and respond to user needs. The integration of AI with user experience (UX) has resulted in more personalized, adaptive, and reciprocal interactions between humans and machines [14]. Human-AI collaboration is no longer limited to command-based interfaces; it is becoming more intuitive and context-aware, fostering smoother and more meaningful engagements [15]. This transformation is deeply rooted in advancements in machine learning and data analytics, where there has been a succession of evolution in big data analytics, starting from descriptive analytics to diagnostic analytics to predictive analytics, and excelling towards prescriptive analytics out of which three (descriptive, predictive and prescriptive analytics) are dominant [16]. These stages enable AI systems to move from explaining what happened and why, to predicting future outcomes and, importantly, recommending the best actions to take. These functions transform AI into a proactive decision-support partner.

Recommender systems are a direct result of these developments, employing prescriptive analytics and behavioural data to suggest content, products, or actions tailored to individual preferences [17, 18]. These systems support personalisation in diverse areas, from entertainment and marketing to education and healthcare, and play a vital role in what constitutes modern Human-AI interactions. Furthermore, affective and multimodal computing enhances these interactions by allowing systems to interpret emotional and contextual signals. Approaches that integrate text, audio, and visual cues have become increasingly sophisticated in recognizing both individual and group emotional states [19–21]. Innovative applications such as smartphone-based sentiment mapping [22] show how AI systems can dynamically adjust responses based on emotional feedback, contributing to more engaging and human-centric interactions. These advancements reflect the shift toward emotionally intelligent and socially aware AI systems, marking a critical stage in the ongoing evolution of artificial intelligence and smart cities.

The Smart City Intelligence Ecosystem model integrates AI across six domains, namely governance, environment, living, people, mobility, and economy, to support sustainable and citizen-centric services [23]. These systems depend on interconnected infrastructures, particularly IoT, whose integration is recognized as essential for realizing the full potential of intelligent urban solutions [24]. This stage in the evolution of AI, focused on intelligent information systems and applications, demonstrates the transition from algorithmic research to operational deployment in real-world systems across diverse domains [25, 26]. These implementations prioritize functionality, personalization, explainability, and adaptability as core attributes of modern AI. In the context of smart cities, intelligent systems support critical infrastructure by enabling data-driven management, automation, and user-centred services. Applications span domains such as energy, health, transportation, and emergency communication, where AI technologies improve efficiency and resource allocation.

Fire emergency alert systems have adopted fuzzy logic to generate personalized notifications. These systems account for user-specific conditions and environmental inputs to optimize alert relevance and responsiveness [27].

Home energy management within smart cities relies on explainable AI and interoperable architectures to coordinate energy consumption across devices and households [28–30]. These systems contribute to overall energy optimization and support the integration of renewable sources.

Smart homes also apply AI for personalization through virtual agents and behaviour modelling. Studies indicate that end users respond positively to adaptive systems that infer preferences and automate domestic tasks [31, 32].

In health-focused applications, AI enables in-home safety monitoring. For instance, fire prevention systems detect cognitive errors in cooking behaviour among users with memory impairment, reducing accident risks [33, 34].

AI in education includes intelligent tutoring systems, adaptive learning environments, and chatbot evaluation models. These systems use cognitive and fuzzy logic approaches to adjust content delivery and assess pedagogical efficiency [35–37].

In business, AI supports model innovation by aligning technological capabilities with managerial roles and decision processes. Research identifies strategic integration as a factor in effective AI implementation [38].

In medicine, AI applications include LLMs for diagnostics, semi-supervised models for chronic disease prediction, and explainability frameworks. These tools enhance diagnostic precision and improve trust in clinical decision support [39–43].

Public sector use cases involve knowledge systems and labour compliance tools. AI models have been developed to detect undeclared work through supervised learning, accounting for class imbalance and overlap [44–46].

In software engineering, AI is integrated into development processes through prompt engineering techniques. A taxonomy of prompt patterns has been proposed to structure interactions with generative models, improving software task automation [47]. Leadership efforts in standardization also support the alignment of AI tools with development goals [48]. Further work highlights the role of intelligent systems in knowledge-based software engineering [49].

In the energy sector, AI supports prediction, optimization, and explainability. Machine learning algorithms have been applied in the oil and gas industry for operational management [50]. Explainable decision-support frameworks adapt AI outputs to user needs [51], and clustering methods have been introduced for residential load analysis [52]. Broader frameworks enable flexible, intelligent energy management across systems [53]. Theoretical contributions, such as the free energy principle, inform the modelling of inference under uncertainty in intelligent systems [54].

In automated driving, safety and risk evaluation methods have been developed for deep neural network-based systems. These include techniques for analysing hazards and validating model behaviour in real-time control contexts [55].

Behavioural and social dimensions are also critical in smart city systems. Trust, empowerment, and perceived risk influence user engagement with AI-based services, highlighting the need for human-centred design and trust-aware systems [56]. Urban development also benefits from AI-based assessment tools that support goal tracking and inform planning strategies through measurable indicators [57].

AI-driven decision support systems further enhance smart cities by enabling adaptive reasoning in complex environments. These systems assist in strategic planning and operational decisions through context-aware models and explainable outputs [58]. Additionally, models of trust dynamics in human-AI collaboration, such as VIRTSI, offer insights into how perceived reliability affects usage intentions and decision confidence [59].

The effective deployment of such systems also requires increased AI literacy. Studies emphasize that user competence in understanding and interacting with AI systems influences acceptance and correct usage, suggesting the need for structured learning approaches [60].

Human-centred design is essential for building trust in AI systems deployed within smart cities. As AI becomes embedded in energy, mobility, and governance infrastructures, it must address the diverse expectations and levels of expertise among citizens. Personalized interaction models have been introduced to assess how users, both experts and non-experts, engage with AI tools. Findings suggest that transparency, perceived usefulness, and adaptive communication are key to fostering trust and effective use [61].

Further work has emphasized that personalization in AI interaction improves trust and acceptance, particularly in complex urban systems where decision outcomes affect daily life. The evaluation of citizen-AI interaction through tailored frameworks shows that explainable recommendations and contextualized outputs enhance perceived system credibility [62].

Explainability is also a core requirement in decision-support applications, especially in energy management. AI frameworks that provide interpretable reasoning and traceable outputs are more likely to gain acceptance among stakeholders with varied technical backgrounds [30]. Additionally, AI models that improve performance reliability, such as ensemble methods for baseline energy estimation, further contribute to building confidence in automated urban services [63].

1.3 Types of Human-Smart City Interactions and Factors in User-Citizen Experience

Interactions between citizens and smart city technologies occur across multiple domains, including:

- **Direct Interaction**: Citizens using mobile apps for public transport, paying bills online, or reporting issues through smart platforms.
- **Indirect Interaction**: Systems that respond to environmental or behavioral data, such as adaptive traffic lights or pollution monitors.
- **Participatory Interaction**: Platforms that allow citizen input in decision-making, such as open data dashboards and digital forums.

On the other hand, relevant human factors to be considered include citizen engagement and trust, privacy and data protection, social equity, service personalization, and the broader influence of technology on shaping cities to better align with human needs and preferences.

Accordingly, current related research spans along the lines of:

- Environmentally sustainable urban transportation
- Eco-friendly neighborhoods and sustainable built environments
- Integration of infrastructure and operations across energy, ICT, and transport systems
- Approaches centered on citizen needs and engagement
- Regulatory frameworks and public policy development
- Coordinated urban planning and management strategies
- Sharing of knowledge and cross-sector collaboration
- Establishment of benchmarks, performance metrics, and evaluation indicators
- Governance of open data and responsible data use
- Adoption of standards and system interoperability
- Innovative business models, funding mechanisms, and procurement strategies

1.4 Book Structure

The book at hand aims at updating the relevant research communities, including professors, researchers, scientists, engineers and students, as well as general readers, on the most recent advances in *Human-Smart City Interactions and User-Citizen Experiences*.

More specifically, the book consists of this editorial chapter and additional twelve (12) chapters. Each chapter is self-contained, featuring an abstract and a list of references to supplementary or more advanced resources. All chapters in the book were invited from authors who work in the corresponding chapter topic and are recognized for their significant research contributions. In more detail, the chapters in the book are organized into five parts, as follows.

The *first part* of the book consists of two chapters devoted to *Sustainable Urban Planning*.

Specifically, Chap. 2, authored by Dimitrios Panagoulias, Elissaios Sarmas, Vangelis Marinakis and George A. Tsihrintzis, is entitled "Extreme-Value-Aware Synthetic Data Generation in Electric Vehicle Charging: A GAN Framework with EVA-Based Conditioning and Evaluation". The authors propose and test a novel framework that integrates Extreme Value Analysis (EVA) with Generative Adversarial Networks (GANs) to produce synthetic data that not only captures the bulk distribution of real-world observations but also accounts for rare and seasonal extremes.

Chapter 3, authored by Ioannis Papagiannopoulos, Eleni Kanellou and Vangelis Marinakis, is entitled "Energy Poverty Meets Disability: A Critical Analysis of Financial Support Gaps in Europe". The authors shed light on the gap between financial policies supporting people with disabilities and those addressing energy poverty relief.

The second part of the book consists of four chapters devoted to *Data-driven Governance*.

Specifically, Chap. 4, authored by Stamatios Theocharis and George A. Tsihrintzis, is entitled "Security of personal data processing when using artificial intelligence systems in smart cities: Processes and tools to address GDPR compliance challenges". The authors examine the critical issues and concerns in two seemingly incompatible concepts, namely artificial intelligence systems and the protection of personal data during the use of systems that serve the everyday lives of people and businesses and propose specific procedures and tools to address them.

Chapter 5, authored by Dimitrios C. Gkikas and Marios C. Gkikas, is entitled "AI in Citizen-Centric Smart Cities. Exploring Data Privacy, Algorithmic Transparency and Trustworthiness Through Regulation". The authors examine how the development of Artificial Intelligence has a huge impact on smart cities by transforming urban governance, decision making, security, and public services.

Chapter 6, authored by Aristea Kontogianni and Efthimios Alepis, is entitled "Empowering Smart Tourism with Large Language Models". The authors analyze how Large Language Models enhance smart tourism concepts and identify the opportunities and challenges associated with their implementation. Additionally, they discuss the limitations of existing systems, ethical considerations such as privacy and bias, and potential future directions for optimizing Artificial Intelligence-driven smart tourism services.

Chapter 7, authored by Kirti, Divya, Sukhjot Kaur, Shahla Gufran, Divyanshu Singh, and Dev Divyansh Mishra, is entitled "Sentiment Analysis Using LSTM and BERT: A Comparative Analysis using Natural Language Processing Algorithms". The authors explore sentiment analysis using Bidirectional Encoder Representations from Transformers (BERT) and Long Short-Term Memory (LSTM) models, assessing their performance for accurately capturing emotional nuances in social media text.

The *third part* of the book consists of one chapter devoted to *Smart Mobility Systems.*

Specifically, Chap. 8, authored by Ana Vigario, Joana Oliveira, Rodrigo Fernandes, Tiago Pinto, Arsenio Pinto, Tania Rocha, and Joao Barroso, is entitled "Adaptive User Interface for Electric Vehicle Route Information in Urban Mobility Services". The authors present effective, intuitive, and adaptive interfaces between the citizen and mobility systems which have the potential to facilitate mobility and sustainability in smart cities.

The *fourth part* of the book consists of two chapters devoted to *IoT-enabled Infrastructure.*

Specifically, Chap. 9, authored by Ioannis Gkini, Griseld Gkini and Dimitrios Tsipianitis, is entitled "Design and Implementation of a Smart Recycling Bin Using Multi-Sensor Integration and Machine Learning". The authors present a novel Smart Recycling Bin that integrates cameras, audio detection, and various sensors with machine learning to optimize waste sorting.

Chapter 10, authored by Francisco V. Cipolla Ficarra is entitled "Smart Sensors and Social Sensibility". The author presents a new quality metric called "4S" (standing for smart sensors and social sensibility) to measure the gap between technological reality and the sensation perceived by users in smart cities.

Finally, the *fifth part* of the book consists of three chapters devoted to *Digital Public Services.*

Specifically, Chap. 11, authored by Chairi Kiourt, Zoi Lygizou, Anestis Koutsoudis and Dimitris Kalles, is entitled "A Framework for Autonomous Crowd Management through Reinforcement Learning and Digital Twins". The authors propose an innovative framework for autonomous crowd management that integrates Reinforcement Learning, Digital Twins, Multi-Agent Systems, and Intelligent Internet of Things devices.

Chapter 12, authored by Konstantinos Liagkouras and Konstantinos Metaxiotis, is entitled "A Reinforcement Learning System that Learns from Human-Smart City Interaction for providing Better Quality of Services". The authors propose a Reinforcement Learning System that learns from human-smart city interaction and is especially addressed to citizens of megacities.

Finally, Chap. 13, authored by Evgenia Psarra and Dimitrios Apostolou, is entitled "A Diagnostic Clinical Decision Support System which leverages the Medical Information of Patients suffering from Pneumonia". The authors analyze patients' health metrics to determine if they are affected by pneumonia, utilizing a Convolutional Neural Network-based algorithm.

1.5 Conclusions

The effectiveness of smart cities relies not only on technological advancement but also on their ability to respond to the evolving needs of their citizens. While digital innovation offers significant opportunities for improving urban life, several challenges continue to hinder inclusive and sustainable development. One major issue is the digital divide, as not all citizens have equal access to digital infrastructure or possess the digital literacy required to benefit from smart city services. Ethical concerns are also pressing, particularly in relation to data ownership, user consent, and the transparency of algorithmic processes. In addition, the increasing complexity of smart city systems can result in poor usability, leading to reduced citizen engagement. Over time, citizens may also experience engagement fatigue if their participation in consultation processes does not lead to tangible outcomes or improvements.

In response to these challenges, promoting meaningful interaction between citizens and smart city technologies requires a comprehensive approach that prioritizes both functional performance and social responsibility. Accessibility must be central to system design, ensuring that digital services are intuitive and usable by all individuals, including those who are digitally or socially marginalized. Trust and privacy also play a critical role, as citizens must feel confident that their data is handled securely and used ethically. Furthermore, smart city services must offer practical benefits to users by improving convenience, enhancing efficiency, and delivering timely and reliable information. Civic engagement should not be limited to token participation but should enable citizens to contribute meaningfully to decision-making and policy development. Equally important is the principle of equity, ensuring that all communities benefit fairly from smart technologies, rather than reinforcing existing social and economic inequalities.

To support this vision, future research and development in smart cities should adopt human-centred design approaches that emphasize usability, inclusiveness, and responsiveness. Transparent governance is also essential, requiring openness in how data is collected, analyzed, and used in public decision-making. Continuous feedback mechanisms should be built into systems to enable ongoing evaluation and improvement based on citizen input. Moreover, it is necessary to invest in capacity building, including digital literacy programs and equitable access to technology, to ensure that all citizens can participate fully in the smart city environment.

This book aims to contribute to this ongoing discourse by presenting a range of perspectives and research findings focused on Human–Smart City Interactions and User–Citizen Experiences. Through its five-part structure, it explores the technical, social, and policy dimensions of smart city development. The editors aim to ensure that the book provides useful insights and practical guidance to researchers, professionals, and policymakers working in the field. At the same time, the editors recognize that both societal demands and technological capabilities are evolving rapidly. As new challenges and opportunities emerge, future volumes will be needed to further investigate the complex and dynamic relationship between people and smart urban systems.

References

1. https://commission.europa.eu/eu-regional-and-urban-development/topics/cities-and-urban-development/city-initiatives/smart-cities_en
2. https://www.mckinsey.com/capabilities/operations/our-insights/smart-cities-digital-solutions-for-a-more-livable-future#part2
3. A. Sharifi, M. Amirzadeh, A.R. Khavarian-Garmsir, The metaverse as a future form of smart cities: a systematic literature review of co-benefits and trade-offs for sustainable development goals. Cities **161**, 105879 (2025)
4. M. Virvou, T. Nakamura (eds.), *Knowledge-based Software Engineering: Proceedings of the Eighth Joint Conference on Knowledge-Based Software Engineering*, vol. 180 (Ios Press, 2008)
5. G.A. Tsihrintzis, L.C. Jain, *Multimedia Services in Intelligent Environments: Advanced Tools and Methodologies* (Springer, Berlin, 2008)
6. R. Walshe, K. Casey, J. Kernan, D. Fitzpatrick, AI and big data standardization: contributing to United Nations sustainable development goals. J. ICT Stand. **8**(2), 77–106 (2020)
7. Y. Roh, G. Heo, S.E. Whang, A survey on data collection for machine learning: a big data-ai integration perspective. IEEE Trans. Knowl. Data Eng. **33**(4), 1328–1347 (2019)
8. G.A. Tsihrintzis, M. Virvou, G. Phillips-Wren, Surveys in artificial intelligence-based technologies. Intell. Decis. Technol. **13**(4), 393–394 (2019)
9. G.A. Tsihrintzis, M. Virvou, E. Sakkopoulos, L.C. Jain, Applications of learning and analytics in intelligent systems, in *Machine Learning Paradigms: Applications of Learning and Analytics in Intelligent Systems*. (Springer International Publishing, Cham, 2019), pp.1–6
10. D.N. Sotiropoulos, D.G. Giatzitzoglou, G.A. Tsihrintzis, Adaptive boundary-aware artificial immune recognition system for data classification. Inf. Sci. **689**, 121500 (2025)
11. H. Takeuchi, J.H. Husen, H.T. Tun, H. Washizaki, N. Yoshioka, Enterprise architecture-based metamodel for machine learning projects and its management. Futur. Gener. Comput. Syst. **161**, 135–145 (2024)
12. K. Ogawa, T. Nakatani, Supporting code review by a neural network using program images, in *Joint Conference on Knowledge-Based Software Engineering* (Springer International Publishing, Cham, 2022), pp. 69–81
13. Y.W. Chen, L.C. Jain, Deep learning in healthcare, in *Paradigms and Applications* (Springer, Berlin/Heidelberg, Germany, 2020)
14. M. Virvou, Artificial Intelligence and user experience in reciprocity: contributions and state of the art. Intell. Decis. Technol. **17**(1), 73–125 (2023)
15. M. Virvou, The emerging era of human-AI interaction: keynote address, in *2022 13th International Conference on Information, Intelligence, Systems & Applications (IISA)* (IEEE, 2022), pp. 1–10
16. M.G. Kibria, K. Nguyen, G.P. Villardi, O. Zhao, K. Ishizu, F. Kojima, Big data analytics, machine learning, and artificial intelligence in next-generation wireless networks. IEEE Access **6**, 32328–32338 (2018)
17. G.A. Tsihrintzis, M. Virvou, L.C. Jain, Multimedia services in intelligent environments: recommendation services, in *Multimedia Services in Intelligent Environments: Recommendation Services*. (Springer International Publishing, Heidelberg, 2013), pp.1–5

18. J. Aguilar, P. Valdiviezo-Díaz, G. Riofrio, A general framework for intelligent recommender systems. Appl. Comput. Inform. **13**(2), 147–160 (2017)
19. H.F.T. Alsaadawi, B. Das, R. Das, A systematic review of trimodal affective computing approaches: text, audio, and visual integration in emotion recognition and sentiment analysis. Expert Syst. Appl. **255**, 124852 (2024)
20. A.M. Triantafyllou, G.A. Tsihrintzis, GRoup AFFEct Recognition: the side of heartbeats, in *2024 15th International Conference on Information, Intelligence, Systems & Applications (IISA)* (IEEE, 2024), pp. 1–6.
21. A.M. Triantafyllou, G.A. Tsihrintzis, Group affect recognition: optimization of automatic classification, in *Joint Conference on Knowledge-Based Software Engineering* (Springer International Publishing, Cham, 2018), pp. 189–196
22. A. Fasoulis, M. Virvou, G. Tsihrintzis, C. Patsakis, E. Alepis, Sensus vox: sentiment mapping through smartphone multi-sensory crowdsourcing, in *2018 IEEE 30th International Conference on Tools with Artificial Intelligence (ICTAI)* (IEEE, 2018), pp. 438–444
23. J. John, R. David Amar Raj, M. Karimi, R. Nazari, R.M.R. Yanamala, A. Pallakonda, Artificial Intelligence for smart cities: a comprehensive review across six pillars and global case studies. Urban Sci. **9**(7), 249 (2025)
24. W.M. Dahmane, S. Ouchani, H. Bouarfa, Smart cities services and solutions: a systematic review. Data Inf. Manage. **9**(2), 100087 (2025)
25. N. Bourbakis, G.A. Tsihrintzis, M. Virvou, L.C. Jain (eds.), *Extended Selected Papers of the 14th International Conference on Information, Intelligence, Systems, and Applications (IISA 2023)* (Springer, 2024). https://doi.org/10.1007/978-3-031-52858-3
26. N.G. Bourbakis, G.A. Tsihrintzis, M. Virvou (eds.), *15th International Conference on Information, Intelligence, Systems & Applications, IISA 2024*, Chania Crete, Greece, July 17–19, 2024 (IEEE, 2024). ISBN 979-8-3503-6883-3
27. N. Alonistioti, E.A. Tsichrintzi, K. Chrysafiadi, E. Alepis, Requirements for fuzzy logic in personalisation of fire emergency alerts, in *2023 14th International Conference on Information, Intelligence, Systems & Applications (IISA)* (IEEE, 2023), pp. 1–8
28. G. Kormpakis, A. Lekidis, E. Sarmas, I. Papias, F. Serepas, G. Stravodimos, V. Marinakis, Home energy management systems: challenges, heterogeneity & integration architecture towards a smart city ecosystem, in *2024 IEEE International Conference on Engineering, Technology, and Innovation (ICE/ITMC)* (IEEE, 2024), pp. 1–10
29. G. Kormpakis, A. Lekidis, I. Papias, N. Dimitropoulos, F. Serepas, V. Marinakis, Holistic framework for household energy management services, in *2025 10th International Conference on Smart and Sustainable Technologies (SpliTech)* (IEEE, 2025), pp. 1–10
30. E. Sarmas, D.P. Panagoulias, G.A. Tsihrintzis, V. Marinakis, H. Doukas, An explainable AI-based framework for supporting decisions in energy management, in *Machine Learning Applications for Intelligent Energy Management: Invited Chapters from Experts on the Energy Field* (Springer Nature Switzerland, Cham, 2024), pp. 1–27
31. M. Hubert, S. Kazemi, M. Hubert, A. Carugati, M.M. Mariani, The exploration of users' perceived value from personalization and virtual conversational agents to enable a smart home assemblage—a mixed method approach. Int. J. Inf. Manage. **80**, 102850 (2025)
32. X. Chen, S. Chen, Z. Jin, H. Bian, Z. Chen, H. Li, Expressing the needs in smart home: what is the end users' favorite way. ACM Trans. Comput.-Hum. Interact. **32**(2), 1–38 (2025)
33. K. Chrysafiadi, E.A. Tsichrintzi, E. Sakkopoulos, A smart home fuzzy logic-based cooking fire prevention system addressing human errors due to distraction or minor medical memory lapses. Procedia Comput. Sci. **246**, 3390–3399 (2024)
34. J. Blott, Smart homes for the future of dementia care. Lancet Neurol. **20**(4), 264 (2021)
35. M. Virvou, G.A. Tsihrintzis, Is ChatGPT beneficial to education? A holistic evaluation framework based on intelligent tutoring systems, in *2023 14th International Conference on Information, Intelligence, Systems & Applications (IISA)* (IEEE, 2023), pp. 1–8
36. K. Chrysafiadi, S. Papadimitriou, M. Virvou, Cognitive-based adaptive scenarios in educational games using fuzzy reasoning. Knowl.-Based Syst. **250**, 109111 (2022)

37. M. Virvou, G.A. Tsihrintzis, Artificial intelligence chatbots and social robots in education: FEPER framework for efficiency, pedagogical and ethical requirements, in *2024 15th International Conference on Information, Intelligence, Systems & Applications (IISA)* (IEEE, 2024), pp. 1–8
38. P. Jorzik, A. Yigit, D.K. Kanbach, S. Kraus, M. Dabić, Artificial intelligence-enabled business model innovation: competencies and roles of top management. IEEE Trans. Eng. Manage. **71**, 7044–7056 (2023)
39. D.P. Panagoulias, A. Palamidas, M. Virvou, G.A. Tsihrintzis, A framework for evaluation and requirement extraction for fine-tuning of Large Language Models in multimodal medical diagnosis. Knowl.-Based Syst. 113975 (2025)
40. D.P. Panagoulias, G.A. Tsihrintzis, M. Virvou, *Artificial Intelligence-Empowered Bio-medical Applications: Challenges, Solutions and Development Guidelines*, vol. 49 (Springer Nature, 2025)
41. D.P. Panagoulias, M. Virvou, G.A. Tsihrintzis, Augmenting large language models with rules for enhanced domain-specific interactions: the case of medical diagnosis. Electronics **13**(2), 320 (2024)
42. D.P. Panagoulias, M. Virvou, G.A. Tsihrintzis, A novel framework for artificial intelligence explainability via the technology acceptance model and rapid estimate of adult literacy in medicine using machine learning. Expert Syst. Appl. **248**, 123375 (2024)
43. S. Kotsiantis, G. Melagraki, V. Verykios, A. Sakagianni, J. Matsoukas, Semi-supervised learning for predicting multiple sclerosis. J. Person Med. **15**(5), 167 (2025)
44. S.A. Theocharis, G.A. Tsihrintzis, Knowledge management systems in the public sector: critical issues, in *Lecture Notes on Software Engineering*, vol. 4, no. 1 (2016), p. 59
45. E. Alogogianni, M. Virvou, Association rules and machine learning for enhancing undeclared work detection, in *2020 11th International Conference on Information, Intelligence, Systems and Applications (IISA)* (IEEE, 2020), pp. 1–8
46. E. Alogogianni, M. Virvou, Undeclared work prediction using machine learning: dealing with the class imbalance and class overlap problems, in *2022 13th International Conference on Information, Intelligence, Systems & Applications (IISA)*. (IEEE, 2022), pp. 1–8
47. Y. Sasaki, H. Washizaki, J. Li, N. Yoshioka, N. Ubayashi, Y. Fukazawa, Landscape and taxonomy of prompt engineering patterns in software engineering. IT Professional **27**(1), 41–49 (2025)
48. H. Washizaki, Leadership and engagement: IEEE Computer Society 2025 key strategies. Computer **58**(1), 4–8 (2025)
49. G.A. Tsihrintzis, M. Virvou, T. Saruwatari, Special collection of extended selected papers on *Novel Research Results Presented at the 14th International Joint Conference on Knowledge-based Software Engineering (JCKBSE2022)*, 22–24 August 2022, vol. 16, no. 4 (Intelligent Decision Technologies, Larnaca, Cyprus, 2022), pp. 715–716
50. J. Li, Y. Guo, Z. Fu, X. Zhang, F. Shen, An intelligent energy management information system with machine learning algorithms in oil and gas industry. Wirel. Commun. Mob. Comput. **2023**(1), 3385453 (2023)
51. D.P. Panagoulias, E. Sarmas, V. Marinakis, M. Virvou, G.A. Tsihrintzis, H. Doukas, Intelligent decision support for energy management: a methodology for tailored explainability of artificial intelligence analytics. Electronics **12**(21), 4430 (2023)
52. V. Michalakopoulos, I. Papias, E. Sarantinopoulos, E. Sarmas, V. Marinakis, D. Askounis, A hyperparameter-space clustering methodology of residential electricity loads. Appl. Soft Comput. 113497 (2025)
53. E. Sarmas, V. Marinakis, H. Doukas, Artificial intelligence for energy systems: driving intelligent, flexible and optimal energy management. Learn. Anal. Intell. Syst. **46**, 1–266 (2025), Springer
54. D. Collerton, I. Tsuda, S. Nara, Episodic visual hallucinations, inference and free energy. Entropy **26**(7), 557 (2024)
55. T. Kaneko, Y. Takahashi, S. Yamaguchi, J. Hashimoto, N. Yoshioka, Safety and risk analysis and evaluation methods for DNN systems in automated driving, in *Proceedings of the 14th*

International Joint Conference on Knowledge-Based Software Engineering (JCKBSE 2022), Larnaca, Cyprus, August 22–24, 2022. (Springer, 2023), pp. 83–96
56. J.P.F. dos Santos, C.A. de Matos, A. Groznik, The role of artificial intelligence in smart city systems usage: drivers, barriers, and behavioural outcomes. Technol. Soc. **81**, 102867 (2025)
57. H. Ahvenniemi, A. Huovila, I. Pinto-Seppä, M. Airaksinen, What are the differences between sustainable and smart cities? Cities **60**, 234–245 (2017)
58. G.A. Tsihrintzis, M. Virvou, H. Doukas, L.C. Jain, Introduction to advances in artificial intelligence-empowered decision support systems, in *Advances in Artificial Intelligence-Empowered Decision Support Systems* (Springer Nature Switzerland, Cham, 2024), pp. 1–11
59. M. Virvou, G.A. Tsihrintzis, E.A. Tsichrintzi, VIRTSI: a novel trust dynamics model enhancing Artificial Intelligence collaboration with human users—insights from a ChatGPT evaluation study. Inf. Sci. **675**, 120759 (2024)
60. M. Pinski, A. Benlian, *AI literacy for users—a comprehensive review and future research directions of learning methods, components, and effects*. Comput. Hum. Behav. Artif. Hum. 100062 (2024)
61. G.A. Tsihrintzis, E. Sarmas, V. Marinakis, D. Panagoulias, E.A. Tsichrintzi, M. Virvou, Evaluating non-expert stakeholder interaction with artificial intelligence on energy urban domain using VIRTSI: the case of ChatGPT, in *2024 15th International Conference on Information, Intelligence, Systems & Applications (IISA)* (IEEE, 2024), pp. 1–10
62. G.A. Tsihrintzis, E. Sarmas, V. Marinakis, D.P. Panagoulias, E.A. Tsichrintzi, M. Virvou, Energy urban domain: personalized evaluation of expert and non-expert stakeholder interaction with artificial intelligence through ChatGPT using the VIRTSI model. IEEE Access (2025)
63. E. Sarmas, A. Forouli, V. Marinakis, H. Doukas, Baseline energy modeling for improved measurement and verification through the use of ensemble artificial intelligence models. Inf. Sci. **654**, 119879 (2024)

Part I
Sustainable Urban Planning

Chapter 2
Extreme-Value-Aware Synthetic Data Generation in Electric Vehicle Charging: A GAN Framework with EVA-Based Conditioning and Evaluation

Dimitrios P. Panagoulias, Elissaios Sarmas, Vangelis Marinakis, and George A. Tsihrintzis

Abstract Synthetic data generation has emerged as a powerful technique for enhancing data-driven systems where real-world observations are sparse, privacy-sensitive, or highly imbalanced. This chapter presents a novel framework that integrates Extreme Value Analysis (EVA) with Generative Adversarial Networks (GANs) to produce synthetic data that not only captures the bulk distribution of real-world observations but also accounts for rare and seasonal extremes. The proposed methodology includes conditioning the GAN on EVA-derived temporal features such as monthly seasonality and the presence of extreme events, as well as evaluating the synthetic outputs using return level plots and generalized Pareto-based return period analysis. As a case study, we apply this method to Electric Vehicle (EV) charging data provided by DEI Blue (Greece), illustrating how synthetic charging scenarios can be realistically reproduced and evaluated. Results demonstrate that while the GAN successfully learns seasonality and general usage patterns, EVA-based evaluation reveals limitations in capturing tail-end extremes—offering a novel benchmark for synthetic data quality assessment.

Keywords Generative adversarial networks · e-mobility · Extreme value analysis · Seasonality · Weather conditions · Climate · Commercial and business planning · Charger speed · Embedding EVA in analytics workflow

---

D. P. Panagoulias (✉) · G. A. Tsihrintzis
Department of Informatics, University of Piraeus, Piraeus, Greece
e-mail: PANAGOULIAS_D@yahoo.gr

G. A. Tsihrintzis
e-mail: geoatsi@unipi.gr

E. Sarmas · V. Marinakis
Decision Support Systems Lab, School of Electrical and Computer Engineering, National Technical University of Athens, Zografou-Athens, Greece
e-mail: esarmas@epu.ntua.gr

V. Marinakis
e-mail: vmarinakis@epu.ntua.gr

G. A. Tsihrintzis et al. (eds.), *Human-Smart City Interactions and User-Citizen Experiences*, Learning and Analytics in Intelligent Systems 58,
https://doi.org/10.1007/978-3-032-06364-9_2

2.1 Introduction

The *Paris Agreement* [1], adopted by 196 parties during the 2015 United Nations Climate Change Conference in Paris, marked a pivotal global commitment to combating climate change. One of the most impactful contributions towards achieving its climate targets involves the decarbonization of the transportation sector through the widespread adoption of electric vehicles (EVs), effectively replacing traditional fossil fuel-powered cars [2, 3]. This transition is strongly supported by numerous national policies and incentives around the world [4]. However, its full-scale implementation requires overcoming several challenges [5–7], such as the optimal planning and deployment of EV charging infrastructure [8, 9].

In this context, the role of national initiatives becomes critical. DEI Blue, an e-mobility service by the Public Power Corporation (PPC-DEI) in Greece, exemplifies such a proactive strategy [10]. This initiative delivers integrated e-mobility solutions to individuals, enterprises, and public organizations. Leveraging PPC-DEI's extensive electricity distribution network, DEI Blue aims to establish the country's most extensive EV charging infrastructure. Its core objectives are to make electric mobility economically viable, ensure its wide accessibility, and democratize the environmental and economic benefits of clean transportation.

In a recent work of ours [9], we applied extreme value analysis (EVA) techniques on sets of EV charging data with the goal to extract future trend indications regarding the demand for EV charging. However, the available datasets were rather limited, as they covered only a period of twelve months. Our analysis still revealed trends in the data that indicated a probable seasonality, which indicated that a more detailed analysis was required, especially in view of the fast-rising number of EVs in use. This analysis could be effected perhaps with the use of *synthetic data*.

Synthetic data generation is increasingly vital in domains where privacy, scarcity, or cost constrain real-world data availability. However, most generative models, including Generative Adversarial Networks (GANs), optimize for overall distribution fit and may underrepresent rare but critical events such as peaks, failures, or extremes. This is particularly problematic in domains like climate modeling, grid stress testing, or electric vehicle (EV) infrastructure planning, where extreme events are pivotal for design and risk mitigation [11].

This chapter proposes a generalizable framework that incorporates EVA into both the conditioning and evaluation phases of synthetic data generation via Generative adversarial networks (GANs). By embedding seasonality and EVA-informed thresholds and by benchmarking the generated data through EVA return level estimation, we offer a path forward for trustworthy synthetic data in critical applications.

More specifically, the remainder of this work is organized as follows: Sect. 2.2 summarizes the key concepts and theoretical results of EVA and GAN as they are applied in this work. In Sect. 2.3, we describe the datasets used in our analysis and outline our methodology for analyzing them. In Sect. 2.4, we present the results of our analysis. Finally, Sect. 2.5 summarizes this work, draws conclusions and outline some open questions that need to be addressed in the future.

2.2 Theory and Application Highlights of Extreme Value Analysis and Generative Adversarial Networks

Extreme Value Analysis (EVA), or Extreme Value Theory (EVT), is a statistical framework for modeling rare events that deviate significantly from the median of probability distributions [12]. It relies on analyzing *ordered samples* to estimate the likelihood of extreme, previously unobserved occurrences. Two principal methodologies are employed in EVA: (1) the block maxima/ minima approach, which extracts extreme annual values to form the Annual Maxima or Minima Series (AMS) and is grounded in the Fisher-Tippett-Gnedenko theorem [13, 14]; and (2) the Peak-Over-Threshold (POT) method [15], which models exceedances over a specified threshold using the Generalized Pareto Distribution (GPD) and Poisson processes, as supported by the Pickands-Balkema-De Haan theorem [16].

The second principal method in EVA is the Peak-Over-Threshold (POT) approach [15]. This technique involves sorting a dataset and isolating values that exceed a specified threshold. The POT method typically fits two separate types of distributions: (1) one describing the frequency of threshold exceedances over time or space, and (2) another capturing the magnitude of these exceedances. According to the Pickands-Balkema-De Haan theorem [16], the distribution of exceedances converges asymptotically to the Generalized Pareto Distribution (GPD), while a Poisson distribution is employed to model the count of exceedance events. From the fitted models, the return value (RV) is derived–representing a level that is expected to be exceeded, on average, once every T units of time or space, with a probability of $1/T$.

As EVA assesses ordered samples of a random variable, L-moments and L-moment ratios are used to calculate and summarize the shape of its probability distribution. While conventional moments are linear combinations of independent samples of a random variable, L-moments consist of linear combinations of *order statistics* (L-statistics) of the statistical sample. Specifically, let $X_i, i = 1, 2, \ldots, n$, be a set of independent samples of a random variable X. Then, its i-th order statistic $Y_i, i = 1, 2, \ldots, n$, is its i-th smallest sample and $Y_1 < Y_2 < \cdots < Y_n$. Subsequently, L-moments and L-moment ratios are quantities analogous to mean (or location), standard deviation, skewness and kurtosis and are, thus, termed L-mean, L-scale, L-skewness and L-kurtosis [12].

In the realm of energy production and consumption, especially concerning renewable sources like solar and wind energy, EVA plays a pivotal role in energy demand management. Research efforts, such as [17–19], have concentrated on employing the POT methodology. This approach helps to estimate the occurrence frequency of peaks surpassing a defined limit and to determine the distribution of the magnitude of these peaks. In our previous related work we have applied EVA in various contexts and domains [20–22].

Generative Adversarial Networks (GANs), introduced by Goodfellow et al. [23], have become a cornerstone method for synthetic data generation, particularly in domains where real data are limited, imbalanced, or costly to obtain. GANs operate through a game-theoretic framework involving two neural networks–the generator and the discriminator–trained adversarially to produce data samples that approximate the target distribution. Over recent years, GANs have been increasingly employed for data augmentation and synthetic sampling in a wide range of applications, including image generation [24], medical data synthesis [25], and time-series modeling [26]. These methods have shown promise in improving generalization, mitigating class imbalance, and enabling robust model evaluation under scarce data conditions. In the context of extreme event modeling, GANs have emerged as a valuable tool for simulating rare but critical phenomena, such as power system failures, financial crashes [27], or severe weather events [28]. Their ability to learn high-dimensional distributions from relatively small datasets makes them particularly attractive in domains where extreme value theory (EVT) is applied but suffers from data sparsity or tail uncertainty.

Specifically, GAN-driven data generation has been proposed as a means to augment EVT-based risk estimation frameworks [29], offering a data-centric approach to stress-testing and robustness analysis. However, evaluating the fidelity of synthetic data in the tails remains a challenge. Here, extreme value analysis (EVA) can serve as a diagnostic tool to benchmark GAN outputs under rare-event conditions–effectively enabling a probabilistic audit of GAN performance with respect to extremal behavior.

2.3 Materials and Methods

Our proposed methodology integrates Extreme Value Analysis (EVA) and Generative Adversarial Networks (GANs) in a hybrid framework designed to both synthesize and evaluate data under extreme conditions. The motivation behind incorporating EVA-derived conditions into the GAN framework stems from the limitations observed in unconditional generative modeling approaches. When GANs or other generative models are trained without contextual guidance–such as temporal indicators or extremity flags–they tend to focus on replicating the most common patterns in the data, often at the expense of statistical fidelity in the tails of the distribution. This leads to model collapse, underrepresentation of rare but critical events, and synthetic samples that fail to capture the heterogeneity inherent in real-world EV charging scenarios, particularly those influenced by seasonal spikes or high-demand anomalies. By contrast, our conditional approach aims to steer the generative process toward structure-aware synthesis, ensuring that the model not only learns the bulk distribution but also faithfully simulates the edge cases that are pivotal for risk assessment, planning, and stress-testing applications. This approach unfolds in three main stages as can also be seen in Fig. 2.1:

- EVA-Informed Feature Extraction: Statistical analysis is performed using EVA to uncover seasonal trends and identify extreme charging demand events. These are encoded as conditioning features for GAN training.

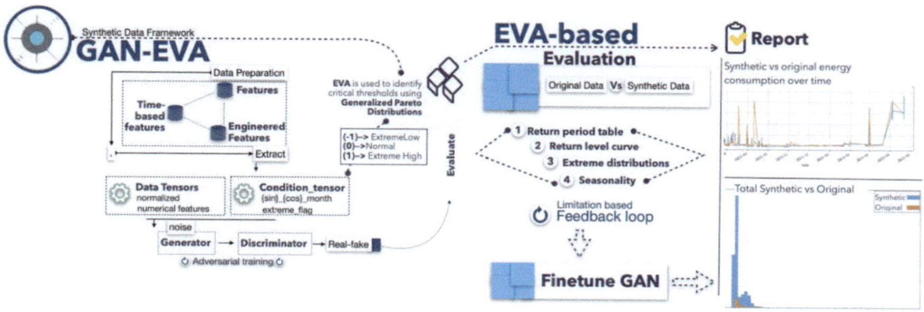

Fig. 2.1 Overview of the proposed GAN-EVA framework for seasonality-aware and extreme-value-consistent synthetic data generation

- Conditional GAN Training: A Conditional GAN (cGAN) is trained on the real dataset, guided by EVA-derived condition vectors to capture both ordinary and extreme behaviors, as well as their seasonal context.
- Post-Generation EVA-Based Quality Evaluation: After synthetic data are generated, they are subjected to the same EVA procedures to assess whether the tail behavior, return levels, and seasonal clustering are faithfully reproduced.

This pipeline begins with raw time-based data, from which engineered features such as monthly seasonality (via sine and cosine encodings) and extreme-value flags (via Peaks Over Threshold EVA) are extracted. These features are embedded into conditional tensors used to train a GAN model. The Generator receives random noise along with EVA-based conditions to produce synthetic samples, while the Discriminator learns to distinguish real from fake data using the same conditional context. After generation, the synthetic data are evaluated using EVA metrics including return level curves, return period tables, and seasonality clustering. This EVA-based evaluation supports a feedback loop, guiding refinement of the GAN model for better extreme value representation.

2.3.1 EVA

To identify and model rare events in EV charging demand, we apply Extreme Value Analysis using the Peaks Over Threshold (POT) method. This enables us to focus on the tail of the distribution and estimate risk under low-frequency, high-impact scenarios. The specific implementation includes:

- Threshold Selection: i.e. the 75th percentile of daily energy consumption is selected as the threshold for defining "extreme" events.
- Distribution Fitting: A Generalized Pareto Distribution (GPD) is fitted to the exceedances to model the severity of rare events.

- Return Periods: Return levels are computed for periods ranging from 1 to 1000 years, enabling long-term capacity planning.
- Seasonality Detection: Monthly clustering of extremes is observed (e.g., in July–August), indicating seasonal effects

The EVA results are used to generate condition vectors for downstream GAN conditioning: sin_month, cos_month, represent cyclic temporal behavior across the 12 months. $extreme\_flag$, a binary indicator marking whether a data point exceeds the extreme threshold. These features are critical for steering the GAN to simulate data across different temporal and extremal contexts.

2.3.2 GAN

We implement a Conditional GAN (cGAN) tailored for structured tabular data. The architecture integrates EVA-derived metadata to enforce contextual learning of extremes. The design is as follows:

- Generator Input: A concatenated vector of:
  - Random noise vector (latent space)
  - sin_month, cos_month (seasonality)
  - $extreme\_flag$ (extremal condition)
- Discriminator Input: A tuple consisting of:
  - Real or synthetic feature triplet: [energy, duration, max energy/session]
  - The same conditioning vector used in the generator

2.3.3 Evaluation

To assess the fidelity and robustness of the synthetic data, particularly under extremal regimes, we employ both visual and statistical metrics:

- EVA Return Level Curves: Comparison of real versus synthetic return levels across 1–1000 year periods.
- Return Period Tables: Tabulated RV values with confidence intervals to quantify differences.
- Seasonal Overlays: Visualization of monthly data distributions to verify whether seasonal clustering of extremes is preserved.
- Distribution Plots: Kernel density and histogram comparisons between real and synthetic samples for each feature.

This dual use of EVA–as both a generative conditioning strategy and an evaluation framework–creates a closed-loop system that emphasizes explainability and realism in the generation of rare event data.

2.4 Application of Methods

The GAN-EVA synthesis framework is applied to real electric vehicle charging data provided by DEI Blue (Greece), demonstrating its ability to generate realistic seasonal consumption profiles with quantified limitations in extreme event reproduction. The model is trained using adversarial loss, with the objective of making synthetic data indistinguishable from real observations across both normal and extreme ranges.

2.4.1 Data Analysis and EVA on DEI Blue Dataset

The data analyzed are daily summarizations from mid-2021 to mid-2022 of charging stations in different locations in Greece. The total chargings examined are approximately 79.000. The total datapoints summarised and analysed are 319. Null values have been removed. The descriptive statistics of our analysis can be seen in Table 2.1.

From the analysis, the 75th percentile of daily energy consumption–approximately 2021 kWh–is identified as the threshold for defining extreme events. This value serves as the cutoff for applying the Peaks Over Threshold (POT) method, which forms the foundation of our Extreme Value Analysis (EVA).

Figure 2.2 illustrates the extreme values that exceed the defined threshold. A noticeable concentration of these extremes during the summer months–particularly in July and August–points to a potential seasonal pattern in charging demand. This trend may be attributed to increased travel activity during summer holidays and the impact of elevated temperatures on battery performance, both of which can drive higher

Table 2.1 Summary statistics of energy consumed

Description	Energy consumed (kWh)
Count	318.00
Mean	2462.02
Standard deviation	5849.52
Minimum	0.000000
25th percentile	265.38
50th percentile	820.36
75th percentile	2021.38
Maximum	52630.31

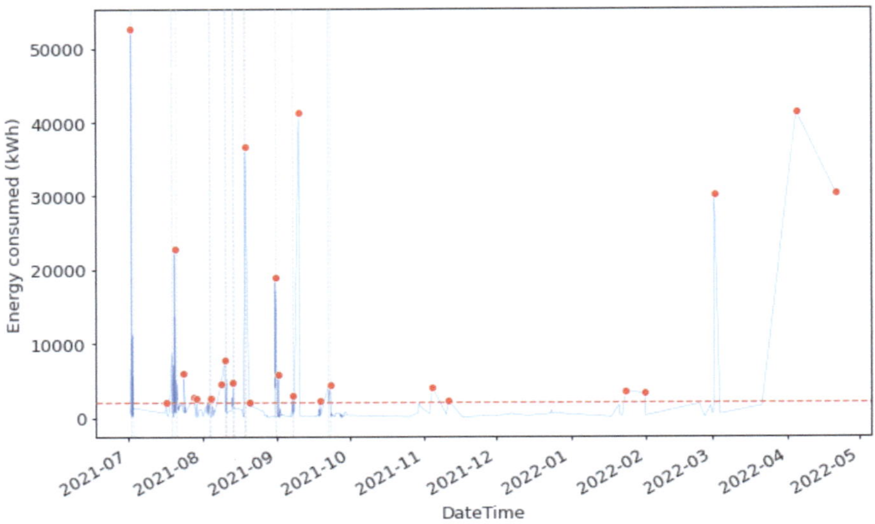

Fig. 2.2 EVA on DEI charging

charging frequency. While these observations are suggestive, additional data over extended periods are needed to confirm long-term patterns and support more robust conclusions. Furthermore, the spikes observed in April and May of 2022 may reflect shifts in user behavior, the commissioning of new infrastructure, or the market entry of additional electric vehicle models and seasonality. **Seasonality** in EV charging demand can be influenced by multiple factors, notably affecting both the frequency and volume of charging sessions. In the context of Greece, two primary contributors to the observed extremes–particularly those highlighted in Fig. 2.2–are the intense summer heat and the peak travel period during the summer holidays. Recent trends suggest that both the duration of high temperatures and the length of vacation periods have increased, likely as a consequence of climate change. Furthermore, Greece has experienced growing popularity as a travel destination, attracting both domestic and international tourists in greater numbers [30]. These dynamics help explain the sustained occurrence of extreme charging events well into late September.

Figure 2.3 presents four diagnostic plots that support the application of Extreme Value Theory (EVT) to the EV charging data. In the upper left plot, the relationship between the return period and energy consumption is illustrated. The x-axis denotes the return period–defined as the expected interval between events exceeding a given threshold–while the y-axis represents the associated energy consumption levels. The red curve corresponds to the fitted Generalized Extreme Value (GEV) distribution, with blue dotted lines denoting the confidence intervals, which are quantitatively detailed in Table 2.2.

The upper right plot depicts the probability density function (PDF) derived from the fitted model. Here, the red line indicates the theoretical density, while the blue dotted lines highlight the alignment between observed data and the fitted distribution.

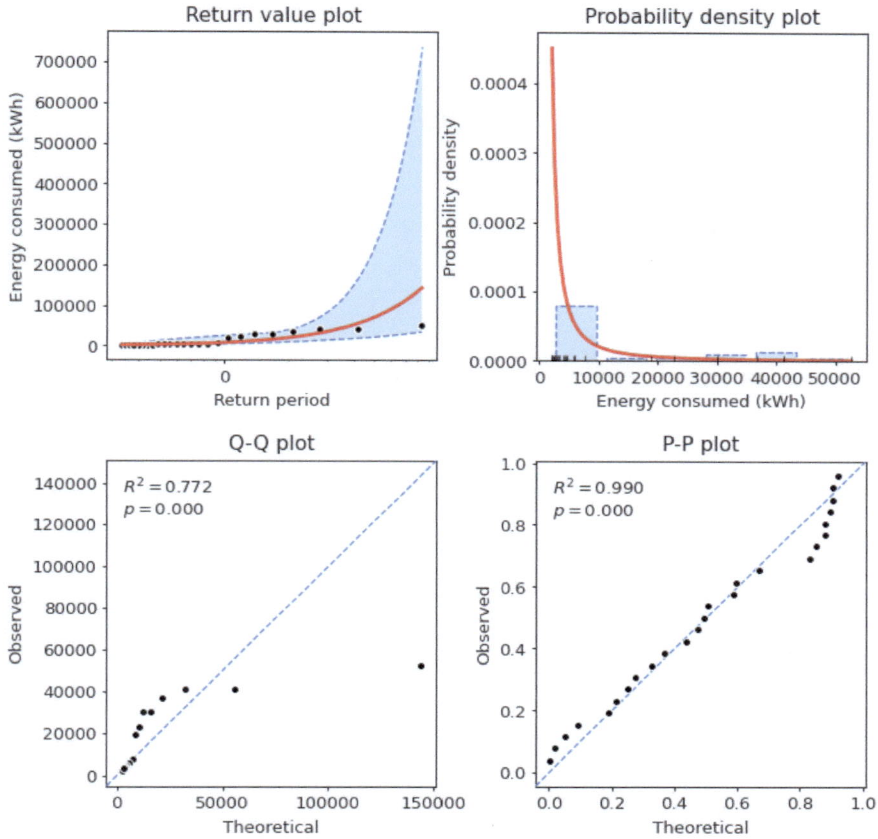

Fig. 2.3 DEI probability plots and distributions

Each return period corresponds to an energy consumption level that is expected to be equaled or exceeded, on average, once every T years, as shown in Table 2.2. In the lower left panel, the Quantile-Quantile (Q-Q) plot evaluates the agreement between the empirical and theoretical distributions. Points aligning with the 45-degree blue dashed line suggest a strong fit. Although the coefficient of determination is moderate ($R^2 = 0.0772$), the associated p-value (p = 0.000) confirms the statistical significance of the fit. Finally, the lower right panel shows the Probability-Probability (P-P) plot, which compares cumulative probabilities of observed versus theoretical values. This plot reveals a high degree of conformity, with a strong correlation coefficient ($R^2 =$ 0.988) and a statistically significant p-value, indicating that the fitted model captures the empirical distribution well (Tables 2.3 and 2.4).

Table 2.2 Return period analysis

Return period	Return value	Confidence interval (Lower CI—Upper CI)
1.0	1.846811×10^5	$43527.149724 - 1.053251 \times 10^6$
2.0	4.803443×10^5	$53571.509594 - 5.020378 \times 10^6$
5.0	1.701623×10^6	$64104.540721 - 3.946676 \times 10^7$
10.0	4.431416×10^6	$71789.935886 - 1.946282 \times 10^8$
25.0	1.570721×10^7	$81673.709858 - 1.621686 \times 10^9$
50.0	4.091080×10^7	$88949.279407 - 7.965103 \times 10^9$
100.0	1.065568×10^8	$96057.427345 - 3.742108 \times 10^{1}0$
250.0	3.777169×10^8	$105205.324435 - 2.893417 \times 10^{1}1$
500.0	9.838122×10^8	$111943.340801 - 1.436680 \times 10^{1}2$
1000.0	2.562466×10^9	$118529.302621 - 7.298954 \times 10^{1}2$

Table 2.3 Descriptive statistics during extreme energy consumption

Statistic	Energy consumed (kWh)	Charging time (h)	Max energy consumed in one session (kWh)
Mean	14278.03	1296.92	72.19
Std	16173.50	1255.29	14.97
Minimum	2027.67	231.12	24.85
25%	2856.24	512.36	66.41
50%	4519.38	685.79	74.48
75%	26471.28	1739.28	79.16
Max	52630.31	5055.67	101.12

2.4.2 Baseline GAN Data Synthesis on DEI Blue Dataset

Initially, we trained a baseline GAN model on the original dataset without incorporating any conditional inputs or temporal logic. In this configuration, the GAN operates purely in an unsupervised manner, learning to reproduce the global statistical distribution of the input features. The data synthesis process is fully stochastic, driven by the latent noise vector input to the Generator, and does not account for

Table 2.4 Descriptive statistics during normal energy consumption

Statistic	Energy consumed (kWh)	Charging time (h)	Max energy consumed in one session (kWh)
Mean	666.24	145.28	54.40
Std	554.79	219.72	19.30
Min	0.00	0.00	0.00
25%	205.65	31.70	40.17
50%	522.57	83.07	57.48
75%	1004.10	190.86	68.76
Max	2002.51	2020.09	101.12

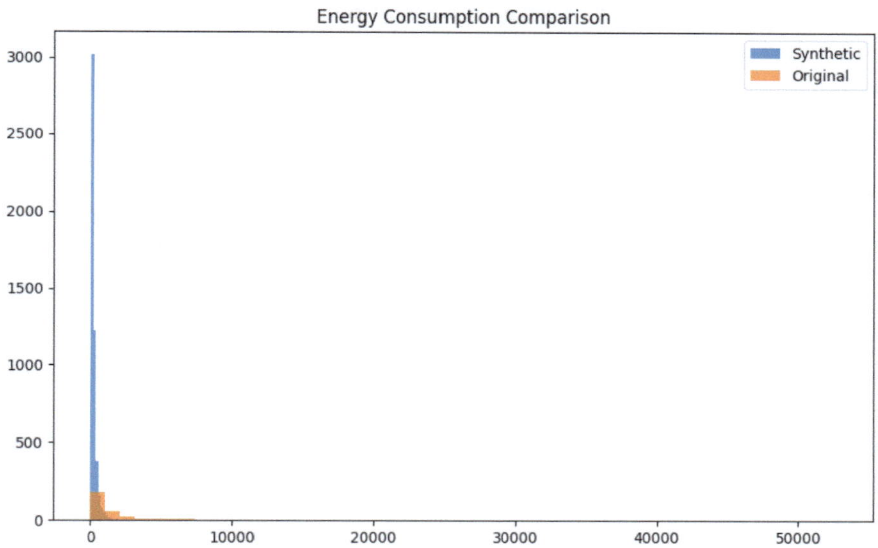

Fig. 2.4 Baseline GAN data synthesis comparison with initial dataset

seasonal patterns, date alignment, or the occurrence of extreme values. As a result, the generated data reflect a randomized distribution centered around the mode of the training data, with limited capacity to reproduce rare or time-dependent phenomena. This baseline experiment serves as a point of comparison for evaluating the added value of the EVA-informed conditioning introduced in the extended model.

Figure 2.4 presents a histogram comparison between the real and synthetic distributions of energy consumption (kWh) (the variable to be used for the EVA) generated by the baseline GAN model. In this configuration, the GAN was trained solely on normalized numerical inputs without incorporating any conditional information, such as seasonality or extreme-value labels. As a result, the synthetic data distribution exhibits a pronounced central bias, with the vast majority of generated samples concentrated near the lower end of the energy spectrum. The synthetic distribution shows a sharp decay beyond approximately 2,000 kWh, with almost no representation of higher-magnitude values. In contrast, the original data distribution contains a broader range, including multiple high-energy extreme events exceeding 10,000 and even 50,000 kWh. This highlights a key limitation of vanilla GANs trained on real-world imbalanced datasets: the generator tends to mode-collapse around frequent, low-energy samples, while rare but critical high-energy events are significantly underrepresented.

2.4.3 Conditional GAN-EVA Data Synthesis on DEI Blue Dataset

To address the issues identified in the previous subsection we apply the GAN-EVA framework as depicted in Fig. 2.1. We apply EVA using the Peaks Over Threshold (POT) method, which focuses on identifying and modeling the most extreme observations in a dataset–specifically those that exceed a certain high threshold. In our case, the threshold is set at the 75th percentile of daily energy consumption, corresponding to approximately 2021.38 kWh and also discussed in Sect. 2.4.1. All data points surpassing this threshold are considered extreme and are subsequently modeled using the Generalized Pareto Distribution (GPD), which is well-suited for capturing the tail behavior of such excesses. The outcome of this analysis forms the basis for constructing conditioning vectors that are used in the training of our generative model. Specifically, the variables sin_month and cos_month are computed to encode the cyclic nature of seasonality across the 12-month calendar year. In parallel, an $extreme\_flag$ is introduced as a binary indicator that distinguishes between normal and extreme consumption days based on the defined threshold.

To synthesize realistic charging data that reflects both typical and extreme usage patterns, we implement a Conditional Generative Adversarial Network (cGAN). In this architecture, both the Generator and the Discriminator are explicitly conditioned on the EVA-derived context features described above. The Generator receives as input a concatenated vector consisting of a latent noise vector, along with the values of sin_month, cos_month -synthesizing the datapoints within the real date ranges of the original dataset- and the $extreme\_flag$. This input guides the Generator to produce synthetic data that align with specific temporal and extremal contexts.

Simultaneously, the Discriminator is trained to differentiate between real and synthetic data samples by evaluating both the generated or observed data and their associated condition vectors. It receives, as input, a tuple that includes either a real or generated sample–comprising energy consumed, charging duration, and maximum per-session energy–together with the corresponding sin_month, cos_month, and $extreme\_flag$ values. The GAN is trained exclusively to generate three continuous-valued outputs: daily energy consumption (kWh), charging duration (hours), and

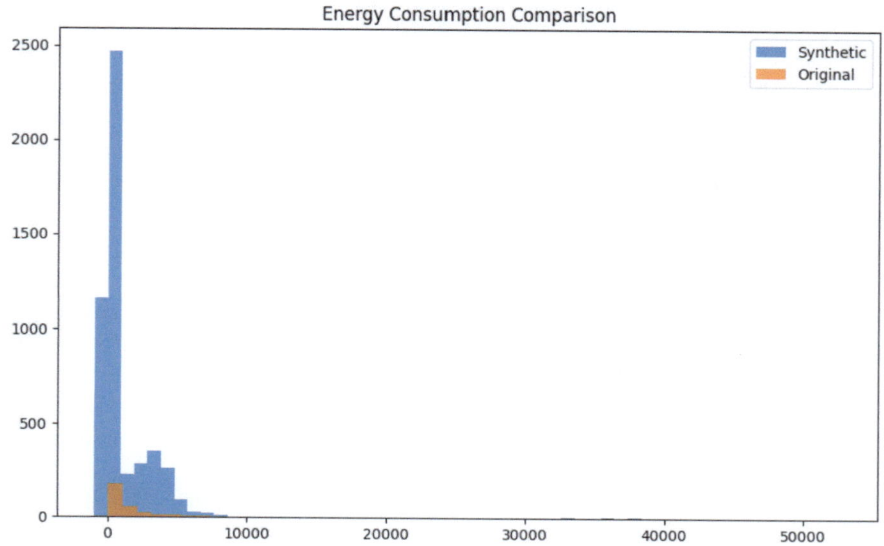

Fig. 2.5 GAN-EVA data synthesis comparison with initial dataset

maximum energy consumed per session (kWh). All other information (e.g., timestamps, charger IDs) is used only for conditioning and evaluation purposes, not for generation.

To validate and evaluate the process we first check the seasonality by comparing the synthesized with the original. Figure 2.5 illustrates the distributional comparison between real (orange) and synthetic (blue) energy consumption values generated using the GAN-EVA framework, which includes explicit conditioning on seasonal features and an extreme-value indicator derived via POT analysis. In contrast to the baseline GAN (Fig. 2.4), which exhibits severe mode collapse and restricts nearly all outputs to low consumption ranges (0–2,000 kWh), the GAN-EVA model demonstrates significantly enhanced diversity in its synthetic output. Notably, the GAN-EVA histogram captures:

- A broader spread of energy consumption values, including multiple clusters in mid- and high-kWh ranges,
- A more continuous tail, extending beyond 10,000 kWh and reaching as far as 50,000 kWh,
- Improved distributional overlap with the original data, particularly in the upper tail where critical extremes reside.

This result confirms that conditioning the GAN on EVA-derived features allows the Generator to learn not only the modal behavior of the system but also the statistical structure of rare events. The inclusion seasonality encodings likely provides the Generator with contextual information to allocate probability mass toward otherwise underrepresented tail regions. Compared to the baseline model, which underrepresents extremes and overproduces low-kWh samples, the GAN-EVA output achieves higher fidelity in both body and tail of the distribution. This validates the usefulness of EVA not only as an evaluation tool but also as an effective conditioning mechanism for guiding GAN training in domains where rare events carry operational significance. To further validate the GAN-EVA approach, we apply Extreme Value Analysis (EVA) to the synthetic data in order to assess how well the model captures, generalizes, and reproduces the statistical properties of extreme events present in the original dataset. This allows us to evaluate the GAN's ability not only to replicate typical consumption patterns, but also to approximate the tail behavior critical for modeling rare, high-impact scenarios. By comparing return levels, return periods, and the distributional structure of extremes between the real and synthetic datasets, we gain insights into the fidelity and limitations of the GAN-EVA synthesis, particularly with respect to its representation of rare but operationally significant outliers.

EVA Evaluation EVA on the original dataset shows extreme return levels ranging up to 5–50 M kWh for high return periods (e.g., 100–1000 years–Table 2.5). In contrast, the synthetic data's return values peak at 400 kWh, following a similar exponential curve shape but failing to capture the true scale of tail risks. This highlights a key finding: GANs capture general structure but under-represent extremes, making EVA an effective benchmark for quality assessment in synthetic data.

Figure 2.6 presents a time-series overlay comparing the original and GAN-EVA-generated energy consumption values across the observed period (July 2021 to May 2022). The plot illustrates the temporal alignment and magnitude similarity between real-world extremes (orange) and synthetic outputs (blue).

Notably, the GAN-EVA model reproduces several seasonal trends and temporal spikes, particularly in:

- Late spring and early summer 2022, where high consumption values are present in both real and synthetic series,
- July–September 2021, where multiple mid-scale spikes appear in both timelines.

While the synthetic series does not match the exact magnitude of the most extreme peaks (e.g., >50,000 kWh in early July), **it does successfully reconstruct the general pattern and clustering behavior of energy surges**. The alignment of peaks–albeit with lower intensity–suggests that the EVA-based conditioning enables the

Table 2.5 Comparison of return values for real and GAN-EVA synthetic data

Return period	Real data (kWh)	Synthetic (kWh)
1	205,626	40,807
10	5,900,000	96,196
100	170,000,000	198,194
1000	4,900,000,000	386,025

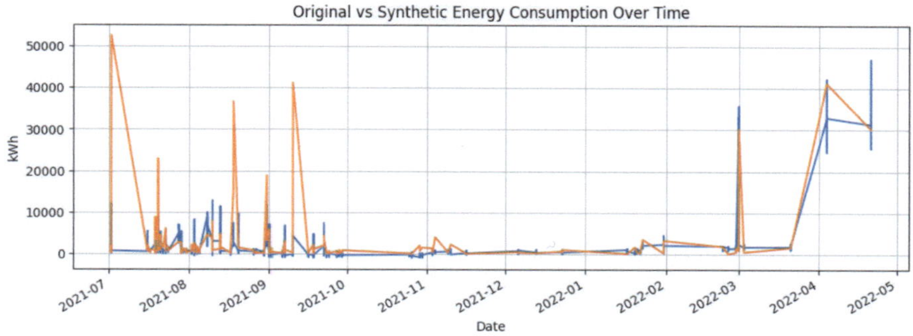

Fig. 2.6 Original versus synthetic energy consumption over time using POT detection based on EVA principles

GAN to recognize periods prone to high demand, even if it remains conservative in amplitude reproduction.

2.5 Conclusions and Future Works

This chapter proposed a general framework for EVA-integrated synthetic data generation, using GANs trained with seasonal and extreme-aware conditioning, and evaluated using classical extreme value theory. The use case of DEI Blue EV charging data illustrated how GANs can reproduce temporal energy patterns and seasonality, while EVA analysis revealed their limitations in reproducing the true magnitude of rare events. This approach provides a blueprint for combining deep learning with statistical extremes, and offers a novel benchmarking method for synthetic data quality. It is particularly relevant for applications involving risk modeling, capacity planning, and resilience testing, where rare events matter most.

In more detail, the GAN-EVA-generated synthetic dataset generated by our proposed framework replicates the seasonal energy demand patterns observed in the real data, with peaks around summer and spring months. Visual inspection shows high alignment between real and synthetic distributions when plotted over time. This finding aligns with EVA-based return level analysis, which showed that the GAN-EVA model underestimates the absolute return values but effectively preserves the exponential structure and seasonality of extremes. Therefore, while further refinement may be required to capture extreme magnitudes, the model demonstrates strong potential in learning when extremes occur, offering value for simulation, scenario planning, and seasonal stress testing.

Abbreviations

The following abbreviations are used in this paper:

EVA	Extreme Value Analysis
GAN	Generative Adversarial Network
cGAN	conditional Generative Adversarial Network
EVT	Extreme Value Theory
EV	Electric Vehicle
AMS	Annual Maxima or Minima Series
POT	Peak-Over-Threshold
GPD	Generalized Pareto Value
RV	Return Value
Q-Q	Quantile-Quantile
AC	Air Conditioning
PPC-DEI	Public Power Corporation

Acknowledgements The authors would like to thank DEI blue for providing the data used in this study and for their continuous support and valuable help.

References

1. Paris agreement, in *Report of the Conference of the Parties to the United Nations Framework Convention on Climate Change* (21st Session, 2015: Paris). Retrieved December, Vol. 4 (HeinOnline, 2015), p. 2017
2. L.M. Fulton, Three revolutions in urban passenger travel. Joule **2**(4), 575–578 (2018)
3. P. Skaloumpakas, E. Spiliotis, E. Sarmas, A. Lekidis, G. Stravodimos, D. Sarigiannis, I. Makarouni, V. Marinakis, J. Psarras, A multi-criteria approach for optimizing the placement of electric vehicle charging stations in highways. Energies **15**(24), 9445 (2022)
4. Intergovernmental panel on climate change. AR6 synthesis report: climate change, in: The IPCC finalized the Synthesis Report for the Sixth Assessment Report during the Panel's 58th Session held in Interlaken, Switzerland from 13–19 March 2023., https://www.ipcc.ch/report/sixth-assessment-report-cycle/
5. D.P. Panagoulias, E. Sarmas, V. Marinakis, M. Virvou, G.A. Tsihrintzis, H. Doukas, Intelligent decision support for energy management: A methodology for tailored explainability of artificial intelligence analytics. Electronics **12**(21), 4430 (2023)
6. G.A. Tsihrintzis, E. Sarmas, V. Marinakis, D.P. Panagoulias, E.-A. Tsichrintzi, M. Virvou, Energy urban domain: Personalized evaluation of expert and non-expert stakeholder interaction with artificial intelligence through chatgpt using the virtsi model. IEEE Access (2025)
7. E. Sarmas, D.P. Panagoulias, G.A. Tsihrintzis, V. Marinakis, H. Doukas, An explainable ai-based framework for supporting decisions in energy management, in *Machine Learning Applications for Intelligent Energy Management: Invited Chapters from Experts on the Energy Field* (Springer, 2024), pp. 1–27
8. J. Wu, S. Powell, Y. Xu, R. Rajagopal, M.C. Gonzalez, Planning charging stations for 2050 to support flexible electric vehicle demand considering individual mobility patterns. Cell Rep. Sustain. **1**, 15 (2024). https://doi.org/10.1016/j.crsus.2023.100006

9. D.P. Panagoulias, E. Sarmas, V. Marinakis, G.A. Tsihrintzis, Circumstance evaluation using extreme value analysis on charging station data: The case of DEI Blue in greece, in *Extended Selected Papers of the 14th International Conference on Information, Intelligence, Systems, and Applications, IISA2023* (Springer, 2024), pp. 121–134
10. DEI blue, https://www.deiblue.gr/
11. H. Doukas, V. Marinakis, E. Sarmas, Machine learning applications for intelligent energy management
12. S. Coles, J. Bawa, L. Trenner, P. Dorazio, An introduction to statistical modeling of extreme values, vol. 208 (Springer, 2001)
13. R.A. Fisher, L.H.C. Tippett, Limiting forms of the frequency distribution of the largest and smallest member of a sample. Proc. Camb. Philos. Soc. **24**(2), 180–190 (1928)
14. B.V. Gnedenko, Sur la distribution limite du terme maximum d'une serie aleatoire [on the limiting distribution(s) of the maximum value of a random series]. Annals of Mathematics (in French) **44**(3), 423–453 (1943)
15. M.R. Leadbetter, On a basis for peaks over threshold modeling. Statist. Probab. Lett. **12**(4), 357–362 (1991)
16. J. Pickands, Statistical inference using extreme order statistics. Ann. Stat. **3**, 119–131 (1975)
17. P. Westerlund, W. Naim, Extreme value analysis of power system data, in *ITISE 2019-International Conference on Time Series and Forecasting*, 25–27 September 2019 Granada (Spain), Vol. 1, 2019, pp. 322–327
18. H. Chen, T. Zhao, Modeling power loss during blackouts in China using non-stationary generalized extreme value distribution. Energy **195**, 117044 (2020)
19. Y. Li, B. Jones, The use of extreme value theory for forecasting long-term substation maximum electricity demand. IEEE Trans. Power Syst. **35**(1), 128–139 (2020)
20. D.P. Panagoulias, D.N. Sotiropoulos, G.A. Tsihrintzis, Extreme value analysis for dietary intake based on weight class., in *2022 13th International Conference on Information, Intelligence, Systems & Applications (IISA)* (IEEE, 2022), pp. 1–7
21. D.P. Panagoulias, D.N. Sotiropoulos, G.A. Tsihrintzis, An extreme value analysis-based systemic approach in healthcare information systems: The case of dietary intake. Electronics **12**(1), 204 (2022)
22. D.P. Panagoulias, G.A. Tsihrintzis, M. Virvou, Extreme value analysis applied in dietary data, in *Artificial Intelligence-Empowered Bio-medical Applications* (Springer, 2025), pp. 83–111
23. I. Goodfellow, J. Pouget-Abadie, M. Mirza, B. Xu, D. Warde-Farley, S. Ozair, A. Courville, Y. Bengio, Generative adversarial nets, in *Advances in Neural Information Processing Systems* (2014), pp. 2672–2680
24. T. Karras, S. Laine, T. Aila, A style-based generator architecture for generative adversarial networks, in *Proceedings of the IEEE/CVF Conference on Computer Vision and Pattern Recognition* (2019), pp. 4401–4410
25. M. Frid-Adar, E. Klang, M. Amitai, J. Goldberger, H. Greenspan, Gan-based synthetic medical image augmentation for increased cnn performance in liver lesion classification. Neurocomputing **321**, 321–331 (2018)
26. J. Yoon, D. Jarrett, M. van der Schaar, Time-series generative adversarial networks, in *Advances in Neural Information Processing Systems*, Vol. 32 (2019)
27. M. Wiese, R. Knobloch, R. Korn, P. Kretschmer, Quant gans: Deep generation of financial time series. Quantitative Finance **20**(9), 1419–1440 (2020)
28. X. Chen, M. Wang, S. Wang, Y. Chen, R. Wang, C. Zhao, X. Hu, Weather radar nowcasting for extreme precipitation prediction based on the temporal and spatial generative adversarial network. Atmosphere **13**(8), 1291 (2022)
29. R.K. Mazumder, G. Modanwal, Y. Li, Synthetic data generation using generative adversarial network for burst failure risk analysis of oil and gas pipelines. ASCE-ASME J. Risk Uncertainty Eng. Syst. Part B: Mech. Eng. **9**(3), 031103 (2023)
30. E. Sarmas, E. Spiliotis, V. Marinakis, G. Tzanes, J.K. Kaldellis, H. Doukas, Ml-based energy management of water pumping systems for the application of peak shaving in small-scale islands. Sustain. Cities Soc. **82**, 103873 (2022)

Chapter 3
Energy Poverty Meets Disability: A Critical Analysis of Financial Support Gaps in Europe

Ioannis Papagiannopoulos, Eleni Kanellou, and Vaggelis Marinakis

Abstract The intersection of disability and energy poverty remains underexplored, despite the disproportionate impact on vulnerable groups. People with disabilities may have specific needs—such as energy-intensive assistive devices, home temperature regulation, and extended periods spent indoors—that make them particularly vulnerable to energy poverty. While a number of financial assistance programs aim to support individuals with disabilities, they are often not coupled with targeted measures to mitigate energy poverty. This study sheds light on the gap between financial policies supporting people with disabilities and those addressing energy poverty relief. Drawing on document review and comparative policy analysis, the study examines the extent to which national frameworks consider the combined challenges of disability and energy poverty. It reviews financial support policies in five EU member states—Greece, Cyprus, Italy, France, and Spain—that face high levels of energy poverty. The findings indicate that most existing policies fail to integrate disability-specific considerations. To ensure equitable access to energy for the most vulnerable households in Europe, inclusive, coordinated, and disability-aware policy solutions are urgently needed.

Keywords Financial measures · Policies · Disabilities · Mapping policy gaps · Energy poverty

---

I. Papagiannopoulos (✉) · E. Kanellou · V. Marinakis
Decision Support Systems Laboratory, School of Electrical and Computer Engineering, National Technical University of Athens, Athens, Greece
e-mail: ipapagiannopoulos@epu.ntua.gr

E. Kanellou
e-mail: ekanellou@epu.ntua.gr

V. Marinakis
e-mail: vmarinakis@epu.ntua.gr

G. A. Tsihrintzis et al. (eds.), *Human-Smart City Interactions and User-Citizen Experiences*, Learning and Analytics in Intelligent Systems 58,
https://doi.org/10.1007/978-3-032-06364-9_3

3.1 Introduction

Energy poverty is a multidimensional issue affecting many households across Europe, with extensive implications for economic stability, social inclusion, and public health [1]. In recent years, the scale of energy poverty has expanded rapidly, with millions of Europeans struggling to keep their homes adequately warm, particularly among low-income households [2]. This problem typically arises from a combination of low income, high energy expenses, and energy-inefficient homes. However, not all members of the population face the same difficulties brought on by energy poverty. Some experience additional challenges due to their particular energy requirements. For example, people with disabilities may have high energy needs, which may be related to longer periods of time spent at home, increased heating or cooling requirements, or energy requirements for the operation of medical equipment [3].

These needs place additional financial strain on households that may already be vulnerable due to lower income levels and barriers to employment. Studies have highlighted that disabled individuals are disproportionately affected by energy poverty, with many unable to afford the energy services necessary for their health and well-being [4–6]. This results in a vicious cycle in which insufficient energy access leads to episodes of energy poverty and worsened health. Although these problems are recognized, current financial policies addressing energy poverty frequently disregard the special needs of disabled people, which are vital to ensuring access to necessary services and enhancing their overall well-being. In addition to being a matter of equity, meeting the unique needs of people with disabilities is crucial to ending the cycle of energy poverty and guaranteeing that everyone has equal access to the resources required for a dignified and healthy life. The purpose of this study is to examine policy gaps, with a particular emphasis on financial mechanisms that tackle the twin problems of energy poverty and disability. By mapping current policies and identifying where they fail to address the overlap of these issues, the paper intends to propose targeted solutions to improve support for disabled households facing energy poverty.

There are four sections to this paper. In Sect. 3.1, the topic of energy poverty in the European context is introduced, pointing out how it is related to people with disabilities, and stressing the importance of financial support systems. With an emphasis on energy-related vulnerabilities and the function of financial aid, Sect. 3.2 investigates the relationship between energy poverty and disability. By mapping national-level programs in Greece, Cyprus, Italy, France, and Spain and critically assessing disability-specific measures, Sect. 3.3 provides a thorough analysis of current financial support policies. Additionally, it highlights important policy gaps where assistance is lacking. Finally, Sect. 3.4 discusses the results through a comparative analysis of national practices, highlighting good practices, shortcomings in support for disabled households, and proposing recommendations for more inclusive and equitable financial policy development in the energy sector.

3.2 Addressing Energy Poverty and Disability Through Financial Support: A Conceptual Framework

3.2.1 Understanding Energy Poverty in the European Context

The uncertainties of the energy market have brought the issue of energy poverty to the forefront, establishing it as a critical European policy priority [7]. This situation is the result of recent crises, including the COVID-19 pandemic, the sharp rise in energy prices, and the Russian invasion of Ukraine in February 2022, all of which have deepened the challenges faced by many Europeans [4, 8]. Recent analyses from 2024 indicate that between 8 and 16% of EU citizens—approximately 35 to 72 million people—experience energy poverty [9]. The European Commission (EC) [10] has defined energy poverty as a set of conditions where 'individuals or households are not able to adequately heat or cool, or provide other required energy services in their homes at an affordable cost'. Energy poverty in the EU is measured through a range of indicators, reflecting the complex and multi-dimensional nature of the phenomenon. Current approaches include expenditure-based indicators, such as the 2M (double the median = high energy burden) and M/2 (half the median = low energy consumption) measures, and consensual, self-reported indicators like the inability to keep homes adequately warm (AW) and arrears on utility bills (UB). A recent JRC (Joint Research Centre) study, using EU-SILC (European Union Statistics on Income and Living Conditions) and HBS (Household Budget Survey) data refined with EUROMOD (European Union Tax-Benefit Microsimulation Model) [11], revealed significant disparities between these indicators, with minimal overlap among them For instance, countries like Greece and Bulgaria report that nearly 30% of their populations are energy poor according to at least two indicators. These indicators include AW and UB, reflecting both self-reported housing conditions and financial constraints. On the other hand, rates below 5% are frequently reported by Western and Northern EU nations [12].

However, rates as low as 0.3% can be obtained using a "intersection approach," which calls for people to fulfil all requirements at once [8]. This glaring discrepancy shows how challenging it is to create a common set of metrics for evaluating and addressing energy poverty in Europe. This complexity is also reflected in the significant variation in energy poverty levels across the EU. Furthermore, the use of subjective markers exacerbates this discrepancy. With AW-poverty rates ranging from about 0% in Sweden and Luxembourg to almost 40% in Bulgaria, for instance, it is clear how reporting trends and national circumstances have a big impact on results [13]. Energy poverty in Europe is a distinct issue that arises from the combination of low income, high energy bills, and energy inefficient homes, even though it frequently coexists with economic poverty, it is mainly a social issue. Its distribution is influenced by gender, housing circumstances, urban–rural disparities, and sociodemographic traits, underscoring the need for comprehensive, cross-sectoral policy initiatives. To effectively combat energy poverty and ensure a fair and inclusive energy transformation for all, coordinated efforts across the energy, social, housing, and health sectors are required [11, 14].

3.2.2 Disability and Energy-Related Vulnerabilities

Meeting fundamental human needs like mobility, communication, education, food security, and health depends on having access to energy. Individuals with disabilities encounter additional obstacles and are especially susceptible to energy poverty, underscoring the pressing necessity for inclusive energy policies and solutions [3]. Their vulnerability is exacerbated by increased energy needs due to mobility challenges, medical conditions, and the use of assistive technologies, as well as lower average household incomes. According to the ASSERT project [15], an initiative supported by the European Union's LIFE Programme. Disabled people are disproportionately affected by energy poverty because of obstacles to obtaining energy-efficient homes, a lack of funds, and the inaccessibility of support systems and public information.

Beyond economic hardship, people with disabilities often face structural and environmental barriers that amplify their energy needs [16]. Inadequately heated or cooled living environments have been shown to cause severe impacts on physical health, particularly on respiratory, circulatory, and cardiovascular systems, while also affecting mental health and overall well-being [17]. The consequences extend beyond individual households to affect broader social and economic outcomes, including development and cohesion [18]. Although many low-income households are energy poor, disability introduces specific needs and barriers not captured by income measures [5]. People with disabilities also face obstacles to employment and education, resulting in lower income and greater reliance on social protection schemes [5]. Additionally, disparities within and between Member States—along urban–rural, socio-demographic, gender, and housing lines—contribute to unequal access [19].

Low incomes, high energy expenses, and inefficient housing are the main causes of energy poverty in Europe. For people with disabilities, these factors are compounded by the lack of inclusive energy solutions [20]. Public policies often fail to recognize the specific energy needs of disabled households. Recent case studies from Greece show how geographic and technical challenges can further limit access to sustainable energy, especially for the most vulnerable [14]. Addressing energy poverty among people with disabilities requires improving energy efficiency, providing accessible information, offering targeted financial support, and designing inclusive services [21]. Policies must take a cross-sectoral approach, integrating energy, health, housing, and social systems, and recognize that disability results from the interaction between personal impairments and external constraints. Ensuring that the green transition includes people with disabilities is essential to achieving a fair and sustainable energy future [22].

Financial limitations are a key factor preventing energy access for disabled individuals. A promising initiative in Greece involves a data-driven decision support tool developed for energy providers to help identify and support at-risk households. These tools enable better targeting and resource allocation [23]. Strong financial support systems tailored to the realities of disabled people are essential to bridge the gap between current energy policy and real-world energy inclusion.

3.2.3 Financial Support Mechanisms: Definitions and Importance

Programs for financial aid are essential for lowering energy poverty, especially for disadvantaged populations like those with impairments. These programs aim to lower the cost of energy use or make energy-efficient dwellings and technology more accessible. They include direct subsidies, grants, refunds, preferential rates, and targeted social transfers. The special needs of households with disabled members necessitate customized, targeted approaches, even though universal schemes guarantee broad coverage and lessen administrative burden. These households frequently require adapted infrastructure, assistive technologies, and higher baseline energy usage for health and mobility needs [21, 24]. At the national and international levels, the significance of such unique methods is becoming more widely acknowledged. While the European Union's energy poverty goals highlight the need to safeguard vulnerable consumers and guarantee equitable access to the advantages of the green transition, the United Nations Sustainable Development Goal 7 asks for universal access to affordable, dependable, sustainable, and contemporary energy. In this regard, financial assistance must be planned to promote long-term gains in energy independence, resilience, and well-being in addition to easing current difficulties [25].

Integrating financial aid within a larger socio-policy framework that is inclusive, grounded in research, and sensitive to the lived reality of energy-poor communities is essential for effective implementation. Particularly for people with disabilities, financial support must go beyond economic relief to address embedded inequalities in housing, infrastructure, and access to services. A multidimensional and participatory strategy, secured in cross-sectoral collaboration can ensure that support mechanisms are not only technically sound, but also socially just and sustainable [26].

3.3 Review of Existing Policies and Identified Gaps

3.3.1 Financial Support Schemes for Energy Poverty in Europe

European countries have implemented a diverse range of financial support mechanisms to address energy poverty, from universal subsidies to targeted assistance for specific vulnerable groups. Understanding this policy landscape helps identify both strengths and structural gaps [27]. These frameworks differ across member states, shaped by national priorities and EU-level guidelines such as the Climate, Energy and Environmental State Aid Guidelines (CEEAG) [28].

While some schemes focus on general affordability, like social tariffs or heating allowances, others, such as capacity mechanisms, are designed to ensure system reliability and often do not consider vulnerable groups [29]. In countries like Finland, Sweden, Germany, and Austria, capacity mechanisms prioritize conventional infrastructure and typically exclude renewables due to their intermittency. This limits long-term sustainability and inclusiveness. Similarly, Portugal and Poland favor conventional systems with minimal social targeting, often neglecting vulnerable consumers such as people with disabilities.

Renewable energy support tools like feed-in premiums and green certificates prioritize decarbonization and market efficiency. However, these are often inaccessible to low-income or disabled households due to high upfront costs or administrative barriers [30]. The design of policies varies across countries—some integrate support into housing policy, others through utility billing or welfare systems—resulting in fragmented and inconsistent access, especially for those facing multiple vulnerabilities [31]. Despite having frameworks to address energy poverty, few countries systematically consider disability-specific needs in their energy support policies. This reflects a broader need for inclusive policy reform across the EU [32].

Findings are summarized in five country-specific tables (Tables 3.1, 3.2, 3.3, 3.4 and 3.5), outlining key characteristics of each support scheme. Although disability-related elements are not explicitly featured, this overview provides an empirical foundation for assessing how inclusive current energy poverty policies are—especially in relation to the needs of people with disabilities, as explored in the next section.

3.3.2 Disability-Specific Financial Measures: An Overview

People with disabilities face a heightened risk of energy poverty due to increased energy needs, such as reliance on medical equipment and climate control, and limited financial resources stemming from lower employment rates and higher healthcare costs. Yet few European programs recognize this link or offer targeted financial

Table 3.1 Overview of energy poverty financial support schemes in Greece [33]

Policy	Description	Beneficiaries
Social tariff scheme—Mitigating the impact of the crisis in the field of energy	Adoption of a preferential electricity sales price for households experiencing energy poverty	Households affected by energy poverty, as determined by the I&IIeq Index and specific income and consumption conditions
Energy upgrading of residential buildings of affected households; and promoting the installation of RES (Renewable Energy Sources) plants	Subsidy for building envelope upgrades, efficient systems, smart devices, and RES for self-consumption	Households affected by energy poverty under Directive (EU) 2023/1791
'Recycling heat switches'	EUR 100 million programme supporting the replacement of water heaters. Beneficiaries: 103,310	Households affected by energy poverty under Directive (EU) 2023/1791
Incentivising existing mechanisms—Just Transition Areas	Measures to increase energy savings units and promote installations in poor-performing residential buildings	Same as above (households affected by energy poverty)
Incentivising existing mechanisms—Energy Efficiency Obligation Scheme	Measures for RES energy use within citizen and renewable energy communities (e.g., "Apollo" program)	Citizen communities and renewable energy communities
Building on the Renewable Energy Communities and Communities Citizens	Implementation of public or blended finance programmes for households in energy poverty	Citizen communities and renewable energy communities
Information and education in the context of Energy Efficiency Obligation Views	Obligated parties provide technical advisors and training to affected households	Households affected by energy poverty under Directive (EU) 2023/1791
Targeted information and education actions	Targeted actions to support households and professionals implementing technical measures	Households affected by energy poverty under Directive (EU) 2023/1791

support addressing their specific circumstances. To ensure equitable access to affordable energy, legislative frameworks must be strengthened to reflect these realities [37]. This section reviews national financial assistance programs for people with disabilities that can impact energy affordability. These include mobility allowances, housing adaptation grants, income supplements, and support for medical energy needs. However, most are not designed with energy use in mind, revealing a lack of integration between disability support and energy policy.

Table 3.2 Overview of energy poverty financial support schemes in Cyprus [34]

Policy	Description	Beneficiaries
Special Public Utility Tariff	Inclusion of vulnerable consumer categories in a special reduced electricity charge arrangement (code 08), based on a Ministerial Decision on Public Utility Obligations	Vulnerable consumers include recipients of welfare benefits, disability or care allowances, and large families with three or more children. Income must not exceed €51,258 annually, plus €5,126 for each child beyond the fourth
Provision of Financial Incentives	Incentives for installing a residential photovoltaic (PV) system of up to 3 kW using net metering (offsetting produced and consumed electricity)	Eligible groups include Minimum Guaranteed Income recipients and large families with three or more dependent children. Income must be up to €51,258, plus €5,126 for each child beyond four
Provision of Financial Incentives—For energy upgrading of homes	Vulnerable groups may receive incentives for energy-saving home upgrades, subject to Ministry of Energy decisions and Managing Authority approval	a. Minimum Guaranteed Income. b. Large or Five-Member Families. **Income criterion**: Up to €51,258, increased by €5,126 for each child beyond four
Installation of PV net-metering systems in homes of vulnerable consumers	Financial support of €900 per installed kW (up to €2,700 total) for installing PV net metering systems in homes of vulnerable consumers	a. Minimum Guaranteed Income. b. Large or Five-Member Families. **Income criterion**: Same as above (€51,258 + €5,126 for each child beyond four)
Measures to protect energy-vulnerable consumers	Electricity disconnection is prevented or reversed for vulnerable consumers with serious health conditions	a. Minimum Guaranteed Income. b. Large or Five-Member Families. **Income criterion**: €51,258 annually, increased by €5,126 for each child beyond four

To capture the current landscape and its limitations, the findings are summarized in five country-specific tables (Tables 3.6, 3.7, 3.8, 3.9 and 3.10), each outlining key financial measures for people with disabilities. These include the type of support, eligibility criteria, and potential impact on energy affordability and independent living. This overview provides a basis for identifying policy gaps and assessing how current disability supports align with energy equity goals.

Table 3.3 Overview of energy poverty financial support schemes in Italy [16]

Policy	Description	Beneficiaries
Social allowances and measures to reduce energy bills	Provision of discounts on electricity and natural gas bills for economically vulnerable households, based on the ISEE (Equivalent Economic Situation Indicator) income criterion	Households with ISEE = (Household Income + 20% × Household Assets)/ Equivalence Scale below: • €9,530 (standard) • €20,000 for families with at least 4 dependent children • €12,000 (2022) or €15,000 (2023), up to €30,000 for large families Coverage: ~30% of annual electricity expenditure and 15% for gas
Zero System Charges for Electricity	Elimination of system charges on electricity bills to reduce costs for all consumers	All electricity consumers
Reduced VAT on Natural Gas	Lowered VAT rates on natural gas supplies to reduce the burden on households	All natural gas consumers
Installment Plans for Energy Bills	Allows households to pay energy bills in installments, easing financial strain during high energy price periods	Households are struggling to pay utility bills
Ecobonus (Energy Renovation Tax Deductions)	Tax deductions for energy efficiency improvements (e.g. insulation, heating systems, RES installations)	Households are investing in energy renovations, incl. low-income and social housing tenants
National Energy Efficiency Fund	Preferential financing for energy efficiency projects in residential and public buildings	Public/private entities incl. low-income households in social housing
Conto Termico Mechanism	Incentives for renewable heating and cooling systems in public and residential buildings	Public institutions and private households incl. those in social housing
Energy Tutors	Personalized guidance and advisory services to help households implement energy efficiency	Low-income households or those at risk of energy poverty
Subsidized Energy Tariffs	Direct energy price reductions for households with very low incomes	Households in the first two quantities of equivalized income
Expanded Eligibility for Bonuses	Temporary increases in ISEE thresholds to expand access to social bonuses	Households with ISEE below: • €12,000 (2022) • €15,000 (2023) • €30,000 for large families

Table 3.4 Overview of energy poverty financial support schemes in France [35]

Policy	Description	Beneficiaries
Energy Vouchers	Small households receive state aid for energy bills or renovations, with vouchers from €48 to €277 based on income and size, plus a €100–€200 exceptional voucher in 2022–2023	Granted automatically based on the Reference Tax Income per Consumption Unit (RFR/UC). Households in the bottom 40% of income distribution qualify
MaPrimeRénov' (MPR)	Grant scheme for energy renovation (insulation, heating system upgrades, ventilation, etc.). Enhanced grants cover up to 90% of renovation costs for the most vulnerable	Small and modest-income households. Income determines eligibility. Higher ceilings for very modest households
Tariff Shields	Caps energy price increases (e.g. electricity increases limited to 10–15% in 2022–2023)	Universal application, with strong impact on low-income households
Winter Truce and Minimum Electricity Supply	Disconnections of electricity or gas are prohibited in winter, with a 60-day minimum supply period for vulnerable users since April 2023	Households are eligible for energy vouchers and the Housing Solidarity Fund

3.3.3 Mapping Policy Gaps: Where Financial Support Falls Short

Despite recent advances in policy efforts targeting energy poverty and broader social protection, a critical gap persists at the intersection of these domains particularly for individuals with disabilities. While financial aid schemes exist within both sectors, they are often developed and implemented in isolation, lacking integration or coordination that would allow them to effectively respond to overlapping vulnerabilities. As a result, people with disabilities frequently fall through the cracks of support systems, facing disproportionately high energy needs without adequate or tailored financial relief [43]. Eligibility criteria for energy support are often income-based, overlooking additional energy costs related to disability, such as medical equipment and climate control. While disability programs may support independent living or mobility, they rarely address the energy burden. Likewise, energy poverty measures seldom include specific provisions for households with disabilities, despite their heightened vulnerability [44]. When energy and disability policy frameworks do not work together, the result is disjointed, unfair support that ignores the increased risks faced by people with disabilities. In order for financial mechanisms to treat energy

Table 3.5 Overview of energy poverty financial support schemes for Spain [36]

Policy	Description	Beneficiaries
Social Electric Bonds (BSE)	Discounts on electricity bills for households identified as vulnerable. ~17% annual bill reduction for vulnerable households and up to 28% for severely vulnerable households	Low-income households. Based on income thresholds and local vulnerability assessments
Vital Minimum Supply	Ensures uninterrupted electricity supply for six months under the precautionary principle, regardless of unpaid bills	Vulnerable households under Royal Decree-Law 17/2021, especially those facing financial hardship
Energy Renovation Programs	Funding for building renovations, reducing primary energy consumption by ≥30%. Includes: Residential area rehabilitation • PREE Programme • PREE5000 Demographic Challenge Programme	Vulnerable or low-income households in urban/rural areas affected by energy poverty
National Energy Efficiency Fund (EENF)	Financial aid and loans for individuals and communities to invest in energy-saving measures	Vulnerable households, low-income communities, and SMEs. Higher aid for energy-poor households
Temporary Measures for Vulnerable Consumers	Includes temporary electricity and thermal social bonds for immediate financial relief	Low-income and energy-poor households. Groups defined under Spain's National Energy Poverty Strategy
Social Climate Fund Measures	Structural and temporary support to reduce energy demand and decarbonize heating/cooling	Vulnerable households with limited access to energy-efficient technology
Energy Performance Renovation Funding	Support for upgrading household appliances and energy infrastructure. Includes partnerships with financial institutions	Homeowners and tenants in vulnerable or energy-inefficient housing

poverty and disability as intersecting aspects of social exclusion that call for inclusive, coordinated solutions, rather than just as separate problems, this disconnect must be addressed [45].

Table 3.6 Disability-specific financial measures related to energy in Greece [38]

Policy	Description	Beneficiaries
Mobility allowance	Provided to individuals with severe mobility disabilities (insured or uninsured)	Eligibility requires severe disability (e.g., ≥80% or limb loss), legal residence in Greece (or equivalent status), no institutional stay over 3M
Nutritional allowance for kidney patients, heart and liver transplants, etc.	Granted to individuals with serious chronic illnesses or transplant recipients, to meet increased nutritional needs	Eligible individuals include those with end-stage renal failure, organ transplants, or on transplant waiting lists, with legal residence in Greece and no similar benefit received
Financial Support Programs for People with Severe Disabilities	Financial aid supports individuals with disabilities, end-stage renal disease, or insulin-dependent diabetes	Disabilities ≥67%0.2. End-stage renal failure ≥80%0.3. Type 1 diabetes (insulin-dependent) ≥50%
Financial Support Program for people with severe intellectual disability	For individuals assessed by a health committee with specific mental or developmental disabilities	Severe/profound mental retardation ≥80%, IQ ≤34.2. Down syndrome with ≥80% disability.3. Autism spectrum disorders ≥80%. Conditions: uninsured/indirectly insured. No other pension >360€ or >399.54€ (2024)
Financial Support Program for paraplegic, quadriplegic and amputees	Assistance for insured/ uninsured individuals with paraplegia, quadriplegia or amputation. Includes public sector and military personnel	1. Medical: ≥67% disability due to paraplegia, quadriplegia, or equivalent/ amputation.2. Residence: Legal permanent residence in Greece, EU/EEA, or Convention countries.3. Conditions: No equal/ greater benefit elsewhere
Financial Support Program for people with congenital hemolytic anemia or bleeding disorders, AIDS	Includes Greek/EU/EEA citizens, refugees, and recognized third-country nationals with relevant permits	(a) Mediterranean anemia, thalassemia, sickle cell anemia (≥67%). (b) Hemophilia A/B, bleeding disorders (≥50%). (c) HIV/ AIDS or antibody immunodeficiency (≥50%)
Financial Support Program for Deaf and Hard of Hearing People	Support for individuals who are deaf or hard of hearing and evaluated by a health committee	(a) Children and elderly (≥67%). (b) Adults 18–65 with deafness and comorbidity. (c) Youth (18–25) in education (≥67%). Applies to uninsured, insured, and retirees

Table 3.7 Disability-specific financial measures related to energy in Cyprus [39]

Policy	Description	Beneficiaries
Severe Motor Disability Allowance Scheme	Financial assistance to cover needs of people with severe motor disability who cannot walk and permanently use a wheelchair	Permanent residents using wheelchairs due to severe mobility impairments
Care Allowance Scheme for Paraplegic & Quadriplegic Persons	Financial support for care services regardless of income or age. Aims to support independence and dignity. €400–€1100 depending on condition and need	Paraplegics/quadriplegics needing a wheelchair or daily assistance
Mobility Allowance	€75–€150/month for mobility support based on disability type and employment/education status	Mobility or vision impaired persons who work or study
Mobility Allowance Scheme for Patients under Regular Dialysis Treatment	€75/month (municipality) or €150/month (community council)	ESRD patients on regular dialysis at public
Allowance for the Provision of a Car to Persons with Disabilities	Financial aid of €3,500–€4,500. Also includes exemption from car registration and road tax	Adults (18–70) with severe limb or vision disabilities under disability schemes
Scheme for Financial Assistance for the Purchase of a Wheelchair	Financial support ranging from €700 to €13,500 depending on wheelchair type	Persons unable to walk or needing assistive devices for short distances
Financial Assistance Scheme for Technical Means, Instruments, and Other Aids	Promotes independence and participation of persons with disabilities through funding for appropriate aids	Cyprus/EU citizens ($\geq$12 months), refugees, with severe disabilities

3.4 Results and Discussion

3.4.1 Identified Gaps in Financial Support for Disabled Households

This section explores the intersection of energy poverty—a condition where households cannot afford adequate energy services—and disability-specific needs. Intersectionality is used as a framework to understand how overlapping social vulnerabilities (such as low income and disability) intensify exclusion from energy access.

While general financial schemes aim to mitigate energy poverty broadly, they often overlook disability-specific challenges. This section focuses on where and how these general frameworks fail to support disabled individuals effectively.

Table 3.8 Disability-specific financial measures related to energy in Italy [40]

Policy	Description	Beneficiaries
"Assegno di inclusione" ("Inclusion Allowance")	Economic support is offered for social and professional inclusion, based on residency, citizenship, and ISEE income criteria. Beneficiaries must join a personalised inclusion program	Support is provided to people with disabilities, minors, seniors (60+)
M5C2 (Component C2—"Social Infrastructures, Family, Community and Third Sector")	Promotes independent living and social inclusion of people with disabilities through accessible housing, personalised care, and community services	Disabled individuals needing support for independent living; families benefiting from support and accessible services
Article 8	Economic support for individuals with disabilities and their families for socio-psycho-pedagogical services, home assistance, and domestic help	Disabled individuals and their families
Article 31	Capital grants to municipalities/cooperatives for housing adaptations for people with disabilities or limited mobility	Individuals with limited mobility; public entities adapting housing
Legge 68/99 (Right to Work for Disabled Persons)	Covers employment incentives, workplace accommodations, and hiring benefits for disabled persons	Disabled jobseekers and employers hiring disabled workers
Legge 118/71 (Civil Invalids Benefits)	Provides pensions, allowances, and health benefits for individuals with civil disabilities	Persons with recognized civil disability
Accompanying allowance for the absolutely blind	Monthly benefit (~€938) for blind individuals unable to walk without assistance or perform daily tasks	Absolutely blind persons who apply, regardless of age or income

Despite existing national energy poverty schemes, most fail to consider the unique needs of people with disabilities. Disability is rarely included as a criterion, and programs focus mainly on income, overlooking higher energy demands from medical devices, climate control, or extended home presence. Thus, a large number of energy-stressed disabled households are left out. None of the schemes under review provide direct assistance for energy-intensive medical equipment or take utility and health costs into account. Inaccessible application procedures also erect additional

Table 3.9 Disability-specific financial measures related to energy in France [41]

Policy	Description	Beneficiaries
AAH—Allowance for Adults with Disabilities	Provides a minimum income (up to €631.63/month) for adults with disabilities, adjusted based on income. Means-tested	Individuals aged 20+ (or 16 if not eligible for family benefits) with ≥80% disability, or 50–79% with major employment restrictions
APA—Personal Independence Benefit	Aims at elderly people losing autonomy, supporting home care or specialized facility accommodation	Individuals aged 60+ with assessed loss of autonomy; aid varies by dependency level
Housing Assistance	Includes: Priority social housing access for disabled persons/dependents; financial aid for home adaptations (e.g., ramps)	Individuals with disabilities or dependents needing accessible housing or adaptation support
Independent Living Supplement	Monthly support (€104.77) for housing and independent living	Persons with ≥80% disability, living independently, receiving housing benefit
ASH—Social Housing and Accommodation Support	Financial aid covering up to 90% of accommodation in care facilities or residences	Individuals with ≥80% permanent disability or major employment restriction. Minimum €271/month retained for personal use
Tax Benefits	Income tax and VAT reductions for disability-related expenses and assistive devices/services	Persons with disabilities qualifying for tax benefits or reduced VAT
Disability Pension	For those with reduced work capacity due to non-occupational illness/injury. Monthly benefit based on salary and disability category	Persons with ≥66% disability and contribution history (≥600 h in past year)
PCRTP—Third-Party Assistance Supplement	Monthly supplement (€563.17–€1,689.58) for those needing third-party help with daily activities	Persons with ≥80% permanent disability and assessed need for daily assistance

obstacles. Disability is addressed in social policy but mainly disregarded in energy assistance frameworks, exposing a systemic gap.

The majority of energy poverty initiatives are created at the federal level, usually by the finance, social affairs, or energy ministries. While the operational design may involve local authorities or welfare agencies in charge of implementation, eligibility criteria are frequently established based on household income thresholds. Nevertheless, there is still a lack of interministerial coordination and little involvement from disability advocacy organizations in program design. As a result, when designing

Table 3.10 Disability-specific financial measures related to energy in Spain [42]

Policy	Description	Beneficiaries
Royal Decree 4/99 (January 8, 1999)	Subsidy increased for hiring workers with disabilities (approx. €3,900)	Employers who hire persons with disabilities; aimed at encouraging labor market inclusion
Legislative Royal Decree 1/2013, of November 29, 2013	Consolidated General Law on Rights of Persons with Disabilities and their Social Inclusion. Ensures rights to equality, accessibility, and non-discrimination	Persons with disabilities protected under this law, benefiting from social inclusion rights and accessibility provisions
Reserve Quota (2%)	Companies with >50 employees must reserve 2% of jobs for persons with disabilities	Companies fulfilling quota may receive benefits (e.g. tax reliefs, subsidies); supports employment for people with disabilities
Funding for Supported Employment	Employment promotion through European funds and foundation donations, even in absence of specific national incentives	Individuals and programs involved in supported employment schemes
Subsidised Employment	Employers hiring disabled individuals may receive substantial subsidies to offset hiring costs	Disabled individuals gaining employment; employers incentivized for inclusive hiring

and implementing policies, the unique needs of disabled populations are usually disregarded.

Beyond financial mechanisms, legal and institutional frameworks, as well as cultural perceptions of disability, play critical roles in shaping access to energy-related support.

Moreover, disability is not a monolithic experience individuals may have physical, sensory, or cognitive disabilities, each with different implications for energy use. For example, some require life-sustaining medical equipment, others need consistent indoor temperatures, and many face barriers related to communication or mobility. These diverse needs must be taken into account when designing support systems.

3.4.2 Towards Inclusive Financial Support: Recommendations

Although none of the five reviewed countries currently offer integrated financial schemes that address both energy poverty and disability, several national programs offer promising entry points for inclusive reform. In France, the *Prestation de*

Compensation du Handicap (PCH) and the *Independent Living Supplement* provide personalized support to people with high care needs—an infrastructure that could be expanded to cover energy costs [46].

In Greece, OPEKA's condition-specific allowances present a replicable model of targeted support that could be extended to include energy-related expenses with minimal structural change. Cyprus also delivers financial assistance through mobility and care allowances, which could be adapted to reflect the energy dimensions of disability [46, 47].

Italy's Law 104/1992 provides support for home modifications and assistive technologies but remains disconnected from energy affordability efforts. However, it offers a legal framework that could be used to support energy-intensive medical or mobility devices [48].

Spain promotes employment among people with disabilities through hiring incentives and quotas, but it fails to recognize the link between disability, home energy use, and remote work needs. Overall, there is a fragmentation between disability and energy policies. Despite acknowledging complex vulnerability, energy poverty strategies often exclude disability from formal definitions. Yet people with disabilities consistently face higher energy needs and are overrepresented in households lacking adequate heating or cooling. The share of Europeans without access to heat rose from 6.9% in 2021 to 10.6% in 2023, especially in Eastern and Southern Europe. Integration may involve coordinated policy planning, inclusive eligibility criteria, or fully joint programs addressing both energy access and disability support. One promising solution is the integration of renewable energy technologies, such as solar photovoltaic (PV) systems, which can lower energy costs and improve energy security for disabled individuals. Smart home energy management systems can also optimize energy use, improving comfort, autonomy, and affordability. Additionally, the installation of smart home energy management systems can optimize energy use, guaranteeing that the unique needs of residents with disabilities are effectively satisfied. These systems support comfort and autonomy while also lowering costs, which is consistent with the larger objectives of inclusive energy policies.

Addressing this issue requires financial frameworks that are inclusive and evidence-based. Disability-related energy needs must be reflected in eligibility criteria, and automatic enrollment based on disability certification could improve access. Application processes should follow universal accessibility standards. Cross-sector coordination, linking energy support to health, housing, and welfare, is also essential. In conclusion, energy poverty cannot be addressed through generic measures alone. Recognizing disability as a structural driver of vulnerability is essential to designing financial support systems that meet the real needs of Europe's most energy-dependent citizens.

Acknowledgements This analysis is conducted in the context of the ASSERT project, co-funded by the EU LIFE Programme, which aims to strengthen inclusive approaches to energy poverty across Europe, with particular attention to people with disabilities.

References

1. E. Kanellou et al., Lessons learnt and policy implications from implementing the POWER-POOR approach to alleviate energy poverty. Sustainability **15**(11), 8854 (2023). https://doi.org/10.3390/SU15118854
2. EPAH Energy Poverty Advisory Hub Handbook 1: A Guide to Energy Poverty Diagnosis | Energy Poverty Advisory Hub. https://energy-poverty.ec.europa.eu/observatory/publications/epah-energy-poverty-advisory-hub-handbook-1-guide-energy-poverty-0. Accessed 23 May 2025
3. R. Gillard, C. Snell, M. Bevan, Advancing an energy justice perspective of fuel poverty: household vulnerability and domestic retrofit policy in the United Kingdom. Energy Res. Soc. Sci. **29**, 53–61 (2017). https://doi.org/10.1016/j.erss.2017.05.012
4. V. Leopatra, View of climate change, water, sanitation and energy insecurity: invisibility of people with disabilities. Can. J. Disab. Stud. https://cjds.uwaterloo.ca/index.php/cjds/article/view/58/79. Accessed 23 May 2025
5. D. Ivanova, L. Middlemiss, Characterizing the energy use of disabled people in the European Union towards inclusion in the energy transition. Nat. Energy **6**(12), 1188–1197 (2021). https://doi.org/10.1038/s41560-021-00932-4
6. M. Pinilla-Roncancio, Multidimensional measures and the extra costs of disability: how are they related? Int. J. Environ. Res. Public Health **20**(3), 2729 (2023). https://doi.org/10.3390/ijerph20032729
7. S. Bouzarovski, H. Thomson, M. Cornelis, Transforming energy poverty policies in the European Union: second annual report of the European Union Energy Poverty Observatory. *ENACT The Energy Action Project* (2019). http://www.coldathome.today/exposed. Accessed 16 May 2025
8. Joint Research Centre, Who's energy poor in the EU? It's more complex than it seems - European Commission. Joint Research Centre. https://joint-research-centre.ec.europa.eu/jrc-news-and-updates/whos-energy-poor-eu-its-more-complex-it-seems-2024-09-25_en. Accessed 16 May 2025
9. C. Marine, What is the status of energy poverty in the European Union? | Heinrich Böll Stiftung | Brussels office—European Union. *Heinrich Böll Stiftung*. https://eu.boell.org/en/2024/11/20/what-status-energy-poverty-european-union?utm_source=chatgpt.com. Accessed 16 May 2025
10. S. Pye, A. Dobbins, C. Baffert, J. Brajkovic, Energy poverty and vulnerable consumers in the energy sector across the EU: analysis of policies and measures. *Energy Institute Hrvoje Požar* (2015). https://www.researchgate.net/publication/282358589_Energy_poverty_and_vulnerable_consumers_in_the_energy_sector_across_the_EU_analysis_of_policies_and_measures. Accessed 16 May 2025
11. Eurostat, Household budget survey—Microdata—Eurostat. *Eurostat*. https://ec.europa.eu/eurostat/web/microdata/household-budget-survey. Accessed 16 May 2025
12. European Commission, HOME | EUROMOD—Tax-benefit microsimulation model for the European Union. *EUROMOD*. https://euromod-web.jrc.ec.europa.eu/. Accessed 16 May 2025
13. G. Notten, C. de Neubourg, Relative or absolute poverty in the US and EU? The battle of the rates. SSRN Electron. J. (2007). https://doi.org/10.2139/ssrn.1095712
14. A. Varo, G. Jiglau, K. Grossmann, R. Guyet, Addressing energy poverty through technological and governance innovation. Energy Sustain. Soc. **12**(1), 1–19 (2022). https://doi.org/10.1186/s13705-022-00377-x
15. European Commission, ASSERT—Large-scale multi-actor training and mentorship programme to assist people with physical impairments in energy poverty | Energy Poverty Advisory Hub. *Energy Poverty Advisory Hub*. https://energy-poverty.ec.europa.eu/discover-community/epah-atlas/assert-large-scale-multi-actor-training-and-mentorship-programme. Accessed 16 May 2025

16. C. Friedman, Unsafe temperatures, going without necessities, and unpayable bills: energy insecurity of people with disabilities in the United States during the COVID-19 pandemic. Energy Res. Soc. Sci. **92**, 102806 (2022). https://doi.org/10.1016/j.erss.2022.102806
17. A.M. Mosadeghrad, P. Isfahani, L. Eslambolchi, M. Zahmatkesh, M. Afshari, Strategies to strengthen a climate-resilient health system: a scoping review. Glob. Health **19**(1), 1–11 (2023). https://doi.org/10.1186/s12992-023-00965-2
18. P. Howden-Chapman, J. Crane, M. Baker, H. Viggers, R. Chapman, C. Cunningham, Health, Well-Being and Housing, in *International Encyclopedia of Housing and Home*, pp. 344–354 (2012). https://doi.org/10.1016/B978-0-08-047163-1.00021-7
19. A. Arsenopoulos, V. Marinakis, K. Koasidis, A. Stavrakaki, J. Psarras, Assessing resilience to energy poverty in Europe through a multi-criteria analysis framework. Sustainability **12**(12), 4899 (2020). https://doi.org/10.3390/su12124899
20. A. Boeri, V. Gianfrate, S.O.M. Boulanger, M. Massari, Future design approaches for energy poverty: users profiling and services for no-vulnerable condition. Energies **13**(8), 2115 (2020). https://doi.org/10.3390/en13082115
21. R.B. Jackson et al., Human well-being and per capita energy use. Ecosphere **13**(4), e3978 (2022). https://doi.org/10.1002/ecs2.3978
22. E. Sarmas, P. Skaloumpakas, N. Kafetzis, E. Spiliotis, A. Lekidis, V. Marinakis, H. Doukas, Optimal site selection of electric vehicle charging stations exploiting multi-criteria decision analysis: the case of Greek municipalities. Tech. Ann. **1**(2) (2023). https://doi.org/10.12681/ta.33632
23. Renewable energy policies in a time of transition. *Australia Policy Online*. https://apo.org.au/node/216701. Accessed 16 May 2025
24. S.P. Jakhanwal, Urban development and exclusion of the poor: an integrated perspective. Soc. Change **44**(4), 605–614 (2014). https://doi.org/10.1177/0049085714548545
25. D.F. Barnes, W.M. Floor, Rural energy in developing countries: a challenge for economic development. Annu. Rev. Energy Env. **21**(1), 497–530 (1996). https://doi.org/10.1146/annurev.energy.21.1.497
26. European Commission, "CEEAG," *European Commission*. https://ec.europa.eu/commission/presscorner/detail/en/qanda_22_566. Accessed 23 May 2025
27. M. Kojima, C. Trimble, Making power affordable for africa and viable for its utilities. World Bank (2016). https://doi.org/10.1596/25091
28. R. King, M. Orloff, T. Virsilas, T. Pande, Confronting the urban housing crisis in the global south: adequate, secure, and affordable housing. *Cities for All*. www.citiesforall.org. Accessed 16 May 2025
29. L.M. Banks, C. Davey, T. Shakespeare, H. Kuper, Disability-inclusive responses to COVID-19: lessons learnt from research on social protection in low- and middle-income countries. World Dev. **137**, 105178 (2021). https://doi.org/10.1016/J.WORLDDEV.2020.105178
30. Hellenic Republic Ministry of Environment and Energy, "TABLE OF CONTENTS." [Unpublished report]
31. European Commission, Commission Recommendation, Assessment (SWD) and Factsheet of the draft updated National Energy and Climate Plan of Cyprus—European Commission. *European Commission*. https://commission.europa.eu/publications/commission-recommendation-assessment-swd-and-factsheet-draft-updated-national-energy-and-climate-10_en. Accessed 16 May 2025
32. Directorate-General for Communication, Italy—Draft Updated NECP 2021–2030—European Commission. *European Commission*. https://commission.europa.eu/publications/italy-draft-updated-necp-2021-2030_en. Accessed 16 May 2025
33. Directorate-General for Communication, France—Final updated NECP 2021–2030 (submitted in 2024)—European Commission, *European Commission*. https://commission.europa.eu/publications/france-final-updated-necp-2021-2030-submitted-2024_en. Accessed 16 May 2025
34. Directorate-General for Communication, Spain—Final updated NECP 2021–2030 (submitted 2024)—European Commission, *European Commission*. https://commission.europa.eu/publications/spain-final-updated-necp-2021-2030-submitted-2024_en. Accessed 16 May 2025

35. Organisation of Welfare Benefits and Social Solidarity (OPEKA), Organisation of Welfare Benefits and Social Solidarity (OPEKA), *OPEKA*. https://opeka.gr/. Accessed 16 May 2025
36. Republic of Cyprus Deputy Ministry of Social Welfare, Severe Motor Disability Allowance Scheme. [Unpublished online information]
37. Ministry of Labour and Social Policies, Inclusion Allowance | Ministry of Labour and Social Policies. *Ministry of Labour and Social Policies*. https://www.lavoro.gov.it/g7-labour/new-measures-inclusion-and-work-access/pagine/inclusion-allowance. Accessed 16 May 2025
38. F.B. Valdes, Social protection for people with disabilities
39. M.A. Verdugo, A. Jiménez, F. Borja, J. De Urríes, Social and employment policies for people with disabilities in Spain
40. T. Lancet, Prioritising disability in universal health coverage. The Lancet **394**(10194), 187 (2019). https://doi.org/10.1016/S0140-6736(19)31638-1
41. L. Puli, N. Layton, D. Bell, A.Z. Shahriar, Financial inclusion for people with disability: a scoping review, Glob. Health Action **17**(1) (2024). https://doi.org/10.1080/16549716.2024.2342634
42. E. Kanellou et al., Lessons learnt and policy implications from implementing the POWERPOOR approach to alleviate energy poverty. Sustainability **15**(11), 8854 (2023). https://doi.org/10.3390/SU15118854
43. European Commission, Cyprus—European Commission. *European Commission*. https://employment-social-affairs.ec.europa.eu/policies-and-activities/moving-working-europe/eu-social-security-coordination/your-rights-country-country/cyprus_en. Accessed 16 May 2025
44. J. Lee, H. Kim, J. Byrne, Operationalising capability thinking in the assessment of energy poverty relief policies: moving from compensation-based to empowerment-focused policy strategies. J. Hum. Dev. Capabil. **22**(2), 292–315 (2021). https://doi.org/10.1080/19452829.2021.1887108
45. R. Spark, Accessibility to work from home for the disabled: the need for a shift in management style, in *Proceedings of the 14th Web for All Conference (W4A 2017)* (2017). https://doi.org/10.1145/3058555.3058577
46. E. Sarmas, V. Marinakis, H. Doukas, Machine Learning–driven energy consumption forecasting for building profiling, in *Artificial Intelligence for Energy Systems, Learning and Analytics in Intelligent Systems*, vol. 46 (Springer, 2025), pp. 129–147. https://link.springer.com/chapter/10.1007/978-3-031-85209-1_5. Accessed 25 May 25, 2025
47. E. Sarmas, V. Marinakis, H. Doukas, *Artificial Intelligence for Energy Systems: Driving Intelligent, Flexible and Optimal Energy Management* (Springer, 2025). https://link.springer.com/book/10.1007/978-3-031-85209-1. Accessed 25 May 2025
48. S. Bouzarovski, S. Petrova, A global perspective on domestic energy deprivation: overcoming the energy poverty–fuel poverty binary. Energy Res. Soc. Sci. **10**, 31–40 (2015). https://doi.org/10.1016/J.ERSS.2015.06.007

Part II
Data-Driven Governance

Chapter 4
Security of Personal Data Processing When Using Artificial Intelligence Systems in Smart Cities: Processes and Tools to Address GDPR Compliance Challenges

Stamatios Theocharis and George A. Tsihrintzis

Abstract In recent years, artificial intelligence has largely penetrated systems and machines that serve the everyday lives of people and businesses. These technologies are dramatically changing the way in which various innovative applications are developed in various fields such as e-banking, e-commerce, e-government, e-health, etc. as well as the way in which users interact with machines and systems. These systems are now provided by both the private and public sectors to citizens and businesses. However, important issues emerge in relation to the protection of personal data of individuals processed by these systems. In this paper, we examine the critical issues and concerns in two seemingly incompatible concepts, namely artificial intelligence systems and the protection of personal data during the use of these systems. We also propose specific procedures and tools to address the challenges that arise when organizations and developers who develop and use AI systems complying with the requirements and restrictions set by personal data protection legislation.

Keywords Smart cities · Personal data protection · AI · GDPR · Risk management

S. Theocharis (✉)
General Director for Transportation and Communication, Prefecture of Attica, Greece
e-mail: stheohar@webmail.unipi.gr

G. A. Tsihrintzis (✉)
Department of Informatics, University of Piraeus, Piraeus, Greece
e-mail: geoatsi@unipi.gr

G. A. Tsihrintzis et al. (eds.), *Human-Smart City Interactions and User-Citizen Experiences*, Learning and Analytics in Intelligent Systems 58,
https://doi.org/10.1007/978-3-032-06364-9_4

4.1 Introduction—Smart Cities and Smart Environment

The rapid development of innovative information systems and devices that use Artificial Intelligence Technologies, combined with the continuous evolution of the characteristics of Information and Communication Technologies (ICT), has a positive impact on the everyday lives of citizens and businesses. This evolution is occurring in both public and private sectors and is related to the transition from traditional electronic services to advanced smart services, mainly in the cases of e-government, e-commerce, e-banking and e-economy in general.

There is a significant impact of technological progress in ICT on the specific sector of Local Government, a sector that is closer to the everyday lives of citizens. In the last decade, in particular, special emphasis has been placed on the development of the so-called "smart" solutions, "smart applications" and "smart cities" to facilitate and improve the everyday lives of citizens and businesses. As a key example, we mention the case of the European Commission, which, participating in the consolidation of this trend, declares its intention to cooperate with smart cities and communities to address local challenges, provide better services to citizens and achieve the objectives of the European Green Deal [1]. Smart cities and communities are expected to use innovative and emerging ICT to become more effective. To achieve this goal, sensors are often used to collect data from which, after processing, information and knowledge are produced to improve the way the city operates.

By developing new innovative applications, a local government organization is expected to be able to, on the one hand, solve more effectively daily problems related to the basic services provided to citizens and, on the other hand, to offer innovative services with added value for the local economy. These applications aim at the personalized provision of services to citizens and businesses, the automation through the digitization of processes for the benefit of efficiency and speed of services, up to the improvement of transparency, accountability and sharing of information and knowledge with the citizens. For example, smart energy meters can contribute to the control of a building, in terms of efficient heating, for the benefit of energy savings. In the field of mobility, the data collected on transport can help us improve road capacity and reduce the impact on the environment. Smart cities also bring direct benefits to citizens, such as safer public spaces and support for the aging population. On the other hand, citizen participation is essential for creating innovative, smart and sustainable cities. Through digital platforms, citizens can directly communicate their needs, concerns and ideas. This strengthens the sense of community participation, leading to more responsive, effective and democratic digital public services.

As a result of technological evolution, a new dynamic of businesses and government organizations has taken place in recent years, making the provision of new services the focus of economic and administrative activity. In this context, new, more effective and faster applications are emerging that use artificial intelligence technologies, and in particular machine learning and genetic artificial intelligence technologies. Open data, linked open data and the information that can result from their free processing and sharing has been an important parameter towards achieving this goal.

In the innovative environment of smart cities and artificial intelligence technologies, it is now easier for a local government organization to solve common problems such as cleanliness, maintenance of urban equipment, reducing the cost of services provided to citizens, distributing students to schools, parking vehicles, even managing traffic and meeting any social needs. The use of artificial intelligence has begun to significantly impact various sectors of society, and local government organizations are no exception. The development of smart cities through artificial intelligence can provide solutions that improve the quality of life of citizens, increase the efficiency of services and promote sustainable development. However, to fully exploit the potential of AI, organizations must collect and manage data effectively and appropriately. The real problem focuses on the fact that organizations do not have the data sets they desire, while often the data they do have is collected and stored in such a way that it is transformed into simple digital garbage.

In this work, we examine how it is possible for two seemingly incompatible concepts, the development and use of trustworthy artificial intelligence systems and the protection of individuals' personal data, to coexist. We focus on the specific area of monitoring compliance with the basic principles and limitations set by the GDPR [2] with the proposal to adopt specific procedures and tools that can be used by scientists and institutions that develop and use artificial intelligence systems. We take into account both the benefits of the development and use of artificial intelligence systems and the challenges and opportunities for individuals and society, as well as the ways in which risks can be addressed and opportunities enabled through the institutional framework and developing technology. In this work, we first present relevant work in the field of interest, while then the main part is developed in three parts. The first part makes a thorough reference to the issues and challenges that emerge during the development and use of artificial intelligence systems regarding the security of the fundamental rights of individuals and in particular with personal data and the privacy of individuals. The following is a report on the proposed procedures and tools that can be used when checking compliance with the basic principles and limitations set by the GDPR in combination with the Regulation on artificial intelligence. In the next section, we make an assessment of the proposed measures by reporting the advantages and expected results from their adoption. Finally, we record the main conclusions while referring to our future work.

4.2 Related Work

The rapid development and use of systems that use artificial intelligence in cases related to serving the everyday lives of citizens, both by public sector bodies and by businesses, have highlighted as important the seemingly incompatible issue of ensuring the fundamental rights of people of the universal Declaration of Human Rights and in particular, the right to protect the privacy and personal data of the subjects. In this field, both the European Union and International Organizations and the scientific community have been involved in recent years, publishing studies and works concerning the recording of emerging issues as well as proposed solutions in order to bring together the seemingly conflicting fields. Indicatively, we mention below key relevant works.

We refer to the EU act on AI [3] to establish harmonized and horizontal rules on artificial intelligence for all Member States. The interest is already in the recitals of the act where, among other things, the risks and the need to take protective measures for the rights of individuals that must be taken into account by the operators and scientists developing AI systems, are pointed out. Also, our interest is focused on the consideration and classification of artificial intelligence applications in four levels according to the risk involved in their use, starting from the impermissible risk, where applications that pose a risk of this level should not be allowed to be used until the level of tolerable risk, where applications that pose such a risk will be used with some minimal safeguards in place.

The study in [4] addresses the relationship between the General Data Protection Regulation (GDPR) and artificial intelligence (AI). Authors, introducing some basic concepts of AI and also review the state of the art in AI technologies and focus on the application of AI to personal data. Authors also consider challenges and opportunities for individuals and society, and the ways in which risks can be countered and opportunities enabled through law and technology. The study then provides an analysis of how AI is regulated in the GDPR and examines the extent to which AI fits into the GDPR conceptual framework. It discusses the tensions and proximities between AI and data protection principles, such as, in particular, purpose limitation and data minimisation. It examines the legal bases for AI applications to personal data and considers duties of information concerning AI systems, especially those involving profiling and automated decision-making. It reviews data subjects' rights, such as the rights to access, erasure, portability and object. The study carries out a thorough analysis of automated decision-making, considering the extent to which automated decisions are admissible, the safeguard measures to be adopted, and whether data subjects have a right to individual explanations. It then addresses the extent to which the GDPR provides for a preventive risk-based approach, focusing on data protection by design and by default. The possibility to use AI for statistical purposes, in a way that is consistent with the GDPR, is also considered.

Authors in [5] mention that as Artificial intelligence is not a new concept for DPOs and data protection professionals instead of Generative AI. Nowadays, with ChatGPT and other familiar applications in the hands of over 100 m users globally, and many other providers such as Google Bard and Anthropic's Claude entering the market, it has become an operational reality, and necessity, for data protection professionals to deal with the consequences of generative AI tools being rapidly utilised within organisations. Whether these tools are adopted simpliciter or are fine-tuned by organisations using their own data sets, novel and as-yet unexamined data protection implications exist, all of which data protection professionals must rapidly come to terms with. So, the aim of this work is to guide data protection professionals through the maze of issues that are unfolding as these technologies gain rapid adoption in organisations. Amongst other key issues, authors mention the need of looking at data-sharing risks, accuracy of personal data, conducting DPIAs on generative AI tools, implementing data protection by design, selecting a lawful basis for training generative AI systems, optimising organisational structures, applying privacy-enhancing techniques, and handling data subject rights in the context of these technologies. Authors, refer that there will be no future without generative AI, and with data playing such a pivotal role in the training and operating of these systems, DPOs will play a central role in ensuring that both data protection and data governance standards are at the heart of these technologies.

The UK Information Commissioner's Office in [6] states that the use of AI has the potential to make a significant difference to society. However, there must be confidence that AI is being developed appropriately and lawfully, in order for individuals to realise the benefits. They suggest that, as the use of AI crosses regulatory remits, regulators should consider the lawful and appropriate use of personal data and AI systems, as well as the opportunities and risks involved. This approach will assist regulators when assessing the use of AI in the sector or sectors they regulate. It will also help regulators who use AI as part of their regulatory mission to do so appropriately and legally. This is achieved through specific advice given through this work, including a risk-based approach to the development of AI technologies, the explanations given to individuals about how these systems make decisions to how GDPR principles are applied in the development and use of artificial intelligence systems.

The General Secretariat of the Data Protection Authority of Belgium published an information brochure [7]. The brochure outlines the complex interplay between the General Data Protection Regulation and the Artificial Intelligence in the context of AI system development. Author emphasizes the importance of aligning AI systems processing personal data with data protection principles while addressing the unique challenges posed by AI technologies. Key points of the work include among other the GDPR and AI Act alignment, AI system definitions, the basics data protection principles and the crucial role of human oversight in AI system development and operation. By providing insights into the legal framework and practical guidance, this information brochure aims to empower legal professionals, data protection officers, technical stakeholders, including controllers and processors, to understand and comply with the GDPR and AI Act requirements when developing and deploying AI systems.

4.3 Data Collection

We indicate the main sources from which, in the smart city environment, large amounts of primary data can be collected, including citizens' personal data.

1. Sensors and Internet of Things Devices: Organizations can install sensors in various infrastructures, such as street lights, trash cans and water systems. These sensors collect real-time data on traffic, energy consumption, and other critical parameters.
2. Data from Cameras and Surveillance Systems: Security cameras can collect images and audio that can be analyzed through artificial intelligence algorithms to detect traffic problems, prevent crime, and improve public safety.
3. Social Networks and Citizen Applications: Information drawn from social networks and special applications developed to communicate with citizens can often provide data on citizens' needs and preferences.
4. Data from Public Services: Information collected from services, such as applications for permits, tax payments, and health records, can be used to analyze trends and predict needs.
5. Decisions of Administrative Authorities Legal acts of the Central Administration and the bodies of Local Government Organizations as well as their accompanying introductory documents constitute the primary stage and basic source of collection of processable data.

The data collected from the sources mentioned above are processed by the bodies in charge, in order to contribute to the production of useful information. The use of artificial intelligence technologies is expected to contribute to the more effective, more targeted and optimal proposal of solutions in specific sectors in the context of sustainability and economic development:

1. Traffic and Transportation Improvement: By analyzing traffic data and using predictive algorithms, municipalities can optimize the operation of traffic lights and consequently the regulation of flows on city streets with the aim of reducing traffic congestion and saving resources.
2. Waste Management: Data from smart trash cans can be used to optimize collection routes, reducing costs for local government organizations and therefore citizens as well as environmental impacts.
3. Energy Management: Smart meters and energy consumption data can help organizations improve the energy efficiency of buildings and promote the use of renewable energy sources.
4. Healthcare: Data analysis from healthcare services can improve health delivery, predict epidemics, and tailor services to the needs of the population.
5. Urban Planning: Using population and traffic data, organizations can plan infrastructure development, manage land, and improve urban areas.
6. Service Improvement Data analysis of processes and decisions leads to continuous improvement of service efficiency and reduction of time spent on citizen cases.

In order to make the most of the data collected, it is imperative that it is processed systematically and based on standardization. In order for the data collected from the above sources to be used in the best possible way and form the basis for the production of useful information that can be shared, it is desirable that they are converted into open data sets with the aim of developing vocabularies, open ontologies and linked open data.

In particular, the standardization of decisions made by algorithms using artificial intelligence is a critical factor for the effective and reliable operation of AI systems. And this is mainly for the following reasons:

1. Reliability and Consistency Standardization of decisions allows AI systems to make consistent decisions based on defined rules and patterns. This ensures that the system's performance remains stable regardless of the environment or conditions.
2. Transparency and Explainability Standardized models and algorithms help to understand how AI makes decisions. This is particularly important for explainable AI, as it makes it easier for users and regulators to understand.
3. Compatibility and Interoperability The use of standards helps to ensure that different AI systems can work together and exchange data without problems. This is critical for the development of broader AI ecosystems.
4. Security Standardized protocols and procedures help protect against malicious attacks and accidents. They ensure that decisions are made in a way that minimizes risks and potential errors.
5. Performance and Optimization Standardization helps optimize AI algorithms and processes, ensuring that the system operates at its best possible performance. It can be used to fine-tune parameters and improve decision accuracy.
6. Regulatory Compliance Standardized AI systems can ensure compliance with legal and ethical rules. This is particularly important in areas such as healthcare, financial management, and public safety.
7. Ease of Maintenance and Update Standardized systems are easier to maintain and update. Changes and improvements can be incorporated in a way that does not disrupt the overall operation of the system.
8. Improving User Experience The consistency and reliability provided by standardized AI systems improve the user experience, as they ensure that services are predictable and reliable.
9. Research and Scientific Value Standardization facilitates scientific research and knowledge exchange. Researchers can compare results and develop new techniques based on commonly accepted standards.

Overall, decision standardization in Artificial Intelligence is essential for the development of reliable, transparent, and secure AI systems. It allows for the optimization of processes and ensuring that systems operate according to expected rules and standards, while providing reliable and accurate services to users.

In order for organisations to prepare for the development and use of a safe, reliable and trustworthy AI system, specific safeguards must be ensured. AI, and in particular machine learning and genetic AI, pose major challenges for data protection, confidentiality and privacy. The implementation of specific procedures for data security and the maturity of technologies support citizens' trust. In particular, personal data protection measures taken by institutions are an effective way to gain this trust. Furthermore, the development of appropriate policies and regulations for AI is a declared priority for the European Union, as AI significantly increases the opportunities and, at the same time, the risks for the rights of individuals in ways that have significant social and legal implications. AI can enhance human capabilities, improve safety and efficiency, and enable the universal provision of knowledge and skills. However, artificial intelligence can also increase opportunities for control, manipulation, and discrimination. Social interactions can be disrupted and individuals exposed to various harms resulting from technological failures or disregard for individual rights and social values [1, 2].

There is already a debate on specific ethical and legal issues related to AI in various areas, such as civil liability, insurance, data protection, security, contracts and crime [2]. Such issues are gaining greater importance as more and more organizations are withdrawing conventional systems and replacing them with smart applications. A central role and common feature of the innovative services being developed is the collection of data, including personal data of citizens, whether simple or special categories. Usually, the processing of such data concerns a large scale and a wide range of categories of affected citizens. Thus, although the information provided by the data is crucial for the development and optimization of the services provided, an important issue is the safeguarding of people's fundamental rights, and in particular the privacy and security of personal data. It is therefore observed that both citizens and professionals who manage data are now more interested in emerging issues regarding the existing risks to the security and privacy of personal data and whether they can be eliminated or at least mitigated. These issues, however, also existed in the past when using traditional technologies, but are intensifying in the more complex technological environment at the center of which are AI technologies.

The emerging intense interest in the protection of personal data and the seemingly contradictory use of AI systems, especially in the case of European states, can be mainly attributed to the passage and implementation of the General Data Protection Regulation 2016/679 and the recently published Regulation 2024/1689 [1] for the establishment of harmonized rules on artificial intelligence as well as the discussion taking place at all levels of the EU and scientific bodies on the achievement of a trustworthy artificial intelligence and the identification of all the necessary measures to protect the fundamental rights of individuals.

4.4 Issues—Challenges for Securing Personal Data When Developing and Using AI Systems

As artificial intelligence systems enter the daily lives of citizens, businesses and organizations, important issues for rational, efficient and trustworthy artificial intelligence are regulated by international organizations such as the EU (Council and European Commission), OECD, UNESCO, the G7 et al. with relevant instructions or decisions. However, there are also issues such as the symbiosis of artificial intelligence and the protection of personal data and the privacy of individuals which for many are still considered open.

We remind that artificial intelligence is a technology that allows machines to reproduce cognitive functions of humans such as learning, planning and creativity [2]. Functions are based on data while machines perform these functions replacing human intelligence. Artificial intelligence has many applications and benefits, such as improving the daily lives of citizens, healthcare, education, transportation and security. However, artificial intelligence also poses challenges and risks to privacy and personal rights, as it often involves the collection, processing and analysis of large amounts of personal data, according to the European Personal Data Board (EPDB).

The use of personal data by artificial intelligence systems can have various implications for individuals' privacy, including:

- Intrusion: AI systems may collect and process personal data without individuals' knowledge or consent, or in ways that exceed their reasonable expectations. For example, AI systems may use facial recognition.
- Profiling: AI systems can use personal data to create profiles of individuals or groups and make predictions, recommendations or decisions about them. For example, AI systems may use personal data to assess individuals' creditworthiness, employability, health status or personality, or to target them with personalized advertising, offers or services.
- Discrimination: AI systems may use personal data to treat individuals or groups differently or unfairly, based on characteristics such as age, gender, race, ethnicity, religion or disability.

Specific issues related to the application of the principles and requirements of the GDPR during the development and use of IT systems are listed as follows.

- Data protection by design and by default.
  This requirement is described in Article 25 of the GDPR [2] and plays a key role in promoting the protection of individuals' privacy and data. It is therefore imperative that data controllers take this responsibility seriously and implement their obligations in relation to the GDPR when planning processing operations using AI systems. The basic obligation is the implementation of the appropriate measures and the necessary guarantees that provide effective implementation of the principles of data protection and, consequently, the rights and freedoms of the subjects already by design and by definition. Since controllers are required by law to consider this principle early, while planning a new processing operation, they should act accordingly when designing and developing IT systems by providing the relevant safeguards to satisfy this obligation.
- The principle of data minimization.
  It is described in Article 5 of the GDPR [2] and is the crux of the relationship—and apparent incompatibility—of personal data protection with artificial intelligence. The key issue relates to how the minimization principle can coexist with the

optimal functioning of an artificial intelligence system. However, it is a relationship of interdependence. As is commonly accepted [7], data is the driving force behind artificial intelligence. On the AI side, the more data that flows into an AI system, the more accurate results that system can produce. From the point of view of personal data protection, this principle advocates the broader principle of proportionality, which requires the limitation of the scope of the intervention in the privacy of the individual through the use of his data. In practice, this limitation can be achieved by not making it easy to identify the individual, by limiting the amount and nature of the subject's information—which is to feed the AI system. Especially through the use of anonymization or pseudonymization techniques, it becomes difficult to identify the subject and thus the degree of intervention is limited and—at the same time—it is allowed to retain the (not easily identifiable) data for long periods of time, thus satisfying both any operational needs, but also in compliance at the same time with another GDPR principle, that of limiting the storage period. By limiting the data that will be used by the AI we are in effect weakening it and perhaps inadvertently introducing some biases. This is because artificial intelligence learns from all the data it receives. If trained on a small database—which will only reflect a part of the population—then these AI systems will produce a biased and shallow view of the real situation. In this way, however, we notice that there is also a conflict with the principle of fair processing. AI systems need data so that they can deal with any biases. Data about color, ethnicity, gender and other sensitive data can help identify and correct biased AI.

- Discriminating Artificial Intelligence.
  In practice, the application of artificial intelligence has already created conflicting opinions as some believe that artificial intelligence may exacerbate social inequalities, affecting more the weak social groups, the weaker economically and generally bringing with it the prejudices that possess people without any control and limit. Although the purpose of artificial intelligence aims to contribute to the facilitation of human life, the negative effects are linked to the potentially incorrect use or not sufficiently protective of the underlying legislation. Algorithms potentially using seemingly unrelated personal data can autonomously create specific profiles based on social or other criteria and lead to rejection of requests or exclusion from benefit settings. In these cases, human intervention is considered important in order to achieve a substantial control of the automated decisions as well as the criteria used for the results of the algorithms.
- Principle of transparency—The case of the black box phenomenon.
  Often, both systems developed with conventional tools and those using artificial intelligence are characterized as black boxes, as it is impossible for ordinary users to understand the inner workings and decision-making process based on algorithms and any data correlations. This issue concerns even ICT scientists who do not have the expertise in the field of artificial intelligence and therefore even if they have access to the code they are not able to confirm the adherence of the algorithm to the principles of the GDPR. The issue that emerges is related to the incompatibility with the fundamental principle of transparency provided for

in the GDPR and is already of considerable concern to the scientific community and personal data professionals.

- Explanation of processing.
  In order to be able to explain the logic of processing according to GDPR principles, the computational model should by design and by definition be trained to use correct and relevant data, as well as to assess which data is most relevant to the individual purpose of processing. It should be able to deal with the following requests: 1. What were the specific factors that led to this decision? 2. If we changed any factor, would there be a modification of the decision? 3. What is the reason that led to 2 different approaches to similar cases? The essential issue relates to how to audit the algorithm and code developed by the developers so that the auditors are able to confirm the compliance obligation of the developers.
- Making a decision without human intervention—the right of a natural person to intervene.
  The specific right under Article 22 of the GDPR translates as the subject's right not to be subject to a decision made solely on the basis of automated processing, including profiling. Therefore, the data controller must implement appropriate measures protecting at least the right to ensure human intervention, to express an opinion and challenge the decision. In addition, AI increases the ability to monitor and analyze people's daily habits. For example, there is a potential risk that it could be used, in breach of EU data protection and other rules, by government authorities or other entities for mass surveillance and by employers to monitor how their employees behave. By analyzing large amounts of data and identifying links between them, AI can also be used to identify and de-anonymize data about individuals, creating new privacy risks, even in datasets that do not include personal data. The issue relates to the amount of autonomy an AI system is chosen to have while it is up to developers to implement it and document their compliance.
- Obtain consent—Withdraw consent of the subject.
  The very nature of the operation of artificial intelligence systems presents a significant challenge in obtaining the consent of data subjects. According to the GDPR, the processing of personal data, as long as there is no other legal basis for processing, is lawful only if and as long as the data subject has consented to the processing of his personal data for one or more specific purposes. Consent is obtained under specific conditions provided for in the legislation and which mainly concern the clear and transparent information of the subject. The key issue identified in the case of artificial intelligence is whether the controller can clearly and fully inform the data subject of the purposes of the processing, given that AI systems can largely proceed autonomously in creating "invisible" or implicit correlations between the data and in the creation of new data concerning the data subject. At the same time, IT systems continuously reuse data redefining the purposes of processing, making the challenge of properly informing the subject and obtaining consent even greater. Especially with respect to machine learning systems, the involvement of the Controller in the formation of new associations and therefore new data is minimal to non-existent, which reduces the value of consent given the lack of predictability of the AI system's outcome.

An important issue is also identified in the case where the data subject withdraws his consent. The withdrawal of consent does not have retroactive effect, however, it entails the obligation to completely delete the subject's data by the Controller. exercising this right, however, may pose a threat to the development of AI, which relies on data to train itself. In addition, there may be a distortion in the representativeness of the data set, which could lead to a "bad" training of the AI system and therefore a bias in its decision-making. For this reason, it is important to technically ensure that data deletion does not affect the representativeness of the data set. In any case, the complete deletion of the subject's data is not possible, since as mentioned above the data is reused constantly producing new data related in whole or in part to the subject.

- Are the legislative provisions for data protection sufficient?
In the modern era characterized by the development and use of innovative technologies and the so-called big data and data analysis, it would be correct to avoid on the one hand the so-called "techno-optimism" that uncritically accepts every innovation and on the other hand the "techno-pessimism" that opposes the adoption of technology as such and desires its destruction. Let's not forget that artificial intelligence and data analysis are products of the work of expert scientists and professionals. The effectiveness of the algorithms is related to the training methodology and is directly related to the quality of the training data. In fact, the ethics of the scientist are directly reflected in the "ethics" of the algorithm. Another important issue that arises is that of the violation of privacy and personal data as the increasing use of artificial intelligence tools undoubtedly leads to the processing of a huge amount of data. It is clear that the selection, storage, processing and collection of this data in the machines of artificial intelligence expands and inflates the risks highlighted by the use of traditional IT techniques.

On the part of the legislative authority both at the national, European and international level, for decades the conceptual content of "personal data" and measures to protect the fundamental rights of individuals have been defined and updated. It has been, over the years, a complex process for legislators and the respective agencies that has involved a constant change and renewal of the terminology of their protection terms within the legislative texts. A key feature of the legislation is the fact that in the issue of such great importance and risk of the use of artificial intelligence systems and the consequent processing of personal data, and indeed in the field of exercising public authority, according to many, it seems to be content with a rudimentary wording regarding the need to observe transparency and a even more rudimentary obligation to inform persons. At the same time, they include independent authorities for the monitoring and supervision of the implementation of the provisions without, however, proposing specific practical solutions for the aforementioned issues.

It is therefore clear that the regulatory proposals that are submitted and increasingly allow the use of artificial intelligence technologies in everyday applications in combination with the relatively limited conditions that are set as safeguards to ensure the protection of people and their personal data from it, they reveal the constantly observed tendency of postmodern societies to "give in" to the speed of technological development which increasingly undermines social cohesion with the constant new risks that it inherently poses, I often argue when these are not adequately demarcated legally.

4.5 Procedures and Useful Tools for Checking GDPR Compliance

The rapid development, spread of Artificial Intelligence technologies and their use in processes in which personal data processing is carried out, make the need to comply with the GDPR increasingly imperative. A key priority for ICT operators and scientists should be privacy issues, ensuring that appropriate personal data protection and security measures are in place.

Important in the direction of compliance is the cooperation of agencies and ICT developers and scientists with the regulatory authorities of privacy protection and their contribution throughout the process of development and use of new technologies, can help to address concerns and avoid delays or bans. In this context, the definition of a supervisory authority over trusted artificial intelligence is required. According to the EDPS, personal data supervisory authorities already have experience and expertise that can be used when dealing with the impact of IT on the fundamental rights of individuals, in particular the right to protection of personal data, and will therefore should be designated as market surveillance authorities in certain cases. This will ensure better coordination between the various regulatory authorities, enhance legal certainty for all stakeholders and enhance oversight and enforcement of both AI and EU data protection law [7].

The adoption of privacy by design protection measures is considered self-evident on the part of AI operators and developers, in addition to the observance of privacy by default protection measures, and a commitment to comply with the GDPR. The following have dominated as proposed protection measures:

- Development of control and security mechanisms in order to limit the possibility of incorrect decisions, mainly due to system failures, while the subject will feel safe that he is not a burden of automated systems.
- Development of an accountability mechanism, which means that persons developing automated systems must demonstrate that they have organized appropriate procedures to explain how algorithms produce specific results and propose corresponding solutions.
- Human control, both when defining the operating criteria, and later as a quality control of the decisions and criteria used, to ensure the actual smooth operation of the algorithm.

In order to strengthen the security of personal data for citizens, specific actions are required. For example, we mention the following:

- Risk assessment for personal data in the context of the use of emerging ICT and G-cloud. In particular, it is recommended to carry out early risk assessments and start with small processing and then expand its scope and specialization later.
- Establishing the necessity of carrying out a personal data impact assessment study and taking modern technical and organizational measures.
- Carrying out an algorithmic impact assessment according to the provisions of Article 5 of Law 4961/2022 [8]. These studies take into account, among others, the risks that may arise for the rights, freedoms and legal interests of the natural or legal persons concerned or affected by the decision-making by artificial intelligence and the expected benefit derived for society as a whole in relation to possible risks and effects that the use of the system may bring, in particular for racial, ethnic, social or age groups and categories of the population such as people with disabilities or chronic diseases.
- Appeal to obtaining certification of bodies for compliance according to articles 42, 43 of the GDPR and the relevant directives of the supervisory authorities.
- Addressing the Digital Divide. It is important for actors and developers to be aware of the issues associated with the development and control of artificial intelligence, while protecting fundamental rights when using it.
- Collaboration with scientific bodies on critical issues that were reported with the aim of seeking optimal solutions.
- Extensive data protection training, both for staff using the new innovative technologies within the agencies and for developers and ICT scientists who design and develop systems using artificial intelligence technologies.

In particular, specific technical measures are proposed to achieve the goal of protecting personal data already by design and by definition.

- Achieving a high level of security also during the processing of personal data (maintenance, storage, destruction, etc.).
- Integration of security methods at every level and in all systems and in particular methods of encryption, pseudonymization and anonymization of data.
- The configuration of the security architecture results from the determination:
  - Which system resource is at risk and needs to be protected?
  - The required level of protection for the resource at risk.
- Anticipation of dealing with the most important risks from which a P.S. must be secured.
- Dealing with viruses, worms, Trojans.
  - Internal risk—Attacks by employees
  - Destruction of important data due to employee negligence
  - External attacks on the system
  - Attacks that prevent the provision of services
- The integration of security in the information system should be based on a flexible security architecture, which will allow its redefinition and expansion, as security mechanisms can significantly increase the cost of the information system.

- Provision for protection mechanisms such as the use of anti-virus protection applications and regular backups. The protection of the system from internal attacks is achieved:
  - By organizing the architecture of the information system in access zones,
  - By securing the system from physical access,
  - By securing sensitive data from misappropriation or destruction,
  - Data encryption where possible against anonymization or pseudonymization,
  - By defining user groups and corresponding, appropriate access rights,
  - By defining effective password policies to protect system resources or use smart cards,
  - By defining effective policies for workstations (e.g. prohibiting the use of periperal storage units, use of network profiles, turning workstations into blind terminals, etc.),
  - By effectively managing and updating system security,
  - By observing and monitoring the event logs (logging) of the system.
- Provision for protecting the systems from external attacks by partitioning the P.S. in zones, as well as by integrating into the security infrastructure system, which will be constantly checked and updated.
- Provision for protection of "inside" risks, i.e. the internal users of the P.S. which are vulnerable through Trojan attacks etc.
- Provision for protection against an external Denial of Service (DOS) attack. Public bodies are targets of such attacks. For example, one solution is to achieve a level of presentation in the information system with equal distribution of resources. In addition, a good way is to distinguish external users according to their geographical origin and route them to different subsystems of the system or serve them from geographically distributed subsystems.

4.6 Expected Results

In the modern environment in which ICT solutions based on artificial intelligence systems have already been adopted, the taking of specific measures by the entities that use them in order to achieve compliance with the GDPR and, in general, the safe processing of personal data with these systems, have specific advantages while they expect specific results.

Significant help is expected for the agencies that use artificial intelligence systems to form themselves as data controllers a direct and objective judgment on the level of compliance they achieve in relation to what is defined in the legislation for the protection of personal data. This will also help them, in making the decisions for any update or revision of the policies they have adopted while more effectively preventing the loss of their data.

The measures we mentioned above still serve in the field of monitoring and internal control within the bodies, facilitating the internal auditors and the corresponding external control mechanisms through a check list, thus strengthening the control process.

From the point of view of the image towards the citizens, the agencies are expected, through appropriate communication with the citizens, on the one hand, to improve the reputation of the agency in the field of protecting the fundamental rights of individuals, as well as to improve the relationship between citizens and the organizations since the the latter feel more secure and satisfied with the level of services provided.

4.7 Conclusions—Future Work

In conclusion, the intended transparency towards individuals provided for in the GPDR appears to be the biggest challenge in terms of the collection and processing of personal data by artificial intelligence systems. Getting consent right, in particular, is a major gamble that seems to be hard won by Data Controllers, with the risk of affecting the proper training of AI systems.

The relationship between artificial intelligence and personal data is a relationship of interdependence, a feedback relationship, which, however, requires careful handling, flexible and dynamic legislative frameworks that will be able to monitor developments.

It is necessary to ensure subjects for their security and privacy while at the same time innovation and the development of new technologies will be promoted. The very recent act of the E.U. on artificial intelligence sends an auspicious signal for Europe's quick response to the new data, setting criteria based on the dangerousness of artificial intelligence. Organizations must prioritize privacy and work with regulators and data protection professionals to ensure transparency, address concerns and meet the strict requirements of the GDPR.

It is necessary to ensure subjects for their security and privacy while at the same time innovation and the development of new technologies will be promoted. The very recent act of the E.U. on artificial intelligence sends an auspicious signal for Europe's quick response to the new data, setting criteria based on the dangerousness of artificial intelligence. Organizations must prioritize privacy and work with regulators and data protection professionals to ensure transparency, address concerns and meet the strict requirements of the GDPR.

GDPR compliance not only mitigates legal risks, but also builds user trust and helps create a sustainable and privacy-focused AI ecosystem. In conclusion, T.N. and Personal Data are in a relationship of interaction that can lead humanity to leaps in progress or take on uncontrollable dimensions.

Our interest in the future is directed towards the development of a control system for the compliance of entities with the requirements for the protection of personal data, which will take into account the carried out and intended processing, the existing situations as well as the legal restrictions and will make an autonomous assessment risks related to the compliance objective and will recommend specific actions and actions to the agencies.

References

1. European Commission, Shaping Europe's digital future, retrieved by official site at https://digital-strategy.ec.europa.eu/el/policies/smart-cities-and-communities
2. European Union, Regulation (EU) 2016/679 of the European Parliament and the Council of 27 April 2016 on the protection of natural persons with regard to the processing of personal data and on the free movement of such data, and repealing Directive 95/46/EC (General Data Protection Regulation) (Text with EEA relevance), https://eur-lex.europa.eu/legal-content/EN/TXT/PDF/?uri=CELEX:32016R0679
3. European Union, Regulation (EU) 2024/1689 of the European Parliament and of the Council of 13 June 2024 laying down harmonised rules on artificial intelligence and amending Regulations (EC) No 300/2008, (EU) No 167/2013, (EU) No 168/2013, (EU) 2018/858, (EU) 2018/1139 and (EU) 2019/2144 and Directives 2014/90/EU, (EU) 2016/797 and (EU) 2020/1828 (Artificial Intelligence Act) (Text with EEA relevance), https://eur-lex.europa.eu/eli/reg/2024/1689/oj/eng, European Union
4. European Parliamentary Research Service, The impact of the General Data Protection Regulation (GDPR) on artificial intelligence. Scientific Foresight Unit (STOA), PE 641.530, June 2020, retrieved by https://www.europarl.europa.eu/RegData/etudes/STUD/2020/641530/EPRS_STU(2020)641530_EN.pdf
5. Confederation of European Data Protection Organisations AI Working Group, Generative AI: The Data Protection Implications, 16 October 2023, retrieved by https://cedpo.eu/generative-ai-the-data-protection-implications/
6. UK Information Commissioner's Office, How to use AI and personal data appropriately and lawfully, 8 November 2022, retrieved by https://ico.org.uk/media/for-organisations/documents/4022261/how-to-use-ai-and-personal-data.pdf
7. General Secretariat of Data Protection Authority of Belgium, Artificial Intelligence Systems and the GDPR A Data Protection Perspective, December 2024, retrieved by https://www.autoriteprotectiondonnees.be/publications/artificial-intelligence-systems-and-the-gdpr---a-data-protection-perspective.pdf
8. EDPB, Statement—Press Release on 17 July 2024, retrieved by https://www.edpb.europa.eu/news/news/2024/edpb-adopts-statement-dpas-role-ai-act-framework-eu-us-data-privacy-framework-faq_el

Chapter 5
AI in Citizen-Centric Smart Cities: Exploring Data Privacy, Algorithmic Transparency and Trustworthiness Through Regulation

Dimitris C. Gkikas and Marios C. Gkikas

Abstract This chapter examines how the development of Artificial Intelligence (AI) has a huge impact on smart cities by transforming urban governance, decision making, security, and public services. However, the continuous collection of citizen data raises essential ethical concerns, regarding the user and data privacy, user consent, data bias, and civilian surveillance aspects. This chapter attempts to explore the ethical challenges of AI adoption in smart cities and user experience, elaborating how governments handle user sensitive data while ensuring compliance with regulatory frameworks including EU AI Act and the General Data Protection Regulation (GDPR) while approaching the role of AI transparency, citizen evolvement, public trust in AI decision-making processes and trustworthiness. Furthermore, the chapter highlights ethical considerations, regulatory frameworks and AI initiatives which balance technological innovation with human rights, emphasizing transparency, accountability, and the importance of keeping smart cities human centric. To conclude, the chapter recommends policymakers and engineers with some of the best practices for responsible and transparent AI governance in smart cities.

Keywords Smart cities · Artificial Intelligence · Data privacy · Algorithmic transparency · Ethics · Regulation · Trust · GDPR · EU AI Act · Human-centric design

---

D. C. Gkikas (✉) · M. C. Gkikas
Department of Fisheries and Aquaculture, University of Patras, Messolonghi, Greece
e-mail: dgkikas@upatras.gr

M. C. Gkikas
e-mail: mario.gkikas@upatras.gr

G. A. Tsihrintzis et al. (eds.), *Human-Smart City Interactions and User-Citizen Experiences*, Learning and Analytics in Intelligent Systems 58,
https://doi.org/10.1007/978-3-032-06364-9_5

5.1 Introduction

Since 1950, when Alan Turing first questioned the ability of machines to think, Artificial Intelligence (AI) has shown exponential progress in human societies and economic transformation. AI aims to increase productivity revenues, improve life quality and respond to major global challenges, including climate change, energy consumption, natural resources and health crises. However, the adoption of AI leads to concerns regarding human values, privacy, trust, transparency, safety and accountability. As cities adopt AI decision making systems to enhance energy consumption, health services, resource management, and citizen involvement, the frame of smart cities evolves into a data-centric, autonomous and self-sustained system. Along with the benefits of optimized system efficiency and responsiveness, AI also generates a series of profound ethical and legal considerations regarding data privacy, algorithmic transparency, and human autonomy [63].

According to the OECD's group of AI Experts, called AIGO, an AI system is a "machine-based system that can, for a given set of human-defined objectives, make predictions, recommendations or decisions influencing real or virtual environments. It uses machine and/or human-based inputs to perceive real and/or virtual environments; abstract such perceptions into models (in an automated manner e.g. with ML or manually); and use model inference to formulate options for information or action. AI systems are designed to operate with varying levels of autonomy" (Fig. 5.1) [64].

One of the major pillars of the European Union's strategy is to promote "trustworthy AI" systems that are consistent with the legal, ethical, and robust societal values [23–28].

Additional initiatives, such as the recent AI Act regulation published by the European AI office, the European AI Alliance and the Ethics Guidelines for Trustworthy AI by the AI [1], preserving human equality, operations transparency and trustworthiness as foundational AI principles.

AI applications meet a rapid growth in several society sectors including the:

- Healthcare use AI decision making systems provide early warnings and predictions in disease prevention diagnoses and cure with optimized treatments, medicines and monitoring tools.

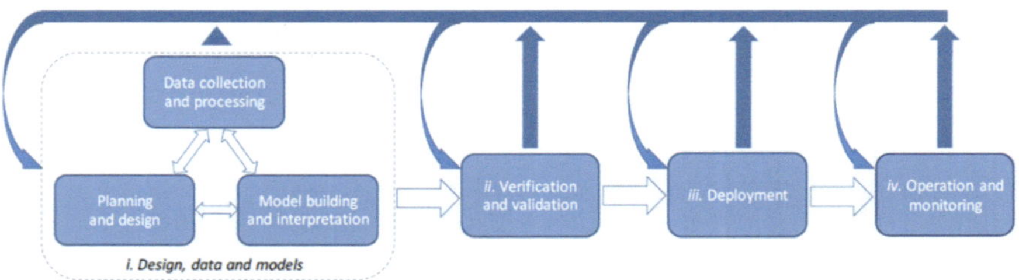

Fig. 5.1 AI system lifecycle

- Transportations leverage AI to fully autonomous vehicles use smart driving systems, optimised routing mitigating the environmental footprint.
- Finance AI is used in fraud detection, credits assessment, customer service costs management and trading automation.
- Justice where AI assists decision-making, case handling and legal research.
- Information systems security where AI supports real time threat detection.
- Agriculture where AI is used to monitor and predict crops, soil and aquaculture health and livestock considering environmental factors.
- Marketing leverage AI data analysis consumers behavior prediction, content curation, user personalization and advertising optimization [64].
- E-governance where AI systems provide high performance data management, e-governance services and optimized and personalized citizen-centric services.

In an AI citizen-centric emerging society the regulatory frameworks act as a counterbalance strategy to the technology-centric world. The adoption of the EU Artificial Intelligence Act sets a precedent for risk-based regulation of AI, with specific relevance for urban implementations [25, 72, 71].

UNESCO's report regarding the Ethics of Artificial Intelligence provides a fundamental ethical framework that promotes the foundations of AI, and it applies to the governance of AI, including smart cities. This report endorses a multidimensional, intercultural, and progressive approach to AI ethics, reinforcing human integrity, well-being, and the harm reduction as primary principles. These values are important in smart city ecosystems, where AI technologies affect governance of urban systems, essential services and facilities management, public service deployment, and citizen engagement. UNESCO's report advocates for ethical monitoring across the pillars of the AI system life cycle starting from research and deployment, oversighting, and use terminations reflects the complexity and interconnectivity of smart cities nature, where AI systems interpret huge volumes of data to generate decisions impacting sectors like transportation, energy, healthcare, education, and public safety. Nevertheless, digital ethical challenges are present more than ever and must be addressed with caution since they refer to algorithmic bias, unequal access to digital technology, surveillance, and the deterioration of privacy are increased in urban environments, requiring solid safety measures to ensure equity, accessibility, universal participation, transparency, trust and accountability [83].

This chapter explores the AI application in citizen-centric smart cities by examining data privacy, algorithmic transparency and trustworthiness through policy and regulatory frameworks. Aiming to assist the co-ordination and consistency of the AI regulations across the global societies, it navigates through the AI regulatory frameworks including the EU AI Act and the General Data Protection Regulation (GDPR), intersecting with ethical standards and concerns [27, 37, 60].

The aim is to explore how smart city technologies can be aligned with these ethical principles and citizen-centric values such as transparency, equity, participation, and trustworthiness, while accounting for international trends to enhance quality of life while safeguarding fundamental rights and democratic values [41, 55, 65, 83].

5.2 Related Works

As AI ethics in public contexts emerged as an interdisciplinary domain applying in computer science, urban development, AI regulation, philosophy, and public administration, it has evolved as an essential area of regulatory, technical, and academic point of reference in smart cities. Recent regulatory frameworks, including these from the European Commission, try to regulate Artificial Intelligence in public services highlight the significance of implementing trustworthy AI systems for public environments. Reports like these showcase the rapid adoption of AI in cities applications including social service optimization, surveillance and behaviors while considering the ethical issues between innovation, privacy, trust, and democratic principles. The need for firm data governance, transparency, and trustworthiness is imperative especially when AI affects public decisions [72].

The European Commission highlights, though the AI Watch reports, that the ethical issues emerging from AI development in e-governance are closely related to smart cities. Reports like this pinpoint the joint role of governments to balance the differences between citizens and regulators prioritizing transparency, equality, and trust. Towards this direction, there are series of key considerations regarding algorithmic obscurity, danger of biased decisions, and risks for citizen privacy through data collection and personalization [36].

Thus, it is important to build regulatory bodies and committees from all stakeholders which will act as the foundational guide for supervising the AI deployments according to human rights to ensure citizen-centric, transparent AI aligned with human values the report serves as a Misuraca and van Noordt [36].

Furthermore, the AI international standards implemented and processed by regulatory bodies like ISO, IEEE, and ETSI are aligned with the proposed EU AI Act's system requirements which also impact smart cities deployment. Also, there are ethical frameworks such as the ECPAIS and the P7000 series which are considered the cornerstones of AI systems implementation within urban ecosystems. Among states human-centric AI principles are imperative to be embedded into the legal, ethical and technical digital foundations of smart cities [52, 72, 16].

The OECD report responds to the need for AI systems trustworthiness, human-oriented, transparency, robustness, and accountability in close alignment with other counties' regulatory frameworks and principles including the EU AI Act and GDPR. OECD urges for inclusivity, global coordination regulations and certain ethical requirements to guarantee data protection in AI systems in e-governance, urban deployment, public services integrity and ensure democratic values. The findings upgrade the level of urgency which is required of adapting ethical safeguards into the design and implementation of smart cities infrastructures [63].

The OECD's Scoping the AI Principles [63, 64] report reveals a globally accepted framework for ethical and trustworthy AI systems, providing important guidance to the development of smart cities. Initially developed by the AIGO group of experts and adopted by the OECD Council, the essential principles in an AI lifecycle refer to citizen-centric values, operations transparency and accountability, procedure robustness, and decision fairness. Values like these are essential for urban AI systems which communicate with citizens, especially in key sectors like energy, safety, mobility and public e-services. AI lifecycle allows stakeholders to predict and respond to potential need for ethical risks mitigation of decision bias, data privacy breech, and lack of explainability. Society actors and state authorities are responsible in shaping and sustaining a trustworthy governance framework [63].

One of the most significant parts of the UNESCO report of recommendation on the ethics of AI there is an effort to specifically bond AI ethics to citizen data protection need under the principles of transparency, accountability and supervision in AI e-governance. It clearly suggests privacy affection evaluation, measures against surveillance, and frameworks that secure the right to access, manage, control, and erase personal data and records. These measures are important for conducting transparently and ethically smart cities governance, where citizens' personal data and digital tracks are often collected through AI services, sensors and platforms without warnings or the citizens' consent (UNESCO 2024).

Jobin et al. [46] researched 84 AI ethics guidelines and principles and discovered that there is a harmonization among principles such as fairness, trust, transparency, and privacy. The essential nature of AI ethical governance was researched and pinpointed by [13] while [54, 61] suggested to be collective when incorporating human rights in AI systems. Eubanks [23] in their report showcase that the predictive decision making, regarding welfare services, disproportionately aim and include the poor, enlarging this way existing differences and inequality. Following the same context, Kitchin [69] has conducted research about the challenges of data management, algorithmic lack of transparency, and accountability in city boards while [42], recommended the term of "smart city citizenship," highlighting the change from infrastructure-based optimization to citizen-centric policy development.

AlgorithmWatch [4] along with and the OECD AI Observatory offered overviews that show that the European municipalities make some or no use of transparency tools and algorithm registries. Following, the Mozilla Foundation [60] examined institutional and practical difficulties towards AI explainability in public-sector. The Data & Society [20] challenged the performance unquestioned narrative around smart technologies, pushing the necessity for equal civic engagement and equity hearing.

The European [33] Research Service has recently published extensive reports on algorithmic regulation, civic overview, and national security issues of AI apps in smart cities. Veale and Zuiderveen Borgesius [85] have been critical against the EU AI Act draft generic definitions, terms, principles and law enforcement mechanisms, while DLA Piper [17] explored its interaction with the GDPR.

The ALTAI framework, the AI Pact by the European Commission [31], and the People-Centered Smart Cities Programme by UN-Habitat [81] assist towards a global change for civic participation, transparency, and inclusiveness AI e-governance in smart cities. The European Commission, Third draft of the general purpose AI code of practice [28], the AI Risk Management Framework [62], and the Montreal AI Ethics Institute reports (2023–2024) proved addition assistance to deploy AI principles in smart cities. Towards that direction there is a series of upcoming tools like the EU Digital Principles, algorithmic transparency registries [4, 50] etc.

5.3 AI to Smart Cities

AI applications in governance, mobility, environment, living, economy, and people refer to the domains that are closely related with each other and define the nature of smart cities (Fig. 5.2) [6].

Moreover, AI application in governance provides predictions, enables data governance and digital strategies while speeding up the processes while shaping policy decisions and mitigating risk forecasting. Furthermore, chatbots applications handle citizen demands and serve common good [38].

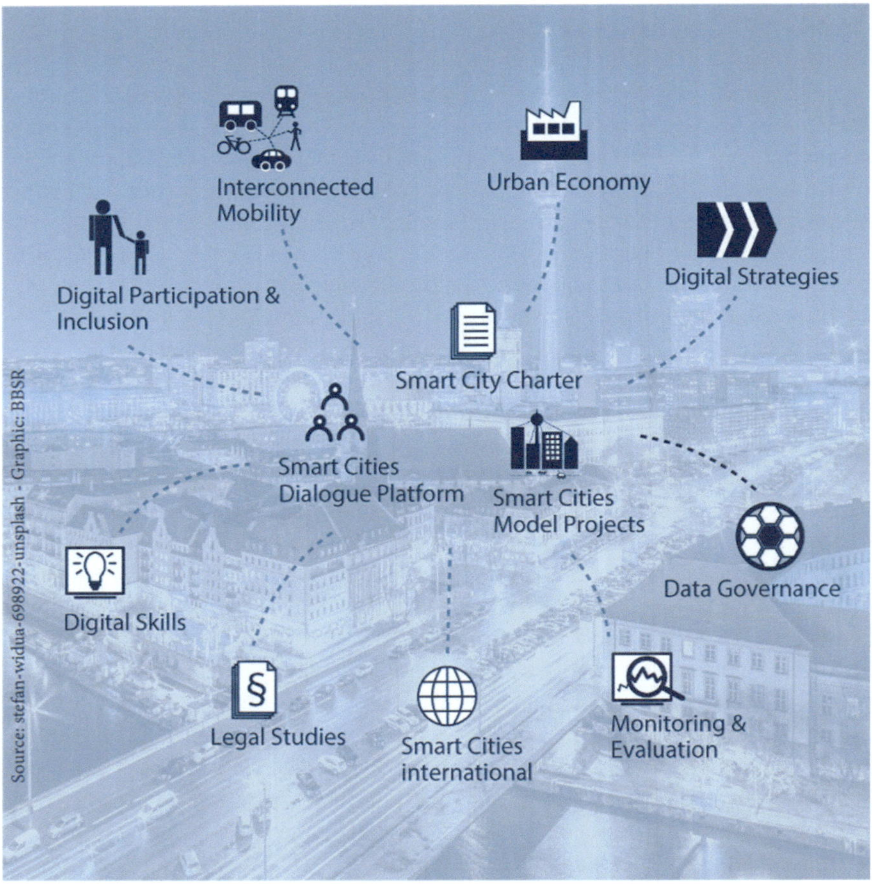

Fig. 5.2 Smart cities research cluster

Regarding interconnected mobility, AI empowers autonomous vehicles, automates transportation, conducts traffic enhancement, and offers a dynamic real-time map routing [12, 49].

AI is a powerful tool for optimizing the peoples' living quality within a city ecosystem by providing legal advice, optimizing emergency systems and smart housing sustainable development.

AI significantly offers solutions regarding environmental protection. Predictive analytics including machine learning, deep learning and data mining models monitor air quality, provide accurate weather forecasts, and predict environmental risks while due to dynamic use of resources helps lowering the carbon emissions, eliminating food waste etc. [81].

Regarding the urban economy, AI applications provide a sustainable model of job finding and role matching predictions in vacancy platforms, enhancing digital skills and allowing innovation and competitiveness while decreasing unemployment rates within a city [55].

Digital participation, inclusion and equality, personalized learning and skills development, smart health monitoring, and personalized e-government services for citizens leverage AI for societal progress [65].

Nevertheless, despite the advantages of such wide AI use there must be a control system for the drawbacks of such technological appliances along with the critical ethical and security concerns that have arisen. The UK Government has highlighted that AI in public systems must work under the principles of fairness, transparency, accountability, and contestability to provide smart, secure and sustainable interconnections among cities allowing places to function more effectively for the benefit of civilians and businesses. However, connectivity among cities also creates cyber threats and vulnerabilities. While empowering local authorities and organizations to purchase and use interconnected places technology, the UK Government tries to have an international consent for a continuous, efficient and consistent plan towards the security of connected cities mitigating the risks posed to businesses, organizations, the public sector and citizens (Fig. 5.3) [39].

The threat of privacy violations, biased decisions, and limited access of minorities or small communities to AI services is concerning [60].

5.4 AI Ethics in Smart Cities

AI involvement in smart cities generates a series of serious considerations regarding the ethical challenges. In particular, in essential society services like policing, healthcare, education, and mobility AI raises concerns. The combination of AI governance and civic environments increases the concerns over critical aspects like surveillance, discrimination, transparency, and consent [73].

Regarding privacy and surveillance, the development and use of biometric recognition technologies like facial and emotional detection, public space monitoring, and real-time individual behavioral pattern tracking and recognition intensify the

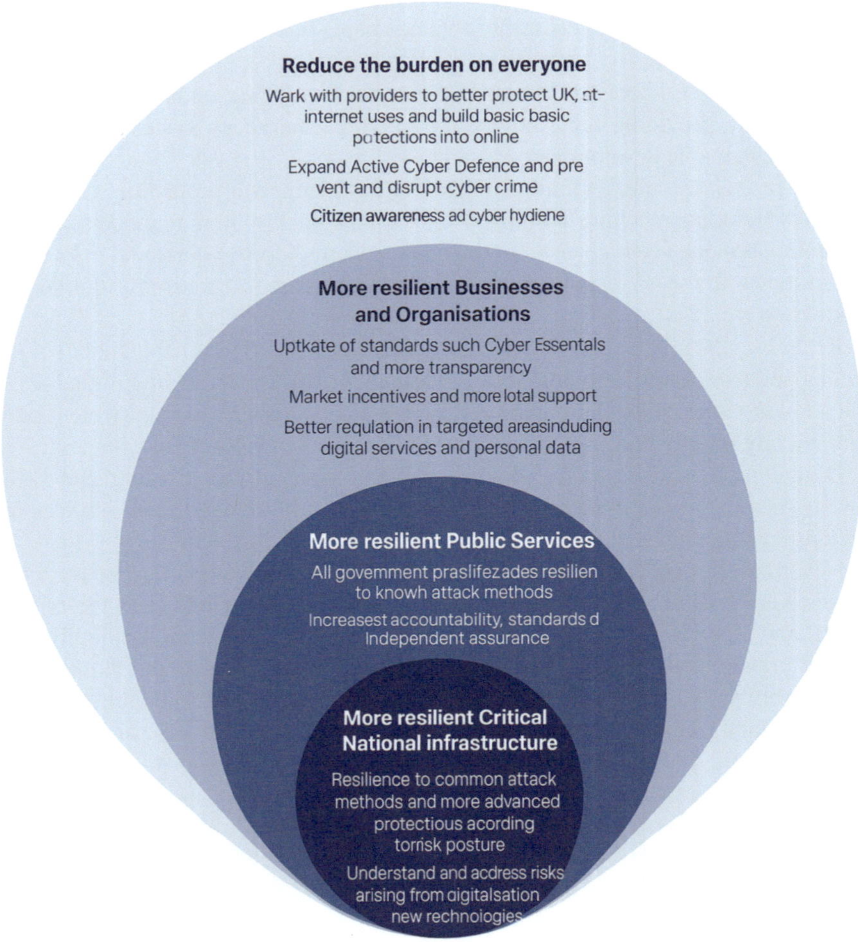

Fig. 5.3 GOV UK Cyber security and resilience framework

concerns about civic surveillance. The Mozilla Foundation [60] highlights that many smart cities concepts have "transparency deficits" that overcast how surveillance data is processed, mined, shared, or monetized.

An [81] report mentions that ethical aspects of AI application like data privacy must be a fundamental requirement instead of an afterthought in human-centric smart cities design.

When it comes to biased decision-making and discrimination, the European Union Agency for Fundamental Rights (FRA), [34] notifies that AI decision systems usually reflect societal behaviors like inequalities and discriminations if they are not properly addressed and audited. AI-based decision systems in important areas like policing, human rights and welfare must decrease systemic biases.

There are AI regulatory frameworks including the ALTAI (Assessment List for Trustworthy AI) and the OECD AI Principles which are responsible for handling cases and conducting tests to eliminate discriminatory recommendations and results in AI-based systems [63, 64].

Transparency and explainability refer to occasions where AI-based systems lack a substantial number of transparency rules and methos, transforming them into difficult cases to be comprehended and controlled by citizens and regulators.

Based on an [4] recent report there are very few European cities keeping records or logs of implemented and embedded algorithms, causing the belief of a "black-box governance" existence.

The Mozilla Foundation [60] report about "AI Transparency in Practice" examines the necessity of technically comprehensible and compiled but also available for public auditing for both computers and institutional explainability bodies.

GDPR Articles 15 and 22 preserve users' rights about autonomy and consent enabling the last to be aware of how their data is used and processed and to how object to AI-based decisions. To this day, most smart cities projects promote the "implied consent" or the "opt-out" choice, undermining this way the actual user control [37].

Following the previews strategy, a report from The Data & Society [20] showcases that many AI implementations present effectiveness as an excuse for public accountability avoidance.

As pinpointed by the [81] and Living-in.eu [51], inclusive AI implementation must requisitely act to eliminate obstacles to equal participation and fairness mitigating vulnerability and digital exclusion.

5.5 Regulation and Legislation

5.5.1 AI Regulation

5.5.1.1 The EU AI Act

Smart cities utilize AI to enhance public services such as traffic control and public security. The AI Act is the first-ever legal framework on AI. It addresses AI risks and positions Europe as a global leader. Established in August 2024, it consists of a legal framework for AI risk management, influencing smart city governance. This chapter takes a closer look at AI Act impact, data protection, and smart cities challenges. The AI Act defines 4 levels of risk for AI systems (Fig. 5.4) [31]:

- Minimal or no risk
- Transparency or limited risk
- High risk
- Unacceptable risk

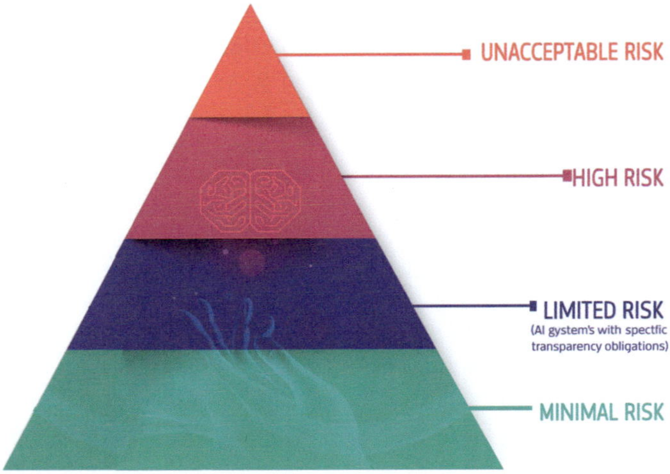

Fig. 5.4 Four levels of risk for AI systems

Minimal or no risk: There are no rules or obligations for AI systems that fall into this category. Systems such as video games, spam filters etc. do not need to be regulated, having minimal or no usage risk.

Transparency or limited risk: AI systems such as Chatbots require transparency. e.g., users need to know that they are chatting with an AI system. Moreover, AI generated content should be mentioned as such, to ensure that users are aware of AI interaction.

High risk: AI systems affecting public health, safety, or fundamental rights are classified as high risk and must be strictly regulated before public use. These systems require thorough documentation, security, accuracy, and high-quality datasets. e.g., public transportation systems, medical assistance, biometric identification, etc.

Unacceptable risk: AI systems that can directly harm public rights are prohibited.

The AI Act prohibits eight practices [31]:

1. Harmful AI-based manipulation and deception.
2. Harmful AI-based exploitation of vulnerabilities.
3. Social scoring.
4. Individual criminal offence risk assessment or prediction.
5. Untargeted scraping of the internet or CCTV material to create or expand facial recognition databases.
6. Emotion recognition in workplaces and education institutions.
7. Biometric categorization to deduce certain protected characteristics.
8. Real-time remote biometric identification for law enforcement purposes in publicly accessible spaces.

5.5.2 General Data Protection Regulation (GDPR)

Another legal framework that enhances the AI Act is General Data Protection Regulation (GDPR). It was established in 2018:

> Aiming at the protection of natural people with regard to the processing of personal data and on the free movement of such data [37].

GDPR protects how the personal data of EU citizens are processed or transferred and applies to manual and automated systems. According to GDPR article 22:

> The data subject shall have the right not to be subject to a decision based solely on automated processing, including profiling, which produces legal effects concerning him or her or similarly significantly affects him or her.

GDPR's emphasis on data minimization and transparency aligns with the AI Act's ethical guidelines. Even though the AI act and the GDPR have different approaches they also overlap [53].

The EU AI act along with the GDPR, establishes a strong framework for AI usage in smart cities. It combines innovation with citizens' rights protection. By aligning these laws, smart cities can encourage trust, transparency and viability (Table 5.1) [53].

5.5.3 The U.S. Approach

In contrast to the European Union's centralized, risk-based AI Act, the U.S adopts a decentralized approach [78].

Instead of a holistic legal framework, U.S. introduces state-level regulations. Each state has introduced AI-specific rules. Hundreds of bills will be added during 2025. For example, the Colorado AI Act which has been effective since May 17, 2024, regulates high-risk AI systems focusing on algorithmic discrimination and consumer protection. Another initiative would be California's AI transparency Act. AI-generated content must be watermarked by providers. Regarding smart cities, this fragmented landscape faces many challenges. Due to variety of laws throughout the states, an alignment through regulations would lead to a more trustworthy and transparent approach to AI systems (Table 5.1) [15]

5.5.4 United Kingdom's AI Regulation

The United Kingdom is developing a framework to regulate AI. The goal is to align new technologies with ethical concerns and public safety (Table 5.1). In March 2023, the UK government introduced a white paper consisting of five principles for responsible AI innovation: Safety, security and robustness, appropriate transparency

Table 5.1 AI Legislation comparison on smart cities

Region	Legislation/policy	Key features	Smart city implications	Alignment with global trends
EU	AI Act (2024), GDPR	Risk-based, bans unacceptable practices	Ensures safety, high compliance costs	Risk-based, digital alignment, collaboration
US	State laws (e.g., Colorado AI Act, 2024)	Decentralized, sector-specific	Innovation,but fragmented compliance	Risk-based (state-level), private-sector
China	Interim Measures on Generative AI (2023)	Centralized, transparency-focused	Rapid deployment, ethical concerns	Risk-based,state-controlled
United Kingdom	Pro-Innovation AI Regulation Framework (2023), AI Safety Institute	Light-touch, sector-led, risk-based	Supports innovation	Risk-based, pro-innovation, international cooperation
Singapore	Model AI Governance Framework (2019)	Ethical, human-centric	Enhances trust, balanced innovation	Ethical principles, sector-specific
Japan	Social Principles of Human-Centric AI (2019)	Human dignity, sustainability	Ethical, sustainable AI use	Ethical principles, emerging risk-based
South Korea	Basic Act on AI (2026)	Trustworthy ecosystem	Balanced innovation and safety	Risk-based, digital alignment
Australia	Voluntary AI Safety Standard (2024)	Flexible, existing frameworks	Rapid adoption, less rigorous	Ethical principles, sector-specific
UAE/ Saudi Arabia	National AI Strategies (2023–2024)	Ethical, responsible AI	Trustworthy,innovative urban applications	Ethical principles, private-sector
The African Union	Continental AI Strategy (2024)	Focus on innovation, infrastructure, regional collaboration	AI for viable development, economic growth, public services	Regional collaboration, ethical AI

and explainability, fairness, accountability and governance, contestability and redress [21].

AI Safety Institute (AISI) founded in November 2023 under the direction of UK Department for Science, Innovation, and technology. AISI through collaborations with AI developers and international partners, organizes evaluations in order to minimize unpredicted risks.

Moreover, UK's national institute for data science and artificial intelligence, the Alan Turing Institute, plays a significant role in shaping UK's approach to AI regulation. Through research and policy making, it provides the public sector with tools to apply ethical principles to AI projects [79].

5.5.5 The Australian AI Systems

The Australian government has proposed 10 barriers considering deployers and developers of high-risk AI systems. These measures are designed to align with existing frameworks with respect to privacy and consumer protection laws (Table 5.1) [22].

5.5.6 China Smart Cities Control

On the other hand, China has become the leader in Smart Cities technologies globally. As of 2020, China has deployed more than 900 pilot smart cities projects. On August 15th, 2023, China introduced Interim Measures on Generative AI services. These measures are a series of regulations applied in mainland China and regulate several aspects of products and services provided. However, ethical concerns arise due to biometric techniques, surveillance practices, and the social scoring system used by the Chinese government (Table 5.1) [80].

5.5.7 The Middle East AI Models

Middle East countries such as UAE and Saudi Arabia lead the way in this geographical area. Saudi Arabia's National Strategy for Data & AI aims to set a high standard in responsible AI utilization. introduce ethical AI practices, ensuring transparency and accountability [74].

UAE's National Strategy for Artificial Intelligence aims to integrate AI systems in government services, transportation and smart cities. Investing in technology, education and infrastructure, UAE's initiatives consist of guidelines to support and enhance AI development [75].

In conclusion, the global landscape of AI legislation and regulation is diverse. While the EU's AI Act stands as the leading framework based on risk classification of AI systems, other regions adopt strategies that align with their vision, culture and ethics (Table 5.1).

5.5.8 *The Asian-Pacific Models*

Singapore's Model AI Governance Framework, released on 23 January 2019, provides a detailed series of regulations aiming to give attention to ethical and governance issues regarding AI applications. Main pillars of this act are transparency, explainability and fairness. AI systems should be anthropocentric. Singapore's innovation in AI regulation and its human-centric approach makes it a global leader in this area (Table 5.1) [76]

Japan was one of the first countries to regulate the use of AI. In 2019, Social Principles of Human-Centric AI was released. It focused on human dignity, diversity and inclusion, sustainability (Table 5.1) [77].

South Korea's Basic Act on AI will be effective from January 2026. It aims to develop AI infrastructure, setting the foundation for a trustworthy and safe AI ecosystem (Table 5.1) [58].

5.5.9 *The African Union's Model*

The African Union's Continental AI Strategy [3] is an African-centric approach, prioritizing the expansion and conversion of AI systems to the African continent. Approved in July 2024 focuses in five key areas: harnessing AI's benefits, minimizing risks, investing in infrastructure, collaborating with international partners, attract investments public and private [3]. The goal is utilizing the technologies for social and economic growth through viable development.

5.6 Transparency and Trust in Automated Decision-Making

Artificial intelligence systems in smart cities are a tool for optimizing the lives of citizens. The decisions made by these systems greatly affect the daily lives of residents, influencing areas such as public transport, safety and health. These areas usually lean on non-transparent algorithms used by neural networks and ensemble methods, which can affect the overall accountability and transparency of those systems [9].

Transparency in the way decisions are made is a key feature of these systems. Citizens will be able, to a certain extent, to understand how these decisions are made. A lack of transparency can lead to suspicion, lack of trust and rejection of these technological advancements.

According to a survey conducted by BCS, The Chartered Institute for IT, 53% of UK adults don't trust organizations to use algorithms when making judgments about them, ranging from education to welfare matters. This chapter explores the need for transparency and explainability in automated AI systems that make decisions for the public, examines guidelines and action plans, analyzing the explainable AI (X-AI) mechanism [7].

Trust and transparency play a crucial role in ensuring AI systems are welcomed by the public. To address this issue, the High-Level Expert Group on AI presented Ethics Guidelines for Trustworthy Artificial Intelligence, published in April 2019. According to the guidelines, for an AI system to be trustworthy, it must meet seven key requirements (Fig. 5.5) [1]:

- Human agency and oversight
- Technical robustness and safety
- Privacy and data governanceresponsible and trustworthy
- Transparency
- Diversity, non-discrimination, and fairness
- Societal and environmental well-being
- Accountability

Moreover, on April 9, 2025, the European Commission launched the AI Continent Action Plan. The AI Continent Action Plan will expand the European Union's capabilities through policies and actions. Investments, infrastructures and collaborations are the goals. Moreover, building trustworthy and anthropocentric AI plays a crucial role in economics and in preserving the fundamental rights of our societies. Aligning with the EU AI Act which categorized the systems based on risk levels, the EU aims to build an environment where AI systems are not only advanced but also responsible and trustworthy aligned with social values [29].

As AI becomes more advanced, there is a growing need for people to trace and understand how AI systems 'think' and extract their results. For that reason, the concept of eXplainable AI, (X-AI) is introduced. These systems are no longer inaccessible and incomprehensible. They will be accessible by everyone, tailored to each person's technical understanding. The mathematical explanation of AI systems can explain the machine learning models used but may not reveal how each model arrived at a specific outcome [45].

This is a deeper issue and there should be an effort to document the algorithms applied for each issue. Complex machine learning systems should be oriented in this direction. Documenting algorithms adds to the concept of XAI. It ensures causality, enables model transferability and builds user confidence in system outputs [9]. The decision-making mechanism should not be a black box, but should be understandable, interpretable, transparent, explainable and accountable [43].

Fig. 5.5 Key requirements for trustworthy AI

There are two types of AI system deployment: inherently transparent systems, which use transparent algorithms (e.g., Decision Trees, Linear Regression), and black-box systems, which use more complex models such as neural networks or ensembles. Thus, there are two ways to track explainability. For the first category it is rather easy to follow the steps until the output. For the second category, post-hoc methods must be applied. These methods do not interfere with algorithms, but they map their procedures. There will be an additional output parallel to the system's output where the procedure will be visualized and explained. All the above methods contribute to explainability of AI systems [9].

With the aim of public interest, efficiency and fairness, registers of algorithms, documentation of systems and data that are processed can be created. This way, citizens will be able to have access to all levels of input, processing and output of decisions.

By applying XAI's philosophy in AI systems, not only is trust built between citizens and technology but also several advantages are established. Fairness detection is now a reality and thus biased content, algorithms or data can be detected. Debugging errors is now easier. Developers trace and correct the root causes of harmful outputs. XAI is aligned with the guidelines of GDPR, where the outputs of automated procedures should be explained to individuals at their request.

Transparency in AI systems is not only a technical negotiation with citizens but promotes democratic handling of issues and citizens' trust in the digital era of smart cities.

5.7 User Behavior in Smart Cities

As AI enters the everyday lives of the citizens, data must be collected to get insights based on behavior. This data consists of how citizens move, interact, consume, and engage with the environment. Machine learning algorithms collect this information and identify patterns. While these data play an important part in behavioral analysis, ethical considerations arise regarding privacy and security.

Smart cities collect behavioral insights from: IoT sensors tracking energy use and mobility, mobile apps and wearables monitoring location and physical metrics, computer vision analyzing movement and digital platforms collecting logs from public services. AI uses these data for personalization and predictive behavior.

Mühlhoff introduces the concept of predictive privacy. Individuals can be affected by AI decisions based on information they did not provide. That means that in smart cities data used by AI systems from one person can influence the output and the decisions made for another person. This raises concerns about consent and autonomy and set the need for data protection collected by AI systems by applying strategies considering the impact of data analytics regarding public safety [70].

Data protection and transparency become visible. By developing AI systems that allow clear access to data analytics processes help build trust between citizens and governance. Data driven outputs should be evaluated regarding usefulness and legitimacy. Thus, accountability and improvement can be the main characteristics of smart city initiatives [48].

Moreover, behavioral data can be high-valued target. Including preferences, personal traits, movement patterns and consumer behavior, these datasets when leaked can be used for manipulation, identity theft and authoritative targeting. Behavioral data is continuously updated. Thus, data exploitation can lead to surveillance and misuse.

For this matter regulatory frameworks such as GDPR and AI Act can be applied. By protecting how user data are transferred and stored, automated procedures can be monitored and evaluated. AI Act by applying risk-based controls over AI systems offers transparency and practicality. Explainable AI adds accountability and explainability, promoting trust between citizens and technology. The Congress of Local and Regional Authorities, Council of Europe [14] underscores the necessity for digital transformation to be grounded in democratic values, transparency and citizen participation. It promotes the protection of fundamental rights, driving cities to apply frameworks that protect privacy, focus on accountability and offer fair access to digital services [14].

5.8 A Human-Centric Smart City

As the concept of smart cities has been evolving for over a decade, it has become a bright expectation for governments and societies. This concept represents the technological advancements applied to cities around the world and highlights ICTs (Information and Communication Technologies) and infrastructures promising to solve many contemporary city problems and accelerate efficiency. This concept often fails to consider the human dimension of digital transformation.

According to UN-Habitat this approach strengthens social inequality and risks public trust as people are treated as data sources rather than active contributors [81]. By placing people at the center of this digital transformation, smart cities' development prioritizes wellbeing, equality and environmental responsibility [82]

Following this principle, the Living-in.EU initiative calls for a citizen-centric approach, an open city that supports socially responsible access of data, active participation of the community, transparency of AI systems, once-only and privacy-by-design principles [50].

To conclude, smart cities should make a turn and be built for the people, not just data-driven systems. Based on the needs of citizens, principles like transparent access, digital equality, privacy, data sharing and active participation should be mandatory to any smart city design.

5.9 AI Initiatives

5.9.1 Strategic Investments and Policy Developments

Recent AI projects' deployments across the globe reflect the rising need for combining ethical regulatory frameworks with strategic investment policies.

The European Commission plans to invest €200 billion in InvestAI initiative to mobilize €200 billion of investment in artificial intelligence "gigafactories," enhancing supercomputing performance infrastructures, and build new Cloud and AI Development Act, considering technology innovation and AI regulation [29, 32].

Similarly, in an effort to enhance education, trust and transparency across the European Union, the EU AI Office has promoted a survey for AI literacy best practices [30].

The Knowledge Centre Data & Society implemented prototypes for human supervision requirements in high-risk AI, translating Article 14 of the EU AI Act into valuable operational and control tools [19, 47, 20].

A policy report published by Data & Society warns against processing speed decrease and cost reduction with performance, calling for solid cost and efficiency AI tools evaluation and auditing in public administration [19].

Canada pinpoints AI governance, talent, and accountability in the new AI strategy for the public administration [41].

5.9.2 Global Ethical AI Programs

Several global program initiatives refer to ethical AI in urban environments:

- The UN-Habitat People-Centered Smart Cities Programme refers to a program focused on inclusiveness and human rights-based smart city models deployment [81].
- The OECD.AI Policy Observatory is a program that provides resources and benchmarks on AI governance schemes [66].
- The European AI Alliance manages to adapt and transform multiple stakeholders' claims and interests on EU AI policy.
- The Montreal AI Ethics Institute publishes reports of open access resources on AI transparency, trustworthiness and e-governance [59].
- The Partnership on AI implements a series of documentation for ethical considerations and regulatory frameworks for responsible and human-based AI applications.

5.9.3 AI Key Operational Suggestions

There is a series of AI Initiatives that apply in AI legislation and human protection:

1. The Algorithmic Registries refer to the responsibility of the municipalities to store, and maintain detailed information for the purpose, the scope, and the performance metrics of deployed algorithms in open registries accessible to the civilians [4].
2. The Institutionalize Ethical Audits refer to a third-party assessment and control of the high-risk AI systems using specific and standardized tools like ALTAI and NIST frameworks [62].
3. The Enhance Public Participation initiative supports civilians' boards and participation platforms for AI policy orientation [81].
4. The Enforce Data Protection by Design integrates GDPR-based articles focused on biometric and behavioral detection-making systems [37].

5. Promote Digital Literacy promotes awareness campaigns and embeds AI ethics into educational guidelines [59].
6. Smart cities classification: criteria and tools for assessing the level of anthropocentric services [18].
7. The Foster Cross-Sector Collaboration refers to the interdepartmental creation of teams of experts who bridge technology, legislation, and social affairs [65].

5.10 Discussion

This chapter explores the multidimensional role of AI in shaping civic-centric smart cities and provides a brief overview of the current interpretation of ethics, law, policy, and public engagement. The current analysis shows that while AI optimizes urban environments efficiency and service provision, it simultaneously introduces important threats and concerns regarding algorithmic bias, personal data erosion, the lack of democratic mechanisms and surveillance risks.

AI-based systems benefits and drawbacks create a tension between innovation and rights preservation in urban governance. As the ubiquity of AI intensifies, the regulatory frameworks are expanding trying to face the challenges of transparent operations, citizen equality and inclusion, and global legislation consistency [8, 61].

While the European Commission tries to provide guidelines or blueprints by publishing the AI Act, it seems that the implementation depends on local capacity, citizen culture, and political will to overcome the barriers. Furthermore, AI trustworthiness is fragile for the public. Personal data manipulation, personalized services without permission, unauthorized facial recognition, flawed predictive legal advice create lack of confidence. Hence, future smart cities must incorporate as perquisites the criteria of inclusiveness, transparency, and responsiveness of AI governance to meet their civic promise requirements [4].

The regulatory frameworks, like the EU AI Act, GDPR, and international ethical initiatives highlight steady progress toward AI accountability institutionalization. Nevertheless, the flaws remain regarding human supervision and the explainability in essential public decisions. As mentioned in the literature and main body of the chapter, trustworthy AI is something that cannot happen only by the regulations or restrictions unless there is an active citizen participation, behavioral limits, and a culture of digital education [37].

The international best practices such as the Canada's strategic AI priorities and the EU's literacy repository enhance the meaning that ethical AI-based governance must be designed by the government, organizations, AI initiatives, AI experts, policymakers, and the citizens. Moreover, the incorporation of behavioral patterns recognition, user profiling and predictive analytics require new rules of actors' engagement that guards citizen dignity.

Finally, this research confirms that the smart city blueprint must consider not only infrastructure enhancements but also the social agreement that combines citizens with digital governance. Trustworthiness, transparency, and participation must be a requirement to urban AI projects design.

5.11 Conclusion

AI technology provides valuable contribution and potential to urban life optimization, but its legitimacy depends on how it protects democratic values, regulations, and ethical principles. Optimizing infrastructure does make a significant contribution to society unless they protect human rights, help citizens, and maintain accountability. AI transparency, human-centric design, and solid regulation frameworks, policymakers can guarantee that AI becomes beneficiary in the urban age. This chapter has explored the interconnections of AI with privacy, transparency, trust, and civic-centric approach in urban environments.

Ethical considerations are present and dangerous. Surveillance, algorithmic bias, unauthorized consent, and mitigated explainability and lack of citizen supervision of AI projects are already forming user experiences.

Therefore, smart city design and governance must embed ethical regulatory frameworks like the EU AI Act and GDPR. Policies should adopt AI transparency tools, citizen participation platforms, and data protection mechanisms. Global AI initiatives, such as the OECD, UN-Habitat, and national strategies of US, UK etc., provide operational holistic models for effective and responsible AI governance.

Ultimately, the legitimacy of AI in smart cities depends on its alignment with democratic values. Ethical AI is a social contract, a mutual agreement that preserves and ensures accountability, inclusiveness, fairness, and fundamental human rights in every stage of AI integration, between government or city infrastructures and the society they serve.

References

1. AI HLEG, Ethics guidelines for trustworthy AI. European Commission (2019), https://ai-watch.ec.europa.eu/topics/trustworthy-ai_en. Accessed 2 Apr 2025
2. AI Security Institute, The AI security institute (2025), https://www.aisi.gov.uk/. Accessed 04 Apr 2025
3. African Union, Continental AI strategy (2024), https://au.int/sites/default/files/documents/44004-doc-EN-_Continental_AI_Strategy_July_2024.pdf. Accessed 06 Apr 2025
4. AlgorithmWatch, AI systems inventory (2024),https://inventory.algorithmwatch.org. Accessed 10 Apr 2025
5. B.C. Stahl, D. Schroeder, R. Rodrigues, *Ethics of Artificial Intelligence* (Springer, Cham, 2022). https://doi.org/10.1007/978-3-031-17040-9
6. BBSR, Research cluster of the federal institute for research on building, urban affairs and spatial development (BBSR) (2023), https://www.bbsr.bund.de/BBSR/EN/research/programs/smart-cities/flyer.pdf;jsessionid=731A4797961199E06991A6477AA3D5BA.live21302?__blob=publicationFile&v=2. Accessed 10 Apr 2025

7. BCS, The public don't trust computer algorithms to make decisions about them, survey finds (2020), https://www.bcs.org/articles-opinion-and-research/the-public-dont-trust-computer-algorithms-to-make-decisions-about-them-survey-finds. Accessed 28 Apr 2025
8. BEUC, EU legislators must close dangerous loophole in AI Act (2023), https://www.beuc.eu. Accessed 10 Apr 2025
9. A. Barredo Arrieta, N. Díaz-Rodríguez, J. Del Ser, A. Bennetot, S. Tabik, A. Barbado, S. Garcia, S. Gil-Lopez, D. Molina, R. Benjamins, R. Chatila, F. Herrera, Explainable Artificial Intelligence (XAI): concepts, taxonomies, opportunities and challenges toward responsible AI. Inform. Fusion **58**, 82–115 (2020). https://doi.org/10.1016/j.inffus.2019.12.012
10. A. Bokolo, S. Sarshar, Enhancing data sovereignty to improve intelligent mobility services in smart cities. Urban Gov. **5**(1), 20–31 (2025). https://doi.org/10.1016/j.ugj.2025.02.002
11. C. Novelli, P. Hacker, J. Morley, J. Trondal, L. Floridi, *A Robust Governance for the AI Act: AI Office* (AI Board, Scientific Panel and National Authorities, 2024). https://doi.org/10.1017/err.2024.57
12. CES, AI in data planning and beyond (2023),https://www.ces.tech. Accessed 12 Apr 2025
13. C. Cath, Governing artificial intelligence: ethical, legal and technical opportunities and challenges. Phil. Trans. R. Soc. A **376**(2133), 20180080 (2018)
14. Congress of Local and Regional Authorities, Council of Europe, European Urban Charter III: Urban living in the era of transformations (2023). Report CG(2023)45-20. https://rm.coe.int/0900001680ad1557. Accessed 25 April 2025
15. Consumer Protections for Artificial Intelligence, SB24-205 (Colorado General Assembly, 2024), https://leg.colorado.gov/bills/sb24-205. Accessed 25 Apr 2025
16. D.C. Sharma, AI Regulation and Standardization in India, in *ETSI AI Conference*, February 2024 (2024), https://docbox.etsi.org/Workshop/2024/02_ETSIAICONFERENCE/S02_REGULATION_GLOBAL/INDIA_SESEI_DINESH_SHARMA.pdf. Accessed 10 Apr 2025
17. DLA Piper, *The EU AI Act's Relationship with Data Protection Law* (2024), https://privacymatters.dlapiper.com. Accessed 10 Apr 2025
18. O. Dashkevych, B.A. Portnov, Human-centric, sustainability-driven approach to ranking smart cities worldwide. Technol. Soc. **74**, 102296 (2023). https://doi.org/10.1016/j.techsoc.2023.102296
19. Data & Society, Dispelling myths of AI and efficiency (2025), https://datasociety.net/library/dispelling-myths-of-ai-and-efficiency. Accessed 10 Apr 2025
20. Data & Society, Research and policy for the public interest (2025), https://datasociety.net/. Accessed 25 Arpil 2025
21. Department for Science, Innovation and Technology, A pro-innovation approach to AI regulation: government response (2023), https://www.gov.uk/government/consultations/ai-regulation-a-pro-innovation-approach-policy-proposals/outcome/a-pro-innovation-approach-to-ai-regulation-government-response. Accessed 20 Apr 2025
22. Department of Industry, Science and Resources, Artificial Intelligence (Government of Australia, 2025), https://www.industry.gov.au/science-technology-and-innovation/technology/artificial-intelligence. Accessed 25 Apr 2025
23. Eubanks V, *Automating Inequality: How High-Tech Tools Profile, Police, and Punish the Poor* (St. Martin's Press, USA, 2018). https://dl.acm.org/doi/10.5555/3208509
24. Eurocities, People-Centered AI in Cities (2023), https://eurocities.eu. Accessed 5 Apr 2025
25. European AI Office, https://digital-strategy.ec.europa.eu/en/policies/ai-office. Accessed 2 Apr 2025
26. European Commission, White paper on artificial intelligence: a European approach to excellence and trust (2020), https://ec.europa.eu/info/publications/white-paper-artificial-intelligence-european-approach-excellence-and-trust_en. Accessed 10 Apr 2025
27. European Commission, Innovation procurement—first findings (2024a), https://research-and-innovation.ec.europa.eu/document/download/0851b090-ba80-490f-abc3-2196f26c3beb_en. Accessed 10 Apr 2025
28. European Commission, Third draft of the general purpose AI code of practice (2024b), https://digital-strategy.ec.europa.eu. Accessed 10 Apr 2025

29. European Commission, Commission sets course for Europe's AI leadership with an ambitious AI continent action plan (2025a), https://ec.europa.eu/commission/presscorner/detail/en/ip_25_1013. Accessed 7 Apr 2025
30. European Commission, New survey to gather practices for the AI literacy living repository (2025b). https://digital-strategy.ec.europa.eu/en/news/new-survey-gather-practices-ai-literacy-living-repository. Accessed 15 Apr 2025
31. European Commission, AI Act (2025c), https://digital-strategy.ec.europa.eu/en/policies/ai-pact#ecl-inpage-What-is-the-AI-Pact? Accessed 1 Apr 2025
32. European Commission, EU launches Invest AI initiative to mobilise €200 billion of investment in artificial intelligence (2025d). https://ec.europa.eu/commission/presscorner/detail/en/ip_25_467. Accessed 1 Apr 2025
33. European Parliament, AI and digital tools in workplace management and evaluation: an assessment of the EU's legal framework (2025). https://www.europarl.europa.eu/stoa/en/document/EPRS_STU(2022)729516. Accessed 1 Apr 2025
34. European Union Agency for Fundamental Rights (FRA), Bias in algorithms and discrimination (2020). https://fra.europa.eu/en/publication/2022/bias-algorithm. Accessed 20 Apr 2025
35. F. Kleiman, S. Meijer, M. Janssen, Behavioral factors influencing the opening of government data by civil servants: initial findings from the literature, in *Proceedings of the 13th International Conference on Theory and Practice of Electronic Governance (ICEGOV '20)* (Association for Computing Machinery, New York, NY, USA, 2020), pp. 529–534. https://doi.org/10.1145/3428502.3428582
36. G. Misuraca, C. van Noordt, *AI Watch, Artificial Intelligence in Public Services—Overview of the Use and Impact of AI in Public Services in the EU* (European commission: Joint Research Centre, Publications Office, 2020), https://data.europa.eu/doi/10.2760/039619. Accessed 10 March 2025
37. GDPR, General data protection regulation—Articles 15 & 22 (2024), https://gdpr-info.eu/art-15-gdpr/. Accessed 3 Apr 2025
38. GOV.UK, *A guide to using artificial intelligence in the public sector* (Government Digital Service, 2020a), https://www.gov.uk/government/publications/a-guide-to-using-artificial-intelligence-in-the-public-sector. Accessed 10 Apr 2025
39. GOV.UK, *Understanding artificial intelligence ethics and safety: a guide for public sector use* (2020b), https://www.gov.uk/guidance/understanding-artificial-intelligence-ethics-and-safety. Accessed 28 Apr 2025
40. GOV.UK, *National Cyber Strategy 2022* (2022), https://www.gov.uk/government/publications/national-cyber-strategy-2022. Accessed 28 Apr 2025
41. Government of Canada, AI strategy for the federal public service 2025–2027: priority areas (2025), https://www.canada.ca/en/government/system/digital-government/digital-government-innovations/responsible-use-ai/gc-ai-strategy-priority-areas.html. Accessed 10 Apr 2025
42. I. Calzada, *Smart City Citizenship* (Elsevier Science Publishing Co Inc., Cambridge, MA, 2021), https://ssrn.com/abstract=3650025
43. IBM, What is explainable AI? (2023), https://www.ibm.com/think/topics/explainable-ai. Accessed 20 Apr 2025
44. J. Mökander, L.A. Floridi, Operationalising AI governance through ethics-based auditing: an industry case study. AI Ethics **3**, 451–468 (2023). https://doi.org/10.1007/s43681-022-00171-7
45. A.R. Javed, W. Ahmed, S. Pandya, P.K.R. Maddikunta, M. Alazab, T.R. Gadekallu, A survey of explainable artificial intelligence for smart cities. Electronics **12**(4), 1020 (2023). https://doi.org/10.3390/electronics12041020
46. A. Jobin, M. Ienca, E. Vayena, The global landscape of AI ethics guidelines. Nat Mach Intell **1**, 389–399 (2019). https://doi.org/10.1038/s42256-019-0088-2
47. Knowledge Centre Data & Society, From policy to practice: prototypes for the human oversight requirement in the EU AI Act (2025), https://data-en-maatschappij.ai/en/publications/rapport-van-beleid-naar-praktijk-prototypes-van-de-menselijk-toezicht-vereiste-in-de-eu-ai-verordening. Accessed 10 Apr 2025

48. L. Bolte, A. Van Wynsberghe, Sustainable AI and the third wave of AI ethics: A structural turn. AI and Ethics. **5**(2), 1733–1742 (2025)
49. H.S.M. Lim, A. Taeihagh, Algorithmic decision-making in AVs: understanding ethical and technical concerns for smart cities. Sustainability **11**(20), 5791 (2019). https://doi.org/10.3390/su11205791
50. Living-in.EU, Declaration on joining forces to boost sustainable digital transformation in cities and communities in the EU (2024b), https://living-in.eu/declaration. Accessed 08 Mar 2025
51. Living-in.eu, AI Act implications for local communities (2024a), https://living-in.eu. Accessed 29 Mar 2025
52. M. Mueck, C. Gaie, D.C. Gkikas, Introduction to the European artificial intelligence act, in *European Digital Regulations. Intelligent Systems Reference Library*, ed. by M. Mueck, C. Gaie, vol 265. Springer, Cham (2025). https://doi.org/10.1007/978-3-031-80809-8_3
53. M. Nisevic, D. Tarasov, C. Arno, D.E. Bruyne Jan, *Explainable AI: Can the AI Act and the GDPR Go Out for a Date?* (2025), https://papers.ssrn.com/sol3/papers.cfm?abstract_id=5056022. Accessed 25 Apr 2025
54. H. Margetts, C. Dorobantu, Rethinking public sector innovation for the AI era. Phil. Trans. R. Soc. A **379**(2196), 20200382 (2021)
55. McKinsey & Company, *Superagency in the workplace* (2024), https://mckinsey.com. Accessed 9 Apr 2025
56. McKinsey & Company, The state of AI: how organizations are rewiring to capture value (2025), https://www.mckinsey.com/~/media/mckinsey/business%20functions/quantumblack/our%20insights/the%20state%20of%20ai/2025/the-state-of-ai-how-organizations-are-rewiring-to-capture-value_final.pdf. Accessed 22 Mar 2025
57. Mehr H, *Artificial Intelligence for Citizen Services and Government* (Harvard Ash Center, 2017), https://ash.harvard.edu/files/ash/files/artificial_intelligence_for_citizen_services.pdf
58. Minister of Science and ICT, A new chapter in the age of AI: basic act on AI passed at the national assembly‘s plenary session (2024), https://www.msit.go.kr/eng/bbs/view.do?sCode=eng&mId=4&mPid=2&pageIndex=&bbsSeqNo=42&nttSeqNo=1071&searchOpt=ALL&searchTxt=. Accessed 25 Apr 2025
59. Montreal AI Ethics Institute, Trustworthy AI whitepaper (2023), https://montrealethics.ai. Accessed 10 Mar 2025
60. Mozilla Foundation, AI transparency in practice (2023), https://foundation.mozilla.org/en/research/library/ai-transparency-in-practice/. Accessed 11 Apr 2025
61. N.A. Smuha, Beyond the individual: AI and the collective dimension of human rights, in *Cambridge Handbook of AI & Human Rights* (Cambridge University Press, 2021)
62. NIST, *AI Risk Management Framework* (2024), https://www.nist.gov/itl/ai-risk-management-framework. Accessed 10 Apr 2025
63. OECD, *Artificial Intelligence in Societ*y (OECD Publishing, Paris, 2019a). https://doi.org/10.1787/eedfee77-en
64. OECD, Scoping principles to foster trust in and adoption of AI—proposal by the expert group on artificial intelligence at the OECD (AIGO) (2019b), https://www.oecd.org/content/dam/oecd/en/publications/reports/2019/11/scoping-the-oecd-ai-principles_71e1b6dc/d62f618a-en.pdf. Accessed 14 Apr 2025
65. OECD, OECD.AI policy observatory—dashboards (2023), https://oecd.ai/en/dashboards/overview. Accessed 10 Apr 2025
66. OECD.AI, National AI policies and strategies dashboard (2022), https://oecd.ai/en/dashboards. Accessed 10 Apr 2025
67. OVIC, *Artificial Intelligence and Privacy: Issues and Best Practices* (Office of the Victorian Information Commissioner, 2022), https://ovic.vic.gov.au/privacy/artificial-intelligence-and-privacy/. Accessed 11 Mar 2025
68. Partnership on AI, Documenting the impacts of foundation models (2024), https://partnershiponai.org. Accessed 13 Mar 2025
69. R. Kitchin, Urban data and city dashboards: six key issues. Cities **110**, 103038 (2021). https://doi.org/10.31235/osf.io/k2epn

70. R. Mühlhoff, Predictive privacy: Collective data protection in the context of artificial intelligence and big data. Big Data Soc. **10**(1) (2023). https://doi.org/10.1177/20539517231166886
71. Regulation (EU) 2024/1689 of the European Parliament and of the Council of 13 June 2024 laying down harmonised rules on artificial intelligence and amending Regulations (EC) No 300/2008, (EU) No 167/2013, (EU) No 168/2013, (EU) 2018/858, (EU) 2018/1139 and (EU) 2019/2144 and Directives 2014/90/EU, (EU) 2016/797 and (EU) 2020/1828 (Artificial Intelligence Act) (Text with EEA relevance), https://eur-lex.europa.eu/legal-content/EN/TXT/?uri=CELEX%3A32024R1689. Accessed 22 Mar 2025
72. S. Nativi, S. De Nigris, AI Watch, *AI Standardisation Landscape State of Play and Link to the EC Proposal for an AI Regulatory Framework* (European Commission: Joint Research Centre Publications Office, 2021), https://data.europa.eu/doi/10.2760/376602
73. S. Ranchordas, Vulnerable citizens and algorithmic decision-making (2020), SSRN https://papers.ssrn.com/sol3/papers.cfm?abstract_id=3938032
74. Saudi Data & AI Authority, *National Strategy for Data and Artificial Intelligence* (2025). https://sdaia.gov.sa/en/SDAIA/SdaiaStrategies/Pages/NationalStrategyForDataAndAI.aspx. Accessed 25 Apr 2025
75. Saudi Data & AI Authority and Vision 2030 (2025), https://sdaia.gov.sa/en/SDAIA/SdaiaStrategies/Pages/sdaiaAnd2030Vision.aspx. Accessed 25 Apr 2025
76. Singapore's Approach to AI Governance (2025), https://www.pdpc.gov.sg/help-and-resources/2020/01/model-ai-governance-framework. Accessed 25 Apr 2025
77. Social Principles of Human-Centric AI (2019), https://www.cas.go.jp/jp/seisaku/jinkouchinou/pdf/humancentricai.pdf. Accessed 25 Apr 2025
78. T. Davtyan, The U.S. approach to AI regulation: Federal laws, policies, and strategies explained (2024), https://papers.ssrn.com/sol3/papers.cfm?abstract_id=4954290. Accessed 25 Apr 2025
79. The Alan Turing Institute, *AI Ethics and Governance in Practice* (2023), https://aiethics.turing.ac.uk. Accessed 12 Apr 2025
80. U.S. China Economic and Security Review Commission, China's smart cities development (Washington, DC, USCC, 2020). https://www.uscc.gov/sites/default/files/China_Smart_Cities_Development.pdf
81. UN-Habitat, *People-Centered Smart Cities Initiative* (2023), https://unhabitat.org/programme/people-centered-smart-cities. Accessed 14 Apr 2025
82. UNECE, *Sustainable and Smart Cities for All Ages* (United Nations Economic Commission for Europe, 2021), https://unece.org/sites/default/files/2021-01/SSC%20nexus_web_opt_ENG_0.pdf. Accessed 10 Mar 2025
83. UNESCO, *Recommendation on the Ethics of Artificial Intelligence* (2021), https://unesdoc.unesco.org/ark:/48223/pf0000381137. Accessed 14 Mar 2025
84. University of Edinburgh, *Emotional AI in Cities: Cross Cultural Lessons from UK and Japan on Designing for An Ethical Life* (2025), https://www.law.ed.ac.uk/research/research-projects/emotional-ai. Accessed 10 Apr 2025
85. M. Veale, F. Zuiderveen Borgesius, Demystifying the draft EU Artificial Intelligence Act. Comput. Law Rev. Int. **22**(4), 97–112 (2021). https://doi.org/10.9785/cri-2021-220402

Chapter 6
Empowering Smart Tourism with Large Language Models

Aristea Kontogianni and Efthimios Alepis

Abstract Smart tourism has emerged as a rapidly evolving field that integrates advanced digital technologies, artificial intelligence, and data-driven solutions to enhance tourism experiences, optimize destination management, and support sustainable development. Large Language Models (LLMs), which can provide real-time assistance, customized recommendations, and improved user engagement, represent a significant development in this field. This chapter investigates the intersection of LLMs with major smart tourism concepts including context awareness, real-time applications, user experience, recommender systems, cultural heritage, and data analytics. Through a structured literature review, we analyze how LLMs enhance these domains and identify the opportunities and challenges associated with their implementation. Additionally, we discuss the limitations of existing systems, ethical considerations such as privacy and bias, and potential future directions for optimizing AI-driven smart tourism services. Our findings suggest that LLMs have the potential to redefine smart tourism by making services more adaptive, user-centric, and efficient. To guarantee their responsible utilization, though, careful fine-tuning, ethical supervision, and integration with human knowledge are absolutely vital. This paper adds insights for academics, business leaders, and legislators trying to use LLMs for more immersive, inclusive, and eventually smart tourism experiences, so contributing to the continuous conversation on AI-driven smart tourism.

Keywords Smart tourism · Large language models · Context awareness · Recommender systems · User experience · Cultural heritage · AI-driven tourism · Data analytics · Privacy and data protection · Real-time assistance · Cyber tourism

---

A. Kontogianni (✉) · E. Alepis
University of Piraeus, Piraeus, Greece
e-mail: akontogianni@unipi.gr

E. Alepis
e-mail: talepis@unipi.gr

G. A. Tsihrintzis et al. (eds.), *Human-Smart City Interactions and User-Citizen Experiences*, Learning and Analytics in Intelligent Systems 58,
https://doi.org/10.1007/978-3-032-06364-9_6

6.1 Introduction

The proliferation of articles and websites devoted to the topic of "Smart Tourism" attests to the concept's maturity. Although the term was initially introduced in the late 1990s [1], it experienced substantial growth approximately a decade ago [2–4]. With hundreds of academic results on Google Scholar and an increasing number of papers in Scopus, interest has been gradually rising since then [5]. With over 34,500 results for the "Smart Tourism" term on Google Scholar and growing yearly publications in Scopus, interest in Smart Tourism has been steadily rising since then [5].

Smart tourism fundamentally utilizes new technology to establish a sophisticated digital ecosystem that advantages industries, visitors, residents, and destinations collectively. Smart tourism amalgamates diverse technologies and approaches through the utilization of sensors, cloud computing, cellphones, RFID, and Wi-Fi [6]. Applications encompass recommender systems that propose attractions according to user profiles [7], location-based advertising [6], and sensor-driven smart tourism platforms [8]. These technologies seek to augment tourism experiences, optimize resource management, and elevate competitiveness.

Actually, in the context of smart tourism, our focus extends well beyond merely enhancing the tourism experience. Through years of research in this field, we have established the following definition of smart tourism:

Smart tourism can be defined as tourism supported by technological infrastructure, algorithms, and Information & Communication Technologies (ICT) in general. It aims to enhance and improve the tourism experience, facilitate tourists' interaction with the environment, increase a destination's competitiveness, preserve tourist attractions, and improve the quality of life for local residents. Additionally, smart tourism provides valuable insights and tools for stakeholders, enabling data-driven decision-making and more efficient management of tourism resources.

Unquestionably, the recent fast expansion of artificial intelligence (AI) has increased the revolutionary possibilities of smart tourism. From personalized recommendations and customer service automation to predictive analytics for infrastructure maintenance [9, 10], artificial intelligence now supports a variety of uses in the aforementioned area of interest. Of these developments, the arrival of Large Language Models marks a major turning point. In smart travel, LLMs–including GPTs–show great promise in terms of real-time information delivery, instantaneous language translations, and custom travel recommendations.

In our previous research, we have identified and analyzed the key concepts that define smart tourism, recognizing that despite its widespread use, the term remains broad and open to multiple interpretations. Through an extensive literature review of academic research articles, conference proceedings, and industry reports, we categorized and evaluated fundamental topics that shape the smart tourism ecosystem. Especially, the following concepts and approaches acknowledged in the field of smart tourism were discovered: privacy and data protection; context awareness; cultural heritage; recommender systems; social media; Internet of Things; user experience;

real-time, user modeling; augmented reality; artificial intelligence; big data; cyber tourism.

We extend this work in the present study by investigating the link between LLMs and each of these significant concepts related to smart tourism. We aimed to find how LLMs connect with and improve components including context awareness, real-time processing, user experience, recommender systems, cultural heritage, privacy and data protection, cyber tourism, and big data analytics. By means of this systematic methodology, we evaluated the prospects and constraints of LLMs in many fields, therefore highlighting places where their influence is still restricted and others where they might be quite beneficial.

Our literature review led us to focus on specific concepts that align closely with LLM applications, while omitting certain areas where further exploration was unnecessary. For instance, since LLMs inherently fall under the broader category of Artificial Intelligence, we did not dedicate a separate section to discussing AI itself. Instead, we concentrated on evaluating how LLMs contribute to various smart tourism applications, identifying knowledge gaps, and proposing frameworks that could bridge existing limitations and optimize their implementation within the field.

Through this analysis, we aim to uncover the broad spectrum of ways in which LLMs can shape the smart tourism industry, offering insights that inspire both further research and practical innovation. Through this holistic approach, we aspire to provide a roadmap for integrating LLMs into smart tourism, highlighting their potential to redefine the industry while addressing critical gaps in current research.

This paper is structured as follows: Introduction 6.1 presents an overview of the study, outlining the significance of integrating Large Language Models (LLMs) into smart tourism. The next Sect. 6.2 details the methodology and research approach, describing the selection process for relevant literature and the key areas of interest examined. Following this, we explore the fundamental concepts of smart tourism and analyze how LLMs interact with essential domains such as context awareness, real-time applications, recommender systems, user experience, and data analytics Sect. 6.3. The subsequent Sect. 6.4 delve into the practical applications and challenges Sect. 6.5 of LLM integration, discussing ethical concerns, privacy issues, and implementation barriers. Finally, the paper concludes Sect. 6.6 with key findings, limitations, and directions for future research, emphasizing the potential of LLMs to shape the future of AI-driven smart tourism.

6.2 Methodology and Research Approach

To explore the integration of Large Language Models (LLMs) in Smart Tourism, a literature review was conducted. Our research aimed to identify relevant academic contributions, analyze key trends, and assess the ways in which LLMs intersect with various aspects of smart tourism.

More specifically, the literature review was carried out using Google Scholar, DBLP, ResearchGate, and Scopus, selected for their extensive coverage of peer-reviewed research papers, conference proceedings, and industry reports, ensuring a high-quality dataset for analysis. The search process followed a structured and iterative approach. Initially, we focused on identifying papers with "Smart Tourism" and "Large Language Models" in the title to establish a core set of relevant literature. To capture broader trends, we then expanded our search to include studies where the terms "tourism" and "LLMs" or "ChatGPT" appeared in the title, abstract, or keywords, acknowledging ChatGPT's prominence in AI-driven tourism research. Additionally, since LLMs are extensively studied beyond tourism, we also examined general AI and LLM-focused papers to gain a broader understanding of their capabilities and challenges in a tourism-related context.

From the reviewed literature, we identified key areas within smart tourism where LLMs could have a significant impact. The primary research themes included:

- Data and Analytics: The role of LLMs in processing and leveraging big data for smart tourism applications.
- User Experience (UX): Enhancing tourist engagement, personalization, and satisfaction.
- Real-Time Applications: Providing live support, recommendations, and event updates.
- Cyber Tourism: AI-driven virtual tourism experiences and interactive content generation.
- Privacy and Data Protection: Addressing security concerns, ethical considerations, and user privacy challenges.
- Context Awareness: Adapting recommendations based on user behavior, location, and preferences.
- Cultural Heritage: AI-powered tools for heritage preservation, museum guidance, and historical site exploration.
- Recommender Systems: Enhancing personalized travel recommendations through intelligent AI-driven approaches.

These thematic areas were chosen as they represent the core pillars of smart tourism and highlight the ways in which LLMs can drive innovation, efficiency, and sustainability. To systematically categorize the literature reviewed, Table 6.1 presents an overview of the selected papers and their primary areas of focus. The classification highlights the intersection of LLMs with various aspects of smart tourism, showcasing how different studies contribute to distinct domains within this field. While numerous papers explore the role of LLMs in tourism, the ones included in this table represent some of the most fundamental contributions that align with the key thematic areas identified in our research.

Table 6.1 Categorization of research papers in smart tourism

Paper	Year	Theoretical contribution	Context awareness	Cultural heritage	Recommender systems	UX	Real time	Data	Privacy and data protection
[11]	2024	x		x			x		x
[12]	2024	x					x	x	
[13]	2024	x							
[14]	2024								x
[15]	2024			x	x	x			
[16]	2024	x							
[17]	2023	x		x					
[18]	2023	x							
[19]	2023	x	x						
[20]	2024						x		
[21]	2023		x	x		x			
[22]	2023			x	x	x			
[23]	2024							x	
[24]	2024		x	x					
[25]	2023			x		x		x	
[26]	2024							x	
[27]	2024		x					x	
[28]	2023	x							
[29]	2024			x					
[30]	2023			x	x				
[31]	2024							x	
[32]	2024	x	x	x	x		x		
[33]	2024	x							
[34]	2024	x	x						
[35]	2024	x							
[36]	2024	x							x
[37]	2023					x	x		
Xiong (2024)	2024	x						x	
[38]	2025	x			x				
[39]	2025			x	x				
[40]	2024			x	x				
[41]	2024		x	x					

After collecting relevant papers, a qualitative content analysis was conducted to identify recurring patterns, methodologies, and findings. Particular attention was given to:

- Current implementations of LLMs in tourism-related applications.
- Potential benefits and improvements offered by LLM integration.
- Challenges and limitations that need to be addressed for effective adoption.

By structuring the research process in this way, we ensured that our study provides a comprehensive overview of the current literature while identifying opportunities and gaps for future research at the intersection of LLMs and smart tourism.

6.3 Intersection of LLMs with Smart Tourism Key Concepts

In this section, we analyze the relationship between Large Language Models (LLMs) and key concepts that shape smart tourism, as illustrated in Fig. 6.1. We specifically investigate how LLMs support Data and Analytics by allowing massive data to be processed and used, hence optimizing tourist applications. By improving visitor involvement, personalization, and general satisfaction, we also investigate how they affect User Experience (UX).Additionally, we discuss their role in Real-Time Applications, where LLMs provide live support, instant recommendations, and event updates. The concept of Cyber Tourism is addressed, highlighting AI-driven virtual tourism experiences and interactive content generation. Moreover, we address privacy and data protection with an eye toward security issues, ethical questions, and user privacy challenges connected to AI-powered travel solutions. Another important feature is context awareness since LLMs can modify recommendations according on user behavior, location, and preferences. We also explore Cultural Heritage, showing how technologies driven by artificial intelligence help to preserve legacy, guide museums, and investigate historical sites. Recommender systems and their connection with LLMs is also explored.

6.3.1 Data

In the realm of smart tourism, "data" is a term that emerges highly, particularly with the prominence of "Big Data" [42]. Indeed, since the early 21st century, Big Data has garnered widespread attention, becoming a transformative buzzword across numerous industries [43]. This so-called "Big Data Revolution" [44] has transformed the way organizations, value, process and utilize vast amounts of information. Despite its ubiquity, it could be said that defining Big Data remains challenging. A commonly cited framework is the "3-V" model, which encapsulates *"high-volume, high-*

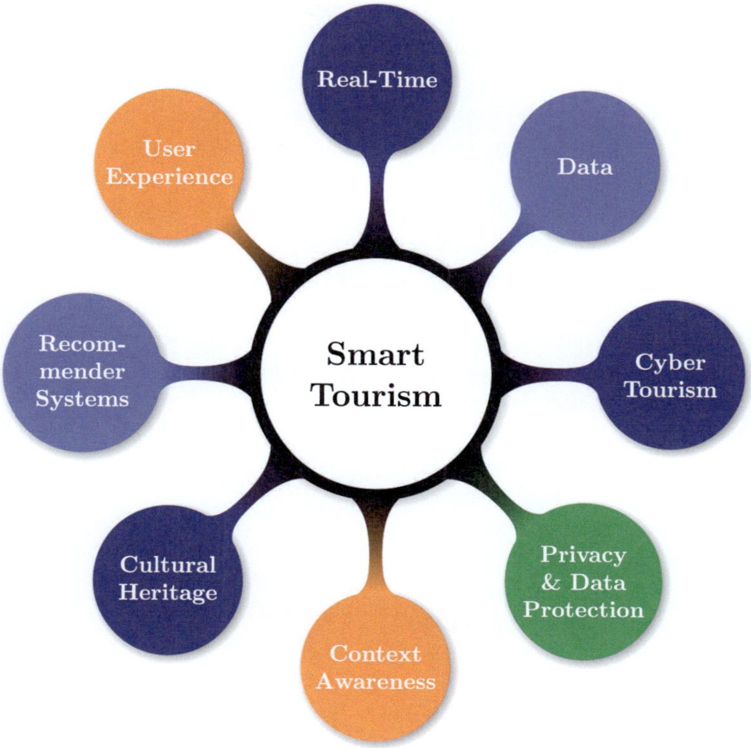

Fig. 6.1 Key dimensions of smart tourism

velocity, and high-variety information assets that necessitate cost-effective, innovative processing to enable enhanced insights and decision-making" [44, 45]. As a tool for improving awareness, predicting outcomes, and solving complex problems, Big Data has inevitably become of great significance in the context of smart tourism, as related literature indicated.

Although widely used, a number of researchers have voiced their disapproval of the phrase "Big Data" due to its lack of clarity, arguing that more specific terms like "data analytics" or "data science" would be more appropriate [44]. Acknowledging that, this subsection provides insights from a literature analysis of the larger idea of data in smart tourism and its intersection with large language models (LLMs). In the context of smart tourism, vast amounts of diverse data originating from various sources such as social networks, mobile applications, and sensors await analysis and utilization [46]. Value creation and innovation in enhancing the smart tourism experience can be built upon this data, which can originate from user activities (whether willingly shared, observed, or inferred) [46] along with factors such as interconnectivity, analytics, and real-time synchronization [43, 47, 48]. That being the case, numerous studies focus on *data* in the smart tourism realm.

At the same time, when it comes to LLMs, data is an essential component as their functionality is totally reliant on it. Books, journals, webpages, and a plethora of other digital sources are among the various kinds of information that are used to train such models. The aforementioned training data serves as the foundation for their learning and ability to recognize language patterns, structures, and links between words and concepts, resulting in the model's ability to offer precise and contextually relevant responses. The quality, amount, and variety of the previously described data actually have a significant influence on their performance [49]. Even after deployment, the necessity to update LLMs remains. The model's flexibility and efficacy in practical implementations are contingent upon this ongoing data integration [50]. Moreover, in numerous cases, feedback data is generated by user interactions, something that could also be employed to enhance the model's performance over time and to result in refined models that user expectations [51].

- Data
  - Data Analysis
    - Analyze trends, tourist behavior &preferences
      - Valuable Insights for decision-making
      - Valuable Insights for stakeholders
  - Tourism Specific Datasets
    - Multilingual &Global Data Integration
      - Improved tourist Experience
      - Promotion of Destination
  - Up to date data
    - Improved tourist Experience
    - Safety

That being the case, we recognize that data, LLMs, and smart tourism are concepts highly tight together, with most research in the field focusing on their interconnectedness in some form. Below, we outline key research areas in this domain, as presented in reviewed literature.

One critical area is the utilization of LLMs to the advantage of tourism stakeholders. To exemplify, authors in [11] emphasize the substantial potential of ChatGPT, a widely recognized LLM, in the aggregation of data from a variety of online sources. As an LLM, ChatGPT excels at analyzing trends, tourist behavior, and preferences, offering valuable insights to tourism stakeholders. Utilizing these data, one may craft marketing tactics that specifically target the needs of the market. Similarly, authors in [23] present an exploratory study leveraging LLMs for analyzing tourist behavior intent and developing a forecasting model using review comments as big data. Use of AI-driven analytics may help stakeholders improve customer engagement and marketing tactics, leading to increased efficiency and profitability in the tourist industry [52, 53]. Another interesting approach is presented in [31] were LLMs are used to generate additional structured data from web content, complementing the information extracted from microdata.

Moreover, another crucial aspect is the exploitation of data that result in the fine-tuning of LLMs. Indeed researchers emphasize the importance of fine-tuning large language models with tourism-specific datasets in order to adjust their functionality to the industry's unique requirements. Historically, tourism has adopted technology developed outside of the industry, resulting in systems that are not entirely aligned with industry requirements. By customizing LLMs based on input from diverse stakeholders–such as residents, employees, tourists, businesses, and policymakers–these models can address key challenges, facilitate evidence-based decision-making, and improve consumer experiences. By integrating multilingual, internationally generated data, such models help break down informational barriers and ensure that varied viewpoints are integrated into initiatives. To ensure accuracy and trustworthiness, comprehensive testing for validation, bias, and explainability is critical, coupled with methods to offer citations for AI-generated outputs [12].

6.3.2 *Privacy and Data Protection*

The fundamental connection between data and smart tourism is undeniable. Extensive research in this domain illustrates that smart tourism systems utilize a plethora of information, including locations, images, preferences, voluntarily shared or observed interests, academic and professional backgrounds, as well as data from smart devices and wearables, and on and on are just a few of the many data sources utilized to provide the so-called smart services. Although the aforementioned data are collected to forecast user preferences, discern behavioral patterns, and ultimately provide personalized technology-driven experiences via smart applications and services, they also result in the creation of precise tourist profiles, raising significant data privacy concerns [46, 54, 55].

That being the case, data management and user privacy, which are often inadequately managed [56], need to be addressed in the smart tourism domain. Thus, it comes as no surprise that this area is gaining more and more attention [6, 46, 54, 57, 58], particularly since systems now need to comply with the EU General Data Protection Regulation (GDPR).

"Fairness and Transparency," "Lawfulness of Processing: Consent, Legitimate Interests, Contract and Public sector," "Purpose Limitation," "Data Minimization: Collection and Retention," and "Accuracy and Up-to-date Processing" are five main approaches to ensuring that smart tourism destinations comply with privacy and data protection principles, as discussed by Masseno et al. [58]. Garanti [59] addresses the inherent challenges in smart tourism that are associated with user privacy. In this research the reliance on personal data for product and service development is emphasized along with the need to implement comprehensive privacy measures to safeguard user information. Another intriguing research is offered by Afolabi et al. [60] who investigate how context influences tourists' privacy concerns when using mobile applications in smart tourism destinations. Their findings reveal that various factors shape users' perceptions of privacy risks. Notably, they observed a negative correlation between privacy concerns and trust in service providers, emphasizing the importance of addressing these issues to build trust within smart tourism environments. In addition, Kovačić et al. [61] underscore the necessity of establishing tourist protection mechanisms, particularly in regions where tourism is a critical component of the economy, such as Mediterranean countries, as inadequate data security can compromise the privacy of tourists.

As already mentioned in previous subsection, LLMs are trained on massive datasets so as to possess the to generate humanlike responses and answer complex queries. Nevertheless, amidst this excitement regarding this new era enhanced by the LLMs' capabilities, concerns about data privacy are getting more and more attention [14]. Specifically, the authors in [14] emphasize two principal privacy risks: privacy leakage, wherein sensitive information may be inadvertently disclosed, and privacy attacks, which capitalize on model weaknesses. At the same time, a substantial percentage of users are familiar with these model and express concerns over data and privacy protection [62].

Thus, it comes as no surprise that research pertaining to smart tourism, large language models, and data privacy and protection is a topic of grave interest. For example, authors in [16] support that businesses should exercise caution when disclosing customers' personal information to ChatGPT, since this data might be utilized for model retraining, which could further compromise customers' privacy. Moreover, the ability of language models to inadvertently capture personal information poses risks of privacy infringements or harm to users [17].

6.3.3 Recommender Systems

When it comes to smart tourism, recommender systems hold a crucial role. Tourists often face an overwhelming amount of information to process, insufficient expertise to make appropriate decisions and in some cases the timing and location are not suitable for the best choices to thrive [42, 63] These systems, hold the key to the aforementioned challenges as their goal is to generate tailored suggestions that enhance the decision-making process [64, 65]. The majority of the early research on recommendation systems relied on reviews and ratings as forms of explicit user input. In order to provide more accurate recommendations, this method has progressed and modern systems are exploiting more implicit user data, such their browsing and purchase history [66, 67].

Due to the plethora of diverse data accessible for processing in the smart tourism industry, it is unsurprising that recommendation systems are extensively employed to customize the tourist experience [7, 63, 68]. Heterogenous data can be utilized to generate recommendations, including the characteristics of available services and products [69], contextual factors such as time, weather, user location, and budget [64], as well as an extensive dataset of options encompassing cultural heritage sites, hotels, restaurants, and more. The primary objective of processing the aforementioned data is to effectively satisfy customers' interests, preferences, and expectations, a task that is more challenging than it appears. Indeed, managing the enormous volumes of accessible data can complicate the recommendation process, a reality that has prompted practitioners and researchers to concentrate on recommender systems to devise efficient solutions.

Inspired by this fact, we have also devoted considerable effort to smart tourism-related recommender systems and frameworks. An example of a crowdsourced recommender system that can conduct image labeling, face and sentiment detection, and is based on the requirement to offer personalized suggestions to travelers is shown in [70]. In [9], we delved further and suggested a smart tourism framework that uses deep learning for picture labeling and predicting user preferences in order to create a smart tourism recommender system. This framework makes use of both explicit and implicit data. Smart tourism and recommender systems have indeed been the subject of many industry studies; these include [7, 8, 64, 71, 71–75].

Given the powerful capabilities of LLMs, various researchers support the potential of these models to revolutionalize recommender systems [76]. By leveraging their ability to process and analyze vast amounts of data, LLMs can significantly comprehend user behaviors and improve the personalization and accuracy of recommendations [77]. Thus, the integration of such models into recommender systems allows them to elevate their suggestions, making them more relevant but and contextually appropriate [77]. Various examples do exist that highlight this possibility. To exemplify, authors in [78] propose Chat-Rec that a system that blends traditional recommender systems with LLMs, exploiting LLMs' ability to learn from context. Another relevant research demonstrating the potential of LLMs to enhance recommendation performance is presented in [79].

That being the case, it comes as no surprise that the interconnection between LLMs, smart tourism, and recommender systems has also become a focal point for researchers in the smart tourism sector with various studies focusing on them. For example, authors in [16] support that tools like ChatGPT can enable "hyperpersonalization" by analyzing customer data, such as past behaviors, preferences, and spending patterns, to create more *to-the-point* marketing efforts. Moreover, authors suggest that integrating ChatGPT into travel websites with real-time data access allows tourists to receive personalized recommendations for hotels, restaurants, attractions, activities as well as other points of touristic interest. This technology could also be a game changer at travel planning by letting users communicate their needs, preferences, and constraints–like budgets or dietary requirements–through natural language, creating a more seamless and personalized experience [16], like having a conversation with an actual travel representative.

Additionally, ChatGPT excels at managing intricate questions and combining information from various sources to provide practical and budget-friendly travel suggestions [16, 17]. It has the ability to analyze what users want, including their favorite places, budgets, and any limitations, to recommend the best choices and even put together comprehensive travel plans that fit personal preferences. ChatGPT showcases its impactful role in smart tourism by creating responses tailored to individual users, providing smooth, context-sensitive solutions that can improve the overall travel experience [18].

6.3.4 Cultural Heritage

Another key research area when it comes to smart tourism is this of *Cultural Heritage*. Cultural tourism, which gained prominence in the early 1990s, is defined as the movement of people to cultural attractions outside their usual environment to fulfill cultural interests [80, 81]. Thus, cultural heritage that encompasses elements such as tradition, monuments, history, language, and folklore [82], and is deeply intertwined with tourism, as cultural immersion often motivates travel [83].

As a driver of economic and social growth [84], cultural tourism has inspired numerous regeneration policies [85]. Information and communication technologies (ICT) have further enhanced cultural tourism by making heritage attractions more engaging and interactive, thereby improving the visitor experience.

For over two decades, personalization in cultural heritage has been a key area of research, particularly with the advent of mobile technologies. Studies have focused on context-aware systems to provide travelers with tailored cultural information [86]. For instance, Nguyen et al. [87] proposed systems that guide tourists to significant nearby venues, while Be Memories [88] introduced a unique communication channel between visitors and cultural sites.

Recently, large language models have demonstrated potential in cultural heritage tourism [11], facilitating enhanced involvement and personalization. In the past two to three years, research on LLMs in the smart tourism sector has emerged. A pertinent study is [19], which examines cultural dimensions like as individualism versus collectivism and their potential impact on ChatGPT's efficacy. In collectivist Eastern cultures, ChatGPT is predominantly utilized for recommendations and information, while technological acceptance in individualistic Western civilizations is influenced by skepticism, privacy apprehensions, and market dynamics. The cultural disparities underscore the imperative for tourist stakeholders to tailor LLM applications to specific regional and cultural contexts, while simultaneously balancing their potential with ethical considerations and sustainability.

ChatGPT has also attracted the interest of cultural heritage tourism organizations due to its capacity to provide real-time, personalized experiences that align with individual preferences and past interactions. For instance, it can be used to create tailored guides or immersive gamified activities, such as role-playing games inspired by historical events, enhancing visitor engagement and learning in cultural heritage settings [11]. Furthermore, ChatGPT supports accessibility by offering features like audio descriptions for visually impaired visitors and sign language translations for those who are hearing impaired [11]. Building on this potential, recent research has explored how LLMs can function as interactive and personalized guides in cultural spaces [11].

Building on this potential, recent research has explored how LLMs can function as interactive and personalized guides in cultural spaces. For instance, in a study involving a virtual museum of contemporary art, researchers utilized GPT-4 to act as a knowledgeable assistant capable of providing tailored navigation and recommendations based on the museum's layout and exhibits [21]. The study focused on assessing the model's ability to understand spatial details and generate accurate guidance, showcasing its application in improving the visitor experience. By integrating LLMs with virtual and physical cultural environments, such research highlights their transformative role in delivering personalized and accessible interactions, as well as enhancing cultural engagement through innovative technologies. Expanding on these applications, systems like ExhibitXplorer [22] demonstrate how personalization in cultural tourism can be further refined through distributed architectures and AI-driven approaches. ExhibitXplorer integrates advanced technologies like geofencing, microservices, and GPT-based APIs to tailor museum content to individual visitor profiles. By combining implicit segmentation (behavioral analysis) and explicit segmentation (user-defined profiles), the system delivers highly personalized content, such as real-time exhibit recommendations and language-independent descriptions.

6.3.5 Real-Time

Real-time is a widely discussed concept in smart tourism-related literature, reflecting its critical role in enhancing tourist experiences and service delivery. Every moment, vast quantities of rich data are generated for the smart tourism industry, including traffic patterns, travel routes, and points of interest, which are readily accessible to travelers. These data streams enable a real-time view of areas of interest and facilitate the on-site provision of smart tourism services, such as dynamic navigation, real-time event updates, and location-based recommendations [89]. Beyond enhancing convenience, real-time data empower stakeholders to respond quickly to changing circumstances, such as crowd management, rerouting in case of traffic congestion, or adapting services based on demand fluctuations.

Moreover, as mentioned above cultural heritage organizations could exploit LLMs for various reasons. A main one could be the fact that these models may offer real-time personalized assistance. Tourists may get immediate, comprehensive answers that take context into account when they ask inquiries about a site's history, artwork, or traditions. Not only does this kind of real-time adaptation improve the experience for visitors, but it also helps tourist organizations promote sustainable practices, manage visitor flow more effectively, and highlight lesser-known locations [11, 22].

Additionally, tourists increasingly rely on real-time in-destination decision-making, enabled by smartphones and the need for immediate responses to emergencies in unfamiliar environments, such as overcrowding, poor weather, or health issues. LLMs significantly enhance real-time assistance by enabling self-service capabilities across a wide range of processes, far surpassing the functionality of single-feature chatbots. With their ability to act as personal assistants in real-time, LLMs offer 24/7 multilingual communication, providing timely and practical advice during medical emergencies or logistical challenges. Their capacity to incorporate past interactions and preferences further refines their real-time adaptability, delivering context-aware support tailored to individual needs By leveraging these advanced real-time capabilities, LLMs empower tourists to maximize their experiences, address unexpected problems efficiently, and plan last-minute trips with ease, ultimately enhancing their overall satisfaction with the destination [17].

6.3.6 User Experience

User experience, a widely used term, refers to the way users interact with and *experience* a platform, application, product, or service. In the global literature, UX is recognized as a multidimensional concept [90, 91], with a significant body of research focusing on it over the last decades [90, 92–95]. UX plays a pivotal role in human-computer interaction, offering a "fresh perspective" [91, 96], with its ultimate goal being a high level of user satisfaction achieved through the usefulness, convenience, and enjoyment derived from human-system interaction [90, 97]. That being said, when it comes to smart tourism, where systems are of grave importance, UX is a term highly researched. Comprehending travelers needs and offering a user experience of high quality via smart tourism systems is crucial so as to maintain user engagement and satisfaction, foster long lasting bonds and attract additional prospective users through positive reviews and recommendations [90, 98].

Recent advancements in human-AI interaction emphasize the role of generative models, such as Large Language Models (LLMs), in enhancing UX across various domains, including smart tourism. As highlighted in numerous researches [20, 21, 26], LLMs can provide personalized, context-aware assistance that adapts dynamically to user input in a more effective way than traditional chatbots [26]. This capability enables seamless interactions that cater to individual preferences, thereby enriching the overall user experience. Moreover, the integration of AI-driven systems into UX design supports real-time decision-making, enhances accessibility, and ensures inclusivity for diverse user groups. The ability of LLMs to deliver assistance in multiple languages also plays a vital role in improving the traveler's experience. By acting as a user's first point of contact, tourism information chatbots powered by LLMs provide valuable information and intelligent responses to inquiries. These systems significantly enhance customer satisfaction by delivering quick, nearly instantaneous answers. Such innovations have the potential to revolutionize smart tourism systems, offering tailored, intuitive experiences that not only meet but exceed user expectations.

Additionally, LLMs are now incorporating features like empathy and persuasive techniques, which further enhance their role in smart tourism. Empathy helps LLMs understand and respond to users' emotions, building trust and making interactions feel more personal [27]. This is particularly useful in areas like hotel bookings, where users value personalized and supportive interactions. At the same time, persuasive techniques such as social proof, scarcity, and personalization make LLMs more effective at engaging users and influencing decisions. For example, an LLM-powered chatbot might recommend a limited-time offer or highlight popular options tailored to a user's preferences. These combined features improve user satisfaction, increase engagement, and make LLMs a powerful tool for delivering adaptive, user-focused experiences [30].

6.3.7 Context Awareness

Context awareness is the capacity of a system to perceive and adapt to its surroundings by means of real-time data including geographical location, time, temperature, user activity [99]. Context-aware technologies have attracted a lot of interest as smart devices–especially in tourism–become essential parts of daily living. These systems provide individualized suggestions and adaptive services by using sensor data from GPS, motion detectors, and other environmental inputs [100]. Context awareness is fundamental in smart tourism in determining tourist decisions, maximizing travel experiences, and enhancing destination management [64]. Context-aware recommender systems have been included into several applications to deliver visitors customized recommendations depending on situational conditions [63, 64]. These solutions guarantee that users get tailored and significant recommendations by means of relevant contextual information, so improving engagement and happiness in smart tourist settings.

Although context-aware technologies are already very important for smart tourism, integrating Large Language Models (LLMs) offers novel opportunities to improve their capabilities. Traditional LLMs are not naturally context-sensitive, which is a major drawback in offering highly pertinent recommendations for use in travel applications. As demonstrated via the creation of context-aware recommender systems for museums, fine-tuning models using location-specific and situational data allows them to give customized solutions that fit individual user needs.

One important use of LLMs in context-aware smart tourism is their dynamic adaptation of recommendations depending on environmental variables. For example, the DualGen Bridge AI system uses fine-tuning approaches in the Tibetan tourism industry to provide real-time tourist viewpoint data depending on location-specific data [32]. Without much effort required, this lets guests get up to date information on surrounding attractions, historical places, or cultural monuments. Research on virtual museum tours also show how LLM-powered guides may offer individualized insights of objects, therefore guaranteeing an immersive and dynamic experience catered to guest preferences.

Furthermore, LLMs' interactive features help them to be adaptive virtual guides. Unlike traditional chatbots, which depend on pre-defined scripts, LLMs can dynamically change their answers depending on user inquiries, therefore providing richer and more significant interactions [20]. Virtual museum applications show how this method lets visitors explore locations more naturally, ask context-aware inquiries, and get answers that change depending on their interests. In smart tourism, where real-time data, cultural sensitivity, and tailored recommendations greatly affect visitor satisfaction, this adaptability is especially important.

LLMs could transform smart travel by improving their capacity for processing and comprehending context. Their incorporation into context-aware systems guarantees that visitors get more relevant, timely, and engaging experiences whether through real-time travel help, personalized itineraries, interactive cultural encounters, or automated content generating. LLMs should become more important in smart tourism as they develop, so closing the distance between artificial intelligence-driven automation and human-centric travel experiences.

6.3.8 Cyber Tourism

Cyber tourism is another concept widely known and highly connected to smart tourism. When addressing cyber tourism, physical, social, and economic constraints can be eliminated, allowing everyone to access an entirely new tourism sphere. At the same time, it has the ability to merge the past, present, and future, something that can only be achieved in the virtual world realm[101]. More specifically, the definition of cyber tourism is, as given by [102]:

> *"Cyber Tourism: Is a developing alternative type of tourism where information technologies, especially virtual reality technology, are widely used. Cyber tourism has led to a change in the traditional space conceptualization, and has become increasingly emphasized in the context of tourism accessibility and smart tourism."*

There are currently numerous researchers and practitioners that examine the prospects of cyber tourism and offer a cyber tourism experience with only a computer or smartphone with internet access required. Numerous studies at the time of writing attempt to empower and expand cyber tourism describing various approaches that incorporate virtual reality with 3D locations, holographies, sensory devices, robotics, etc [103]. A prime example is *Google Arts & Culture* [104], a non-profit initiative that, in collaboration with cultural organizations and artists from around the globe, makes the world's art and culture accessible online to everyone, anywhere, at any time. The Acropolis Museum in Greece offers a website that is worth visiting because its virtual reality 360-degree panoramas enable online viewers to take a panoramic tour of the museum's galleries, providing a *feel* like actual experience.[1] In addition, the collaboration between the Hellenic Ministry of Culture and Sports and Microsoft Corporation has brought the Ancient Site of Olympia to life as never before, creating an exceptional example of cyber tourism. Built on Microsoft's artificial intelligence technology,[2] enables users from across the world to explore the Ancient Site of Olympia in 3D, inspecting various monuments and ancient games. Another noteworthy example is Rickshaw Tours,[3] an innovative platform that connects users with live guides for interactive virtual tours across the globe. By utilizing

[1] https://theacropolismuseum.gr/polymesa/eikoniki-periigisi-sto-moyseio-akropolis.

[2] https://olympiacommongrounds.gr/explore.

[3] https://rickshaw-tours.com/.

video streaming technology, Rickshaw Tours allows users to experience real-time, immersive tours of destinations, cultural landmarks, and natural wonders from the comfort of their homes. This platform emphasizes the human aspect of cyber tourism, enabling tourists to interact with guides, ask questions, and engage in meaningful discussions, thereby creating a truly personalized and enriching experience.

With that in mind, as the attempts to harness the potential of new technology in cyber tourism thrive, some researches have begun to address the integration of Large Language Models (LLMs) into the cyber tourism landscape. Guiding visitors in virtual tours and providing them with desired, tailored content has traditionally been hindered by difficulties in managing large datasets, offering diverse content for different users, and maintaining timely updates, especially for smaller institutions with limited resources [20]. Integrating LLMs into these systems could allow institutions to streamline guidance tailored to individual requests, providing a scalable approach that improves accessibility and engagement.

Recent developments highlight the amazing capacity of LLMs to improve the breadth of multi-modal interactions in virtual tours [27]. One such development is the design of the VirtuWander system By including LLMs for real-time direction, context-aware assistance, and custom experiences, VirtuWander improves conventional virtual tour systems. By use of many bots, the system offers LLM-based chatbots with specialized knowledge, appropriate context, and links to human surroundings, therefore enabling interesting and responsive guidance. All meant to satisfy different user needs and guarantee an interesting and immersive experience, this includes topic tour investigation, individual artwork exploration, and customized personal tours.

User studies have confirmed the versatility and effectiveness of LLMs in virtual tourism systems, showcasing their capacity for personalized communication, expert knowledge delivery, and improved user satisfaction [27]. With features like natural language input, automated wayfinding, and personalized recommendations, LLMs connect what users expect with what virtual tourism platforms can actually do. These systems enhance user experiences by being engaging and adaptive, while also making cultural and historical knowledge more accessible, thus promoting inclusivity in cyber tourism for audiences around the world.

As LLMs keep evolving, they are becoming crucial in cyber tourism, allowing both big and small organizations to offer top-notch virtual experiences that satisfy and surpass what users anticipate. By tackling intricate issues, ranging from data handling to tailoring user experiences, LLMs are leading us into a fresh phase of cyber tourism that emphasizes innovation and puts users first.

6.4 Key Axes of Smart Tourism Enhanced by Large Language Models (LLMs)

In the above sections, we have focused on the integration of smart tourism in several different key areas of interest in the context of smart tourism that we have identified through our research over the last decade. Apart from that, there are numerous key axes in the context of smart tourism that LLMs could enhance, leading to a new era of AI-driven tourism. As illustrated in Fig. 6.2, LLMs have the potential to contribute across multiple dimensions of smart tourism.

More precisely, the necessity to improve the whole smart tourist experience is absolutely vital and maybe the most clearly recognizable aspect of the field. Personalized recommendations and real-time virtual support via chatbots are two areas where the contribution of LLMs might be regarded as substantial. By providing specialized recommendations for itineraries, bespoke travel advice, and identification of unusual events that fit a tourist's expectations, LLMs can alter the whole travel experience [28]. By evaluating user preferences, LLM-powered systems can dynamically provide recommendations for activities, accommodation, and restaurants, therefore ensuring that every vacation experience is especially suitable for the individual. Furthermore, ChatGPT and associated models can provide real-time local attraction insights, advise less-known tourist destinations, and enhance context-awareness by proposing restaurants depending on location and user interests [13]. These features make AI-driven platforms an excellent tool for acquiring and retaining customers, as they substantially improve consumer engagement and satisfaction.

Furthermore improving customer service and concierge features is including LLMs into front-office processes into chatbots. By automating concierge services, hotels and tourism companies using LLM-based chatbots can give visitors real-time help on directions, bookings, and travel-related questions. These chatbots can also create instructions for hotel services, including check-in and check-out processes, facility running hours, emergency protocols, and sustainability policies, so guaranteeing a seamless and instructive visitor experience [16]. Furthermore, LLMs can assist with event-based reservations and bulk bookings, therefore lowering friction in the booking process and providing multimedia materials including photographs, videos, and virtual tours to support guests in their decision-making [13].

Moreover, another key aspect of smart tourism is creating a more authentic and immersive connection between tourists and their environment. This implies not just granting access to natural and cultural sites but also improving the interaction of guests with them, offering a unique experience. LLM-powered systems can play a crucial role in interpreting historical sites, museums, and natural wonders, offering rich, context-aware narratives that deepen the visitor's understanding. Multilingual assistance also guarantees flawless communication, therefore removing language barriers and enabling a worldwide audience to access places. AI-driven solutions can convert passive tourism into a really participatory and interesting experience by closing the distance between visitors and their surroundings.

Another crucial component of the smart tourism concept is destination competitiveness, where LLMs can provide a significant advantage by leveraging data-driven insights to enhance marketing strategies and optimize decision-making. AI-powered analytics via the exploitation of vast amounts of data from tourist reviews [16], social media, travel patterns and platforms, are capable of allowing destinations to "understand" emerging trends and adapt their offerings accordingly. With the term destinations we refer not only to businesses applications but mainly to gov-

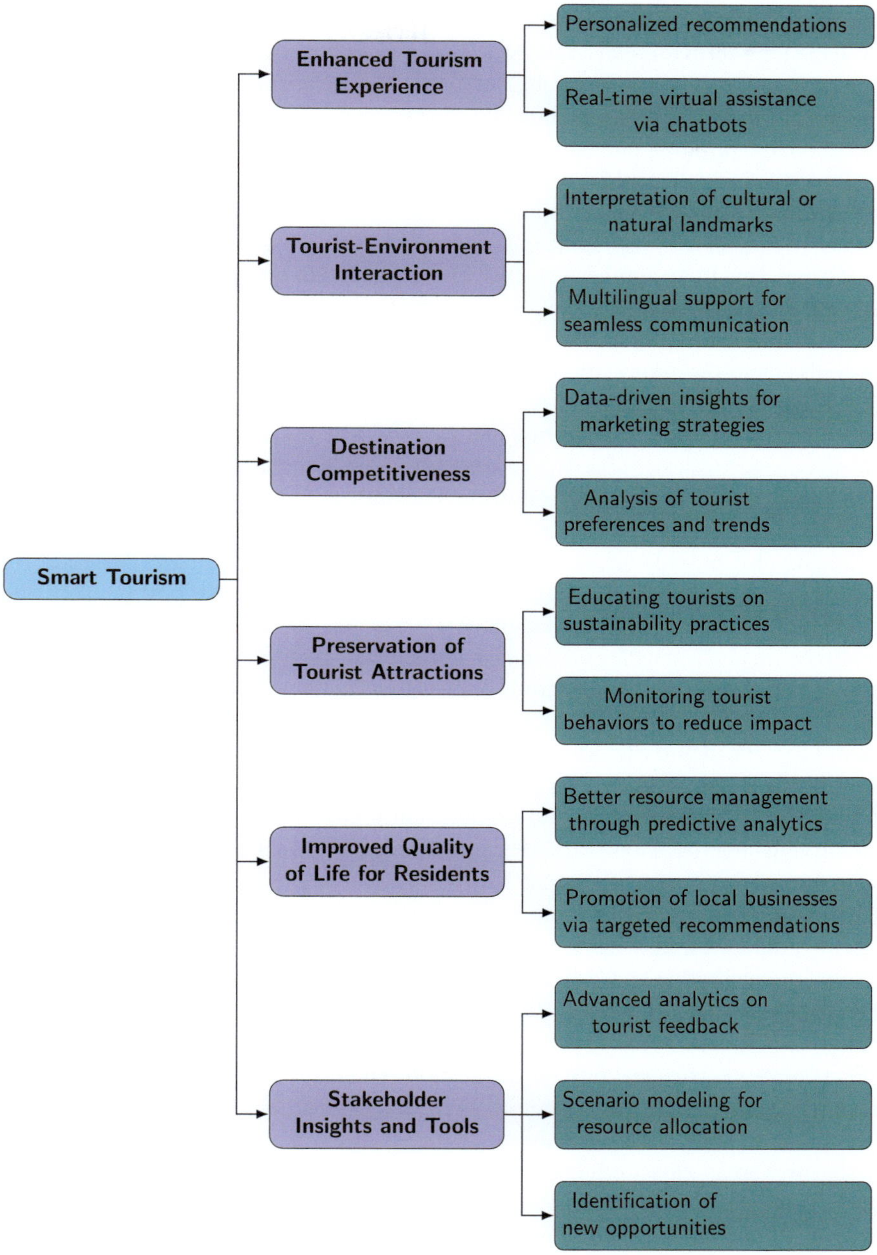

Fig. 6.2 Key axes of LLM-Enhanced smart tourism

ernments and policymakers that could harness the power of LLM-driven insights to shape long-term tourism strategies. By analyzing visitor demographics, spending behaviors, and satisfaction levels, policymakers can make informed decisions on infrastructure investments, sustainable tourism policies, and destination branding. Additionally, if tourist interactions with LLMs are leveraged anonymously and ethically, these insights could further enhance strategic planning. Analyzing such data could identify hidden tourism potential, spot changing travel interests, and more precisely project seasonal demand swings.

At the same time, another key axis where LLMs can provide substantial value is the preservation of tourist attractions, an essential component of sustainable tourism management. Travelers can be taught sustainability techniques by LLM-powered systems, therefore promoting responsible conduct that reduces cultural and environmental damage. Moreover, real-time observation of visitor actions can enable authorities to spot trends of overuse and suggest remedial actions to minimize possible damage to historical sites or delicate ecosystems. This would help authorities allocate resources more effectively, ensuring that tourism hotspots are not overwhelmed, while lesser-known attractions gain visibility.

LLMs also provide advanced analytics on tourist input, scenario modeling for resource allocation, and the identification of novel prospects for development and innovation, so transforming stakeholder insights and decision-making. AI-powered sentiment analysis enables companies and legislators automatically categorize and extract important information from online reviews, therefore guiding their understanding of consumer satisfaction levels and pointing up areas needing development [16]. By identifying prospects for creative tourism products and recognizing developing target customers, LLMs also help businesses grow and expand markets. AI can be used by destinations to examine consumer preferences and behaviors, so identifying underused tourism assets or fresh experiences fit changing market trends. AI-powered recommendations can help restaurant owners in the hotel industry optimize menus, customize recommendations to local culinary tastes, and advise on pricing policies for new markets [13].

Beyond marketing and budget allocation, LLMs are extremely important for policy development since they assist to find gaps in laws, visitor safety issues, and environmental obstacles. They can examine environmental data, evaluate how tourism affects nearby towns, and provide suggestions on how to strike a mix between environmental preservation and economic development. Such insights allow governments to guarantee long-term sustainability by means of proactive policies as visitor caps in sensitive areas, clever pricing methods, or investments in environmentally friendly travel projects. Finally, all the abovementioned advancements contribute directly to improving the quality of life for residents, another critical aspect of smart tourism that should always be kept under consideration.

6.5 Challenges

Although the integration of Large Language Models (LLMs) into smart tourism has numerous, unquestionable benefits, it also raises ethical questions and presents some challenges that need to be identified and addressed. The aforementioned issues are diverse and primarily concerned with privacy, fairness, openness, misleading information, and sustainability–all of which can have an impact on local businesses, visitors, and stakeholder groups. In this section we focus on some of the most pressing concerns.

As already mentioned in Sect. 6.3.2, one of the biggest concerns surrounding LLMs is privacy. Undeniably, these models rely on vast amounts of user data, including personal and behavioral information, to generate personalized recommendations. of this data could result in privacy violations, therefore raising questions of data security and regulatory compliance including GDPR [30]. Ensuring anonymity and ethical data collecting is vital to stop data breaches and illegal profiling since tourism platforms get information via bookings, reviews, and interactions.

Furthermore, LLMs inherit biases from their training data [105], which can distort recommendations and lead to unfair treatment among travelers [28]. This is especially alarming when artificial intelligence-driven systems give particular locations, businesses, or demographics top priority, therefore excluding or unfairly exposing smaller or developing travel markets. Furthermore influencing the accuracy of information could be cultural and policy-related prejudices, so either favoring some narratives over others or misrepresenting areas [19].

Another critical challenge is the lack of transparency in LLM decision-making. These models operate as "black boxes", making it difficult to understand why certain recommendations are generated. The difficulty to track AI-driven decisions causes questions regarding responsibility and credibility for governments and tourism companies [76]. Since consumers should know why a given location, hotel, or activity is being recommended, provide explainability in AI-generated recommendations.

Moreover, a well-documented drawback with LLMs is their propensity to provide false or misleading information, occasionally even creating nonexistent sources or places [12]. In travel, where visitors depend on AI-generated information for local regulations, safety warnings, and itinerary planning, this can especially be risky. Without appropriate validation systems, false information may mislead visitors, therefore influencing their experiences and maybe damaging the reputation of certain locations. In numerous cases could be even dangerous, as activities proposed may be much more risky than the models suggest, something that may be well-know to locals but an unpleasant surprise to tourists.

Another consequence of AI-driven recommendations, that should not be taken lightly is the risk of over-tourism. Since LLMs often highlight popular destinations, this could lead to over-concentration of tourists in already crowded areas while neglecting lesser-known but equally valuable attractions [19]. This disparity can strain infrastructure, aggravate environmental damage, and lower local residents' quality of life as well as visitors'. AI systems should be built to spread tourists more fairly and encourage environmentally friendly travel options in order to help to offset this.

While LLMs greatly improve efficiency and personalization, they also raise concerns about reducing human interaction, a challenge that extends beyond tourism and affects almost every industry. In tourism, overreliance on AI-driven chatbots could mean fewer genuine interactions with local businesses, tour guides, and cultural hosts, potentially diminishing the human connection that makes travel so enriching [106]. At the same time, as AI takes over tasks traditionally handled by people, there are growing concerns about job displacement, particularly in the hospitality and service sectors [28]. While automation can bring convenience, finding the right balance between AI-driven assistance and human engagement is crucial to preserving the authentic and social aspects of tourism–a challenge that many industries face as they adapt to an AI-driven future.

Most general-purpose LLMs, including ChatGPT, have data cut-off limitations and may not reflect recent travel trends, updated regulations, or real-time events [12]. Since retraining large models is extremely costly and time-consuming, tourism stakeholders may struggle to ensure that AI-generated insights remain accurate and relevant. One solution is the development of tourism-specific LLMs that focus on real-time data integration, ensuring more precise and updated recommendations.

6.6 Conclusions and Discussions

The present study investigates the integration of Large Language Models (LLMs) in smart tourism with an eye toward how these AI-driven technologies interact with significant concepts in this field. By means of a study of current research, industrial applications, and case studies, we investigated how LLMs interact with fundamental smart tourism pillars like context awareness, real-time assistance, personalized suggestions, data analytics, user experience enhancement, and cyber tourism. By focusing on these areas, our research emphasizes the possibilities as well as the difficulties of implementing LLMs in the smart tourism realm.

The results of the research show the great possibilities of LLMs in redefining smart tourism by improving efficiency, personalizing, and flexibility to fit changing travel conditions. Their capacity to provide context-aware insights, evaluate big datasets, and interpret natural language searches helps both stakeholders and visitors to make better decisions. Furthermore, LLM-driven systems can support destination management plans, raise visitor involvement, and enhance customer service automation, so generating a more intelligent, smooth, engaging travel experience.

Despite these advantages, challenges remain. Issues such as privacy concerns, data security, algorithmic bias, and ethical considerations require careful regulation and transparency. The reliance of LLMs on large datasets introduces the risk of biased recommendations, and their "black-box" nature raises concerns about explainability and accountability. Furthermore, over-reliance on AI may reduce human interaction, an essential component of tourism experiences.

Future studies should concentrate on optimizing LLMs to more fit regional and cultural tourist environments. Their responsible application depends critically on increasing language support, real-time data integration, and privacy rule compliance. Furthermore, hybrid models combining artificial intelligence with human knowledge could assist to strike a balance between automation and real, human-centered experiences.

Ultimately, this study contributes to the growing field of AI-driven smart tourism by examining both the potential and limitations of LLMs in this domain he ways in which artificial intelligence technologies improve traveler experiences, streamline travel services, and increase accessibility will only become more important as they advance. Dealing with current issues and maximizing their potential will help to guarantee that AI-driven travel solutions provide a more interesting, inclusive, and sustainable travel experience.

References

1. Navruz-zoda Zebiniso Bakhtiyorovna. Three-level smart model of tourist destination. *GWALIOR MANAGEMENT ACADEMY*, page 78, 1998
2. Dimitrios Buhalis and Aditya Amaranggana. Smart tourism destinations. In *Information and communication technologies in tourism 2014*, pages 553–564. Springer, 2013
3. Fuad Mehraliyev, Irene Cheng Chu Chan, Youngjoon Choi, Mehmet Ali Koseoglu, and Rob Law. A state-of-the-art review of smart tourism research. *Journal of Travel & Tourism Marketing*, 37(1):78–91, 2020. https://doi.org/10.1080/10548408.2020.1712309. https://doi.org/10.1080/10548408.2020.1712309
4. Dan Wang, Xiang Li, and Yunpeng li. China's "smart tourism destination" initiative: A taste of the service-dominant logic. *Journal of Destination Marketing & Management*, 2:59–61, 06 2013. https://doi.org/10.1016/j.jdmm.2013.05.004
5. Ulrike Gretzel. From smart destinations to smart tourism regions. *Investigaciones Regionales-Journal of Regional Research*, pages 171–184, 2018
6. Manuel David Masseno and Cristiana Santos. Privacy and data protection issues on smart tourism destinations - A first approach. In *Intelligent Environments 2018 - Workshop Proceedings of the 14th International Conference on Intelligent Environments, Rome, Italy, 25-28 June 2018*, pages 298–307, 2018. https://doi.org/10.3233/978-1-61499-874-7-298. URL https://doi.org/10.3233/978-1-61499-874-7-298
7. Damianos Gavalas, Charalampos Konstantopoulos, Konstantinos Mastakas, and Grammati Pantziou. Mobile recommender systems in tourism. *Journal of Network and Computer Applications*, 39:319 – 333, 2014. ISSN 1084-8045. doi:https://doi.org/10.1016/j.jnca.2013.04.006. URL http://www.sciencedirect.com/science/article/pii/S1084804513001094
8. Michele Nitti, Virginia Pilloni, Daniele D. Giusto, and Vlad Popescu. Iot architecture for a sustainable tourism application in a smart city environment. *Mobile Information Systems*, 2017:9201640:1–9201640:9, 2017. https://doi.org/10.1155/2017/9201640. URL https://doi.org/10.1155/2017/9201640
9. A. Kontogianni, E. Alepis, C. Patsakis, Promoting smart tourism personalised services via a combination of deep learning techniques. Expert Syst. Appl. **187**, 115964 (2022)
10. Aristea Kontogianni, Efthimios Alepis, Maria Virvou, and Constantinos Patsakis. *Smart Tourism-The Impact of Artificial Intelligence and Blockchain*. Springer, 2024b
11. Marianna Sigala, Keng-Boon Ooi, Garry Wei-Han Tan, Eugene Cheng-Xi Aw, Dimitrios Buhalis, Tat-Huei Cham, Meng-Mei Chen, Yogesh K Dwivedi, Ulrike Gretzel, Alessandro

Inversini, et al. Understanding the impact of chatgpt on tourism and hospitality: Trends, prospects and research agenda. *Journal of Hospitality and Tourism Management*, 60:384–390, 2024
12. Cathy HC Hsu, Guoxiong Tan, and Bela Stantic. A fine-tuned tourism-specific generative ai concept. *Annals of Tourism Research*, page 103723, 2024
13. Yogesh K Dwivedi, Neeraj Pandey, Wendy Currie, and Adrian Micu. Leveraging chatgpt and other generative artificial intelligence (ai)-based applications in the hospitality and tourism industry: practices, challenges and research agenda. *International Journal of Contemporary Hospitality Management*, 36(1):1–12, 2024
14. Biwei Yan, Kun Li, Minghui Xu, Yueyan Dong, Yue Zhang, Zhaochun Ren, and Xiuzhen Cheng. On protecting the data privacy of large language models (llms): A survey. *arXiv preprint* arXiv:2403.05156, 2024
15. Hoon-chul Kang, Myeong-Cheol Jwa, and Jeong-Woo Jwa. The task-oriented smart tourism chatbot service. *International Journal of Membrane Science and Technology*, 10(4), 2023
16. I. Carvalho, S. Ivanov, Chatgpt for tourism: applications, benefits and risks. Tourism Review **79**(2), 290–303 (2024)
17. An early glimpse into chatgpt and generative ai, IpKin Anthony Wong, Qi Lilith Lian, and Danni Sun. Autonomous travel decision-making. J. Hosp. Tour. Manag. **56**, 253–263 (2023)
18. Yu. Dogan Gursoy, Li, and Hakjun Song., Chatgpt and the hospitality and tourism industry: an overview of current trends and future research directions. Journal of Hospitality Marketing & Management **32**(5), 579–592 (2023)
19. Sirong Chen, Ke Zhang, Xiaonan Li, Huiyue Ye, Katsy Jiaxin Lin, and Rob Law. Chatgpt: Cross-cultural tourism research imperative. *Journal of Economics and Management*, 45(1):137–146, 2023
20. Iva Vasic, Hans-Georg Fill, Ramona Quattrini, and Roberto Pierdicca. Llm-aided museum guide: Personalized tours based on user preferences. In *International Conference on Extended Reality*, pages 249–262. Springer, 2024
21. G. Trichopoulos, M. Konstantakis, G. Alexandridis, G. Caridakis, Large language models as recommendation systems in museums. Electronics **12**(18), 3829 (2023)
22. R. Ivanov, Exhibitxplorer: Enabling personalized content delivery in museums using contextual geofencing and artificial intelligence. ISPRS Int. J. Geo Inf. **12**(10), 434 (2023)
23. Kaijian He, Linyuan Zheng, Don Wu, and Yingchao Zou. Leveraging large language models for daily tourist demand forecasting. *Current Issues in Tourism*, pages 1–17, 2024
24. Qikai Wei, Mingzhi Yang, Jinqiang Wang, Wenwei Mao, Jiabo Xu, and Huansheng Ning. Tourllm: Enhancing llms with tourism knowledge. *arXiv preprint* arXiv:2407.12791, 2024
25. Jianing Yao. Elevating urban tourism: Data-driven insights and ai-powered personalization with large language models brilliance. In *2023 IEEE 3rd International Conference on Social Sciences and Intelligence Management (SSIM)*, pages 138–143. IEEE, 2023
26. Simone Barandoni, Filippo Chiarello, Lorenzo Cascone, Emiliano Marrale, and Salvatore Puccio. Automating customer needs analysis: A comparative study of large language models in the travel industry. *arXiv preprint* arXiv:2404.17975, 2024
27. Zhan Wang, Lin-Ping Yuan, Liangwei Wang, Bingchuan Jiang, and Wei Zeng. Virtuwander: Enhancing multi-modal interaction for virtual tour guidance through large language models. In *Proceedings of the CHI conference on human factors in computing systems*, pages 1–20, 2024
28. L. Mich, R. Garigliano, Chatgpt for e-tourism: a technological perspective. Information Technology & Tourism **25**(1), 1–12 (2023)
29. Aaswit Kanhed, Pratham Bhagwat, Kunal Palande, Nishant Palav, Sonal Balpande, Kiran Deshpande, and Urjashree Patil. Destinai: Your personalized travel itinerary planner and chat catalyst using generative ai. In *International Conference on Information and Communication Technology for Intelligent Systems*, pages 47–55. Springer, 2024
30. Manolis Remountakis, Konstantinos Kotis, Babis Kourtzis, and George E Tsekouras. Using chatgpt and persuasive technology for personalize d recommendation messages in hotel upselling. *Information*, 14(9):504, 2023

31. Lino Gonzalez-Garcia, Gema González-Carreño, Ana María Rivas Machota, and Juan Padilla Fernández-Vega. Enhancing knowledge graphs with microdata and llms: the case of schema. org and wikidata in touristic information. *The Electronic Library*, 2024
32. Jinhu Qi, Shuai Yan, Wentao Zhang, Yibo Zhang, Zirui Liu, and Ke Wang. Research on tibetan tourism viewpoints information generation system based on llm. In *2024 12th International Conference on Intelligent Computing and Wireless Optical Communications (ICWOC)*, pages 35–41. IEEE, 2024
33. Aristea Kontogianni and Efthimios Alepis. Ai in smart tourism: Llms & gpts leading the way. In *2024 15th International Conference on Information, Intelligence, Systems & Applications (IISA)*, pages 1–8. IEEE, 2024
34. Xuchuan Li, Fei Huang, Jianrong Lv, Zhixiong Xiao, Guolong Li, and Yang Yue. Be more real: Travel diary generation using llm agents and individual profiles. *arXiv preprint* arXiv:2407.18932, 2024
35. Sandra Maria Correia Loureiro, João Guerreiro, Enav Friedmann, Myong Jae Lee, and Heesup Han. Tourists and artificial intelligence-llm interaction: the power of forgiveness. *Current Issues in Tourism*, pages 1–19, 2024
36. Ruiping Ren, Xing Yao, Shu Cole, and Haining Wang. Are large language models ready for travel planning? *arXiv preprint* arXiv:2410.17333, 2024
37. Ommi Balamurali, AM Abhishek Sai, Moturi Karthikeya, and Sruthy Anand. Sentiment analysis for better user experience in tourism chatbot using lstm and llm. In *2023 9th International Conference on Signal Processing and Communication (ICSC)*, pages 456–462. IEEE, 2023
38. Bestoun S Ahmed, Ludwig Otto Baader, Firas Bayram, Siri Jagstedt, and Peter Magnusson. Quality assurance for llm-rag systems: Empirical insights from tourism application testing. *arXiv preprint* arXiv:2502.05782, 2025
39. Angelo Geninatti Cossatin, Noemi Mauro, Fabio Ferrero, and Liliana Ardissono. Tell me more: integrating llms in a cultural heritage website for advanced information exploration support. *Information Technology & Tourism*, pages 1–32, 2025
40. Alexio Cassani, Michele Ruberl, Antonio Salis, Giacomo Giannese, and Gianluca Boanelli. zia: a genai-powered local auntie assists tourists in italy. *arXiv preprint* arXiv:2407.11830, 2024
41. Haoran Lin, Jing Wang, Yutong Zhan, and Yuan Zhang. Tour guide robot based on large language model and user behavior analysis. In *2024 5th International Conference on Artificial Intelligence and Computer Engineering (ICAICE)*, pages 794–798. IEEE, 2024
42. Aristea Kontogianni, Efthimios Alepis, Maria Virvou, and Constantinos Patsakis. Conceptualizing smart tourism. In *Smart Tourism–The Impact of Artificial Intelligence and Blockchain*, pages 7–31. Springer, 2024a
43. Thomas H. Davenport. At the big data crossroads: turning towards a smarter travel experience, 2013. Available at http://amadeusblog.com/wp-content/uploads/Amadeus-Big-Data-Report.pdf, Accessed August 7, 2019
44. Neil M Richards and Jonathan H King. Big data ethics. *Wake Forest L. Rev.*, 49:393, 2014
45. IT Glossary. It glossary: Big data, gartner. Available at https://www.gartner.com/en/information-technology/glossary/big-data ,Accessed April 05, 2022
46. Manuel David Masseno and Cristiana SANTOS, Smart tourism destinations privacy risks on data protection. Revista Eletrônica Sapere Aude **1**(1), 125–149 (2018)
47. Ulrike Gretzel, Marianna Sigala, Zheng Xiang, and Chulmo Koo. Smart tourism: foundations and developments. *Electronic Markets*, 25(3):179–188, Sep 2015. ISSN 1422-8890. https://doi.org/10.1007/s12525-015-0196-8. URL https://doi.org/10.1007/s12525-015-0196-8
48. Q Tu and A Liu. Framework of smart tourism research and related progress in china. *International conference on management and engineering (CME 2014)*, pages 140–146, 01 2014
49. Zhang. Exploring the potential of large language models in radiological imaging systems: Improving user interface design and functional capabilities. *Electronics*, 13(11):2002, 2024. https://doi.org/10.3390/electronics13112002
50. I. Tenney, D. Das, R. Garnett, E. Pavlick, Bert rediscovers the classical nlp pipeline. (2019). https://doi.org/10.18653/v1/p19-1452

51. Serapio-García. Personality traits in large language models, 2023
52. R. Filieri, E. D'Amico, A. Destefanis, E. Paolucci, E. Raguseo, Artificial intelligence (ai) for tourism: an european-based study on successful ai tourism start-ups. Int. J. Contemp. Hosp. Manag. **33**(11), 4099–4125 (2021)
53. R. Pillai, B. Sivathanu, Adoption of ai-based chatbots for hospitality and tourism. Int. J. Contemp. Hosp. Manag. **32**(10), 3199–3226 (2020)
54. Dimitrios Buhalis and Aditya Amaranggana. Smart tourism destinations. In Zheng Xiang and Iis Tussyadiah, editors, *Information and Communication Technologies in Tourism 2014*, pages 553–564, Cham, 2014. Springer International Publishing. isbn:978-3-319-03973-2
55. Manuel Masseno. On the relevance of big data for the formation of contracts regarding package tours or linked travel arrangements, according to the new package travel directive. *Revista Direitos Emergentes na Sociedade Global*, 5(1):56–70, 2017. ISSN 2316-3054. https://doi.org/10.5902/2316305425379. URL https://periodicos.ufsm.br/REDESG/article/view/25379
56. Ulrike Gretzel et al. Guiding principles for good governance of the smart destination. 2020
57. Francisco Femenia-Serra, Athina Ioannou, and Iis P Tussyadiah. Is smart scary? a mixed-methods study on privacy in smart tourism. *Current Issues in Tourism*, pages 1–27, 2021
58. Manuel David Masseno and Cristiana Teixeira Santos, Assuring privacy and data protection within the framework of smart tourism destinations. MediaLaws-Rivista di Diritto dei Media **2**, 251–266 (2018)
59. Z. Garanti, Value co-creation in smart tourism destinations. Worldwide Hospitality and Tourism Themes **15**(5), 468–475 (2023)
60. Olayinka Olasumbo Afolabi, Ali Ozturen, and Mustafa Ilkan. Effects of privacy concern, risk, and information control in a smart tourism destination. *Economic research-Ekonomska istraživanja*, 34(1):3119–3138, 2021
61. M. Kovačić, M. Čičin-Šain, V. Milojica, Cyber security and tourism: bibliometric analysis. Journal of process management and new technologies **10**(3–4), 75–92 (2022)
62. G. Sebastian, Privacy and data protection in chatgpt and other ai chatbots. International Journal of Security and Privacy in Pervasive Computing (2023). https://doi.org/10.4018/ijsppc.325475
63. Matthias Braunhofer and Francesco Ricci. Selective contextual information acquisition in travel recommender systems. *Information Technology & Tourism*, 17(1):5–29, Mar 2017. ISSN 1943-4294. https://doi.org/10.1007/s40558-017-0075-6. URL https://doi.org/10.1007/s40558-017-0075-6
64. J. L. Jorro-Aragoneses, M. B. Díaz Agudo, and J. A. Recio García. Madrid live: A context-aware recomendar system of leisure plans. In *2017 IEEE 29th International Conference on Tools with Artificial Intelligence (ICTAI)*, pages 796–801, Nov 2017. https://doi.org/10.1109/ICTAI.2017.00125
65. Francesco Ricci, Lior Rokach, and Bracha Shapira. Recommender systems: Introduction and challenges. In Francesco Ricci, Lior Rokach, and Bracha Shapira, editors, *Recommender Systems Handbook*, pages 1–34. Springer, 2015. https://doi.org/10.1007/978-1-4899-7637-6_1. URL https://doi.org/10.1007/978-1-4899-7637-6_1
66. C. Ma, L. Ma, Y. Zhang, J. Sun, X. Liu, M. Coates, Memory augmented graph neural networks for sequential recommendation. In Proceedings of the AAAI Conference on Artificial Intelligence **34**, 5045–5052 (2020)
67. Thanh Tran, Xinyue Liu, Kyumin Lee, and Xiangnan Kong. Signed distance-based deep memory recommender. In *The World Wide Web Conference*, pages 1841–1852, 2019
68. Laura Martínez García, Silvana Aciar, Raynel Mendoza, and Juan José Puello. Smart tourism platform based on microservice architecture and recommender services. In *Mobile Web and Intelligent Information Systems - 15th International Conference, MobiWIS 2018, Barcelona, Spain, August 6-8, 2018, Proceedings*, pages 167–180, 2018. https://doi.org/10.1007/978-3-319-97163-6_14. URL https://doi.org/10.1007/978-3-319-97163-6_14
69. J. Bobadilla, F. Ortega, A. Hernando, and A. Gutiérrez. Recommender systems survey. *Knowledge-Based Systems*, 46:109 – 132, 2013. ISSN 0950-7051. doi:https://doi.org/10.1016/j.knosys.2013.03.012. URL http://www.sciencedirect.com/science/article/pii/S0950705113001044

70. Aristea Kontogianni and Efthimios Alepis. Moments of interest: A novel cloud-based crowdsourcing application enhancing smart tourism recommendations. In *2019 11th Computer Science and Electronic Engineering (CEEC)*, pages 144–149. IEEE, 2019
71. Joanna Alvarado-Uribe, Andrea Gómez-Oliva, Germán Molina, Miguel González-Mendoza, María Concepción Parra-Meroño, and Antonio J. Jara. Towards the development of a smart tourism application based on smart POI and recommendation algorithms: Ceutí as a study case. In *Innovative Mobile and Internet Services in Ubiquitous Computing - Proceedings of the 11th International Conference on Innovative Mobile and Internet Services in Ubiquitous Computing (IMIS-2017), Torino, Italy, 10-12 July 2017.*, pages 904–916, 2017. https://doi.org/10.1007/978-3-319-61542-4_92. URL https://doi.org/10.1007/978-3-319-61542-4_92
72. Wen Long Feng, Yu Cong Duan, Meng Xin Huang, Lin Feng Dong, Xiao Yi Zhou, and Tao Hu. A research on smart tourism service mechanism based on context awareness. In *Computer and Information Technology*, volume 519 of *Applied Mechanics and Materials*, pages 752–758. Trans Tech Publications, 5 2014. https://doi.org/10.4028/www.scientific.net/AMM.519-520.752
73. Mickael Figueredo, José Lucas Ribeiro, Nélio Cacho, Antonio Thome, Andréa Cacho, Frederico Lopes, and Valeria Araujo. From photos to travel itinerary: A tourism recommender system for smart tourism destination. In *Fourth IEEE International Conference on Big Data Computing Service and Applications, BigDataService 2018, Bamberg, Germany, March 26-29, 2018*, pages 85–92, 2018. https://doi.org/10.1109/BigDataService.2018.00021. URL https://doi.org/10.1109/BigDataService.2018.00021
74. Hajar Khallouki, Ahmed Abatal, and Mohamed Bahaj. An ontology-based context awareness for smart tourism recommendation system. In *Proceedings of the International Conference on Learning and Optimization Algorithms: Theory and Applications, LOPAL 2018, Rabat, Morocco, May 2-5, 2018.*, pages 43:1–43:5, 2018. https://doi.org/10.1145/3230905.3230935. URL http://doi.acm.org/10.1145/3230905.3230935
75. Alexander Smirnov, Alexey Kashevnik, Andrew Ponomarev, Nikolay Teslya, Maksim Shchekotov, and Sergey I. Balandin. Smart space-based tourist recommendation system. In Sergey Balandin, Sergey Andreev, and Yevgeni Koucheryavy, editors, *Internet of Things, Smart Spaces, and Next Generation Networks and Systems*, pages 40–51, Cham, 2014. Springer International Publishing. isbn:978-3-319-10353-2
76. Zihuai Zhao, Wenqi Fan, Jiatong Li, Yunqing Liu, Xiaowei Mei, Yiqi Wang, Zhen Wen, Fei Wang, Xiangyu Zhao, Jiliang Tang, et al. Recommender systems in the era of large language models (llms). *arXiv preprint* arXiv:2307.02046, 2023
77. Yuyang Ye, Zhi Zheng, Yishan Shen, Tianshu Wang, Hengruo Zhang, Peijun Zhu, Runlong Yu, Kai Zhang, and Hui Xiong. Harnessing multimodal large language models for multimodal sequential recommendation. *arXiv preprint* arXiv:2408.09698, 2024
78. Yunfan Gao, Tao Sheng, Youlin Xiang, Yun Xiong, Haofen Wang, and Jiawei Zhang. Chat-rec: Towards interactive and explainable llms-augmented recommender system. *arXiv preprint* arXiv:2303.14524, 2023
79. Junjie Zhang, Ruobing Xie, Yupeng Hou, Wayne Xin Zhao, Leyu Lin, and Ji-Rong Wen. Recommendation as instruction following: A large language model empowered recommendation approach. *arXiv preprint* arXiv:2305.07001, 2023
80. D. Jovicic, Cultural tourism in the context of relations between mass and alternative tourism. Curr. Issue Tour. **19**(6), 605–612 (2016)
81. G. Richards, Tourism attraction systems: Exploring cultural behavior. Ann. Tour. Res. **29**(4), 1048–1064 (2002)
82. Y. Sun, H. Song, A. J. Jara, and R. Bie. Internet of things and big data analytics for smart and connected communities. *IEEE Access*, 4:766–773, 2016. ISSN 2169-3536. https://doi.org/10.1109/ACCESS.2016.2529723
83. Pam Lee, William Cannon Hunter, and Namho Chung. Smart tourism city: Developments and transformations. *Sustainability*, 12(10):3958, 2020
84. Greg Richards. *Tourism and Culture*, pages 165–178. 01 2000. isbn:978-90-481-5385-5. https://doi.org/10.1007/978-94-015-9584-1_10

85. T. Graziano, Boosting innovation and development: the italian smart tourism, a critical perspective. Eur. J. Geogr. **5**(4), 6–18 (2014)
86. Liliana Ardissono, Tsvi Kuflik, and Daniela Petrelli. Personalization in cultural heritage: the road travelled and the one ahead. *User Modeling and User-Adapted Interaction*, 22(1):73–99, Apr 2012. ISSN 1573-1391. https://doi.org/10.1007/s11257-011-9104-x. URL https://doi.org/10.1007/s11257-011-9104-x
87. Tuong Tri Nguyen, David Camacho, and Jai E. Jung. Identifying and ranking cultural heritage resources on geotagged social media for smart cultural tourism services. *Personal and Ubiquitous Computing*, 21(2):267–279, Apr 2017. ISSN 1617-4917. https://doi.org/10.1007/s00779-016-0992-y. URL https://doi.org/10.1007/s00779-016-0992-y
88. Andrea Gomez-Oliva, Joanna Alvarado-Uribe, Maria Concepcion Parra-Meroño, and Antonio J Jara. Transforming communication channels to the co-creation and diffusion of intangible heritage in smart tourism destination: Creation and testing in ceutí (spain). *Sustainability*, 11(14):3848, 2019
89. Jin Wei, Lei Ma, and Zhongqiu Zhang. A research on smart tourism-oriented big data real-time processing technology. In *2017 29th Chinese Control And Decision Conference (CCDC)*, pages 1848–1851. IEEE, 2017
90. S. Kujala, V. Roto, K. Väänänen-Vainio-Mattila, E. Karapanos, A. Sinnelä, Ux curve: A method for evaluating long-term user experience. Interact. Comput. **23**(5), 473–483 (2011)
91. Christian Schott and Stephen Marshall. Virtual reality for experiential education: A user experience exploration. *Australasian Journal of Educational Technology*, 37(1):96–110, Mar. 2021. https://doi.org/10.14742/ajet.5166. URL https://ajet.org.au/index.php/AJET/article/view/5166
92. T. Chen, W. Guo, X. Gao, Z. Liang, Ai-based self-service technology in public service delivery: User experience and influencing factors. Gov. Inf. Q. **38**(4), 101520 (2021)
93. Jodi Forlizzi and Katja Battarbee. Understanding experience in interactive systems. In *Proceedings of the 5th Conference on Designing Interactive Systems: Processes, Practices, Methods, and Techniques*, DIS '04, pages 261–268, New York, NY, USA, 2004. ACM. isbn:1-58113-787-7. https://doi.org/10.1145/1013115.1013152. URL http://doi.acm.org/10.1145/1013115.1013152
94. Sascha Mahlke and Manfred Thüring. Studying antecedents of emotional experiences in interactive contexts. In *Proceedings of the SIGCHI Conference on Human Factors in Computing Systems*, CHI '07, pages 915–918, New York, NY, USA, 2007. ACM. isbn:978-1-59593-593-9. https://doi.org/10.1145/1240624.1240762. URL http://doi.acm.org/10.1145/1240624.1240762
95. Fedelis Brian Putra Prakasa, Melky Radja, and S Suyoto. User experience based mobile application design for boat loaning at marine tourism in indonesia. *Int. J. Interact. Mob. Technol.*, 14(4):86–102, 2020
96. Susanne Bødker. Third-wave hci, 10 years later—participation and sharing. *interactions*, 22(5):24–31, 2015
97. A. Kontogianni, E. Alepis, Smart tourism: State of the art and literature review for the last six years. Array **6**, 100020 (2020)
98. Shuchih Ernest Chang and Wei-Cheng Shen. Exploring smartphone social networking services for mobile tourism. *IJMC*, 16(1):63–81, 2018. https://doi.org/10.1504/IJMC.2018.10007781. URL https://doi.org/10.1504/IJMC.2018.10007781
99. Melody M Terras and Judith Ramsay. A psychological perspective on mobile learning. In *Encyclopedia of Information Science and Technology, Fourth Edition*, pages 6398–6411. IGI Global, 2018
100. Zahrah A Almusaylim and Noor Zaman. A review on smart home present state and challenges: linked to context-awareness internet of things (iot). *Wireless networks*, 25(6):3193–3204, 2019
101. Graham Dann. *The tourist as a metaphor of the social world*. cabi, 2002
102. Öznur Akgiş İlhan. The rise of digitalization in the tourism industry during covid-19: Cyber space, destinations, and tourist experiences. In *Handbook of Research on the Impacts and Implications of COVID-19 on the Tourism Industry*, pages 843–862. IGI Global, 2021

103. S. Anwar, J. Hamilton, Tourism into the future-towards 2020, and beyond. Tour. Recreat. Res. **30**(3), 77–85 (2005)
104. Google. Google arts & culture. Available at https://about.artsandculture.google.com/ ,Accessed April 11, 2022
105. Miaomiao Li, Hao Chen, Yang Wang, Tingyuan Zhu, Weijia Zhang, Kaijie Zhu, Kam-Fai Wong, and Jindong Wang. Understanding and mitigating the bias inheritance in llm-based data augmentation on downstream tasks. *arXiv preprint* arXiv:2502.04419, 2025
106. M.H. Huang, R.T. Rust, Artificial intelligence in service. J. Serv. Res. **21**(2), 155–172 (2018). Sage Publications

Chapter 7
Sentiment Analysis Using LSTM and BERT: A Comparative Analysis Using Natural Language Processing Algorithms

Kirti, Divya, Sukhjot Kaur, Shahla Gufran, Divyanshu Singh, and Dev Divyansh Mishra

Abstract Social media platforms have become major venues for public expression, making the analysis of sentiments within these data streams increasingly valuable for applications ranging from consumer feedback to public health monitoring. This paper explores sentiment analysis using Bidirectional Encoder Representations from Transformers (BERT) and Long Short-Term Memory (LSTM) models, assessing their performance for accurately capturing emotional nuances in social media text. We conducted an extensive literature review to understand current approaches, identify gaps in the existing methods, and establish research questions focused on improving model efficiency and accuracy in handling diverse social media data. The key findings presented in this chapter highlight the strengths plus the limitations of recent methodologies, offering insights into future research directions. Our study provides a framework for advancing sentiment analysis, aiming to contribute a robust approach to interpreting social media sentiments effectively.

Keywords Sentiment analysis · Natural Language Processing · BERT model · LSTM model · Deep learning models · Emotion classification · Sentiment classes · Supervised learning

7.1 Introduction

In today's digital communication landscape, the rapid increase in user-generated content on social media, product reviews, and online discussion platforms presents an exceptional chance to gain insights into human opinions and emotional states

---

Kirti (✉) · Divya · S. Kaur · S. Gufran · D. Singh · D. D. Mishra
School of Computer of Science and Engineering, Lovely Professional University, Phagwara, Punjab, India
e-mail: kirti.29458@lpu.co.in

G. A. Tsihrintzis et al. (eds.), *Human-Smart City Interactions and User-Citizen Experiences*, Learning and Analytics in Intelligent Systems 58,
https://doi.org/10.1007/978-3-032-06364-9_7

on a large scale [1]. Sentiment analysis, a significant area within natural language processing (NLP), has become an essential tool for identifying and evaluating subjective information from text data. Although traditional machine learning methods have established a foundation for sentiment categorization, the rise of deep learning frameworks has transformed the discipline, providing advanced means to grasp the subtleties of human sentiment.

7.1.1 *Evolution of Sentiment Analysis Methodologies*

The progression of sentiment analysis methodologies has seen crucial developments, ranging from basic lexicon-based techniques to intricate neural network structures [2]. Earlier approaches heavily depended on manually crafted features and predetermined sentiment dictionaries [1], which frequently fell short in recognizing context-dependent sentiments and subtle linguistic distinctions, including sarcasm, irony, and implicit expressions of sentiment. The emergence of deep learning technologies has resolved many of these challenges, allowing models to independently learn hierarchical representations of textual data and understand complex semantic interrelations that were previously challenging to represent.

7.1.2 *Deep Learning Models in Sentiment Analysis: LSTM and BERT*

This paper conducts a thorough examination of two leading deep learning models—Long Short-Term Memory (LSTM) networks and Bidirectional Encoder Representations from Transformers (BERT)—specifically in the context of sentiment analysis tasks. LSTMs have shown impressive success in grasping the contextual flow of text due to their capacity to capture long-term dependencies within sequential data [3]. In contrast, BERT's transformer-based structure, which employs bidirectional training and attention mechanisms, has established new standards in numerous NLP applications, including sentiment analysis [4].

7.1.3 Advancements in Computational Capabilities and Model Efficiency

The swift progress in computational power and the access to extensive datasets have further amplified the capabilities of these architectures [5]. LSTM networks, utilizing advanced gating mechanisms, successfully tackle the vanishing gradient issue that affected previous recurrent neural networks, allowing them to model long-range dependency text sequences. In contrast, BERT signifies a major shift in natural language processing (NLP) due to its approach of pre-training and fine-tuning, which enables it to utilize knowledge from vast text collections and adapt to particular downstream tasks with minimal specialized training data [6].

7.1.4 Study Contributions and Objectives

Our study contributes to the current body of research by:

- Conducting a comprehensive comparison of the LSTM and BERT models for sentiment analysis.
- Analysing performance metrics across various dataset sizes and levels of complexity.
- Exploring the computational demands and training efficiency of both methods.
- Suggesting a hybrid framework that utilizes the strengths of both approaches.
- Assessing the models' resilience across diverse domains and styles of text.

7.1.5 Practical Implications

The importance of this study lies in its practical ramifications for businesses, researchers, and practitioners aiming to establish sentiment analysis systems. By offering an in-depth analysis of the strengths and weaknesses of these architectures, this research intends to aid in selecting suitable models based on specific use-case requirements, computational limitations, and accuracy needs. Additionally, our study addresses the increasing demand for interpretable and efficient sentiment analysis solutions in practical applications, ranging from social media analysis to customer feedback evaluation and market research [7].

7.2 Literature Review

In the current age of digital communication, sentiment analysis is now an essential tool for measuring public sentiment and emotions shared on social media. The earliest techniques heavily relied on lexicon-based approaches and traditional machine learning algorithms, which, while somewhat successful, often struggled to capture the nuances of human emotions. This restriction led to the advancement of more sophisticated methods, especially ones utilizing deep learning, which have demonstrated outstanding effectiveness in processing sequential data.

RNNs and LSTM networks have received a lot of interest in the deep learning field for their ability to effectively handle sequential data. As outlined in the study conducted by Sharma et al. [8], LSTMs are proficient in resolving the problem of diminishing gradients seen in conventional RNNs, enabling them to preserve data over long sequences.

This skill is crucial in analysing sentiments, particularly when the situation changes over a sequence of interactions, like in tweets or other social media entries [8].

Many studies, like the comparative analysis by Alharthi et al. [9], have demonstrated LSTM's capability to effectively interpret sentiment in social media text. LSTMs establish a strong base for analyzing shifts in sentiment over time by capturing the sequential nature of language, improving the comprehension of user opinions and behaviours.

The sentiment analysis field has been significantly transformed by the introduction of Bidirectional Encoder Representations from Transformers (BERT). Unlike LSTMs, BERT is created to grasp context in both directions, allowing it to interpret intricate language structures that unidirectional models frequently miss. The research by Sari et al. [10] emphasizes BERT's superior capacity to analyses nuanced expressions of sentiment, thereby establishing its significance in natural language processing (NLP) tasks.

BERT has a substantial impact on sentiment analysis on social media. Due to its ability to process language in context, BERT has been utilized for analyzing sentiment in tweets and other online posts, revealing insights that traditional models might miss. For instance, Liu et al. [11] elaborate on improvements to BERT that enhance its accuracy in sentiment analysis, showcasing the model's versatility across different datasets and contexts. Moreover, the systematic review conducted by Rojas and colleagues [12] highlights how BERT and other transformer models have transformed the methodologies used in this field.

Recognizing the limitations present in individual models, contemporary research has started to investigate hybrid approaches that merge the strengths of both LSTM and BERT. These models take advantage of BERT's contextual comprehension along with LSTM's efficiency in handling sequential data. The paper by Zadeh et al. [13] examines how these combined methods work together, showing positive results in enhancing the precision and speed of sentiment analysis

Nonetheless, challenges persist in applying deep learning models to sentiment analysis, notably concerning computational requirements and performance variability across different languages and social media platforms. The thorough assessment carried out by Zadah et al. [13] highlights these issues, illustrating how LSTM and BERT-based models may underperform in specific contexts and datasets, underscoring the need for continuous improvement and innovation in this area.

Overall, the literature reflects a clear evolution from conventional sentiment analysis techniques to the incorporation of advanced deep learning methods, particularly LSTM and BERT. Each model offers distinct advantages for analyzing social media sentiment, while the emerging trend of hybrid models indicates a promising trajectory for future research. This literature review not only underscores foundational studies in the field but also pinpoints gaps that the present paper intends to fill by thoroughly examining sentiment analysis in social media utilizing BERT and LSTM (Table 7.1).

7.3 Research Gaps

Due to limited diversity in databases, sentiment analysis has for the most part accentuated a few of the well-known platforms such as Instagram, Twitter, or IMDB [18]. Since over 90% of the research is directed towards these platforms and does not stress having more general knowledge about sentiment in many other different contexts, it is biased. The scope of research should include a much broader range of sources to gain a more in-depth and accurate understanding of sentiment across various businesses and applications. Examples are official documents, such as business reports, research papers, and offline customer feedback, which are generally overlooked but capture user sentiments for most industries [9, 18]. Emotion classification is another gap that is highly significant in sentiment analysis. Most of the studies do not take into account the complexities and subtleties of human emotions binary classifications as positive, negative, and neutral prevail. Such simple classifications can sometimes be enough for the purpose at hand, but they leave out more complex emotions, such as happiness, sadness, or mixed feelings. A more detailed emotional picture of the landscape would be provided, one that is also more accurate for sentiment models and gives a richer insight into user sentiments [10–12]. Polysemy—the fact that one word can have many meanings proves to be nearly as large a barrier as it ever was with models like BERT. It becomes even more obvious in specialized domains, for example, banking or medicine, where terms often have very niche meanings not apparent from their broader context [8, 17]. Models for sentiment analysis are not keen on pinning those context-dependent shades. What is required, therefore, are much more advanced methods that can more accurately capture context-specific and domain-specific language to improve the efficiency of sentiment analysis in those domains and, thereby, come up with much more reliable and meaningful results [13].

Even though their scalability is a problem, state-of-the-art models like LSTM, CNN, and BERT have exploded the boundaries of sentiment analysis. These models

Table 7.1 Key literature on sentiment analysis techniques: LSTM & BERT

References	Year	Focus area	Algorithm	Evaluation metrics	Results
[14]	2010	Sentiment analysis of Facebook data	Naive Bayes, SVM	Accuracy, Precision, Recall	Naive Bayes was more effective than SVM for sentiment analysis on Facebook
[12]	2014	Improvements in sentiment analysis of tweets	Naive Bayes, SVM, MaxEnt	Precision, Recall, F1-score	Combination of machine learning algorithms improved sentiment classification
[15]	2017	Hybrid techniques for sentiment analysis	Hybrid BERT-LSTM	Accuracy, F1-score	Hybrid BERT-LSTM approach improved overall sentiment analysis performance
[9]	2019	Propaganda analysis Using NLP models	BERT	F1-score, Precision	BERT achieved state of the art results in identifying propaganda
[11]	2019	Systematic review of sentiment analysis in social media	Various NLP Models	Accuracy, Precision, Recall	Identified BERT and LSTM as strong contenders for social media analysis
[13]	2019	Comparative study of deep learning methods for sentiment analysis	LSTM, CNN, RNN	Accuracy, F1-score	LSTM performed better than CNN and RNN in sentiment classification
[8]	2020	Fundamentals of RNN and LSTM networks	LSTM	Accuracy, RMSE	LSTM demonstrated superior performance in time series prediction tasks
[16]	2020	Limitations and challenges in deep learning for sentiment analysis	BERT, LSTM	Precision, F1-score	Identified challenges like sarcasm detection and domain adaptation
[13]	2021	Improving BERT performance with self-supervised attention	BERT with Self-Supervised Attention	Accuracy, Precision	Self-supervised attention improved BERT's accuracy in various tasks

(continued)

Table 7.1 (continued)

References	Year	Focus area	Algorithm	Evaluation metrics	Results
[17]	2023	Enhancing LSTM text filtering accuracy	LSTM	Precision, F1-score	LSTM improvements led to better text filtering results in social media data
[18]	2023	Sentiment analysis of Instagram posts	BERT, LSTM	Accuracy, Precision	BERT and LSTM were effective for analysing sentiment in Instagram posts
[10]	2023	Smart grid parameter prediction using ARIMA & LSTM	Bi-LSTM, ARIMA	MSE, MAE	Bi-LSTM Outperformed ARIMA in predicting smart grid parameters

call for a lot of processing power, quite frequently which then burdens their utilization in environments where resources may be scarce. These models should be improved for sentiment analysis to be really meaningful in larger real-life applications. Solving this problem will make it more feasible for sectors with low processing power to use sentiment analysis. Thus, they can use these technologies more besides the power shortages. The main part of sentiment analysis research also relies on fixed datasets, not minding the fact that the sentiment of a social cognition changes over time. Sentiment analysis should focus on temporal dynamics, the concept that acts as a record of how sentiments operate over time, since sentiment alters with the coincidental events. Through the incorporation of time-series analysis, sentiment models would be able to track these shifts, an exercise that is vital in spontaneous environments like social media or the financial markets [10, 14]. More accurate predictions and highlights could follow from implementing this temporal element, which would aid in identifying exactly the different nuances of the mood that people have due to the accurate, and up-to-date information provided by sources that are more transparent [9].

Moreover, even though specialized financial language models such as FinBERT have been developed for the finance sector, their application is only to be utilized within the industry. The possibility of the implementation of these models in the other domains is dictated by the absence of sector-specific models in the USA and the rest of the world in areas such as politics and healthcare. It can be made clear that sentiment analysis in these fields will be driven up by forming mood models that are exclusive to those domains since these models will have the ability to grasp the special language and emotional expressions that go along with each field. Multimodal sentiment analysis is the integration of a range of text, images, videos, and audio data to the recognition of emotions and it is also another area of research which

Table 7.2 Research questions

RQ1	What makes sentiment analysis a valuable tool for modern digital communication, & what roles do LSTM & BERT play in it?
RQ2	In terms of time and resource requirements, how do computational efficiency of LSTM and BERT compare?
RQ3	Which important measures were taken in the research study to assess the efficacy & precision of LSTM & BERT models?
RQ4	What were the primary obstacles encountered when using deep learning models for sentiment analysis tasks, like LSTM and BERT?

can be an interesting field for further studies [12, 16]. Nonverbal hints, such as facial expressions or intonation, can serve as cues that are contextually relevant for emotional analysis. Sentiment analysis might be significantly advanced through synergies of these different data types, thereby providing a more in-depth and accurate understanding of people's emotional gamut in various circumstances. Finally, the lack of evaluation measure for the sentiment analysis models is a major issue that has not been addressed in a large part of the studies. Future work also needs reliable and frequently accessible comparison methods to assess the real performance of the proposed models [13, 17, 18]. The development of sound performance frameworks will further help facilitate such comparisons and support the researchers themselves to better understand the strengths and weaknesses of the respective models to enhance this particular field of study. To further develop the sentiment analysis these gaps have to be closed. Born out of these problems are recommendations for future research to develop the area and make sentiment analysis universally useful in real life situations. To fill them, this will enhance our comprehension of human emotions in a number of settings.

7.4 Research Questions

See Table 7.2.

7.5 Research Methodology

This section describes the methodology used to compare the sentiment classification skills of the Long Short-Term Memory (LSTM) and Bidirectional Encoder Representations from Transformers (BERT) models. The methodology consists of choosing the dataset, preprocessing the data, extracting features, training the model, and evaluating the performance of the two models using evaluation metrics. In order to shed light on the various advantages and disadvantages of transformer-based and recurrent neural network-based designs for sentiment analysis, this study aims to study

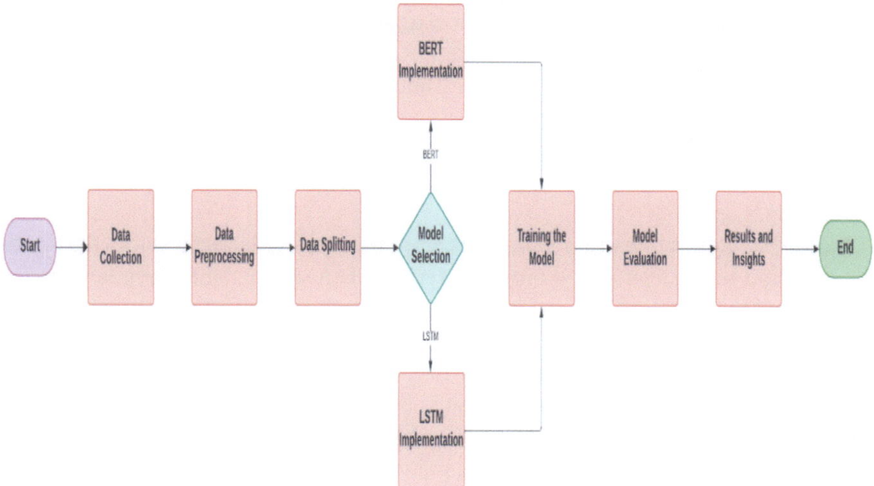

Fig. 7.1 Research methodology flowchart

their effectiveness. The approach has been carefully constructed at every stage to guarantee accurate and repeatable results (Fig. 7.1).

7.5.1 *Data Gathering and Its Significance*

The social media posts that make up the dataset that was chosen for this study span a range of attitudes, including neutral, negative, and positive. This diversity is important because it makes it possible to explore the various ways that various emotions are conveyed in user-generated material. Social media interactions provide a rich background that sheds light on public opinion, feelings, and attitudes, making them a helpful tool for comprehending current affairs.

7.5.2 *Importance of Gathering Data*

- **Illustration of Real-World Circumstances**: Social media platforms provide a window into the attitudes and lookouts of the real world. Researchers can better understand human emotions and activities in a variety of circumstances by gathering data from different platforms, which increases the study's applicability.
- **Model Assessment and Training**: Effective machine learning model training requires high-quality, labeled datasets. Because the dataset includes labeled sentiments, supervised learning techniques where models learn from instances and increase their predictive accuracy—can be used. The dataset's diversity of feelings

makes it easier to evaluate models thoroughly and sheds light on the advantages and disadvantages of various methods.

- **Enabling Comparative Evaluation**: The nature of the dataset makes it possible to compare several sentiment analysis methods, including LSTM and BERT. Researchers may guarantee consistency in their assessments by using the same dataset, which enables a more precise comparison of model efficiency and performance.
- **Handling Robustness and Data Bias**: Getting a varied dataset reduces the biases that could result from using only one type of data source. The study is to assess how well the models handle a range of linguistic styles and emotional expressions by including posts from various people, cultures, and moods.
- **Addition to the Knowledge Base**: The results of this study add to the body of information about sentiment analysis. Businesses, legislators, and researchers can improve decision-making processes by learning about public mood patterns through the study's analysis of social media sentiment.

7.5.3 Data Pre-processing

In order to make sure that the dataset is proper and of high quality for effective model training, data pre-processing is an important phase in the sentiment analysis pipeline. Post ID, Post Content, Sentiment, Label, Likes, Shares, Comments, User Follower Count, Post Date and Time, Post Type, and Language are among the columns that make up the original dataset. But, we will just pay attention to the "Post Content" and "Sentiment" columns for this study. The following preprocessing actions are carried out:

- **Column Selection:** All columns are eliminated except for Sentiment and Post Content. By eliminating unimportant data, this restructuring makes the dataset more manageable and concentrated, which could improve the sentiment analysis.
- **Data Cleaning**: To get rid of any errors or inconsistencies, the remaining data is cleaned. This comprises deleting redundant entries to preserve the accuracy of the data. Depending on the situation, handling missing values may include either discarding those rows or using the proper strategies to fill them in.
- **Text Normalization**: The following procedures are used to normalize the post content: To keep things consistent, all text will be converted to lowercase. Eliminating whitespaces, special characters, user mentions (e.g., @username), hashtags and punctuation marks helps improve the text data's quality.
- **Label Encoding**: Using methods like label mapping, the sentiment labels will be converted into a numbers apt for machine learning models. The sentiment labels will be converted to numerical values for this study, where positive will be represented by 1, negative by -1, and neutral by 0.
- **Tokenization**: Lastly, the preprocessed text input is divided into distinct words or phrases and transformed into a sentiment analysis model-friendly format. In

order to guarantee that the sentiment analysis model can be trained successfully on clear, appropriate, and organized data, these pre-processing steps are essential for preparing the data for further analysis.

7.5.4 Implementation of The BERT Model

In natural language processing (NLP), the application of the BERT (Bidirectional Encoder Representations from Transformers) model is a noteworthy development. This section describes the architecture of BERT, offers implementation-ready pseudocode, and elucidates its operation [19–21].

7.5.4.1 BERT Architecture

The Transformer model, which mostly relies on self-attention processes, is the foundation of BERT's architecture. The main elements that make up BERT's architecture are as follows:

- Transformer Layers: BERT has a number of transformer layers that enable deep contextual interpretation of text, such as 12 for BERT Base and 24 for BERT Large.
- Bidirectional Attention: BERT analyzes text in both directions concurrently, allowing it to comprehend a word's context depending on all other words in the sentence, in contrast to earlier models that read text sequences from left to right or right to left.
- Input Representation: There are three primary embeddings in the input of BERT. Each word in the input sequence is represented by the token embeddings.
- Segment Embeddings: Distinguishes between various sentences in tasks such as answering questions.
- Position Embeddings: These preserve word order by encoding each token's location inside the sequence.
- Output Layer: By including a task-specific output layer (such as a softmax layer for classification), BERT can be improved for a variety of applications, including sentiment analysis. BERT can capture intricate linguistic patterns thanks to its architecture, which also enables it to perform remarkably well on a range of NLP tasks (Fig. 7.2).

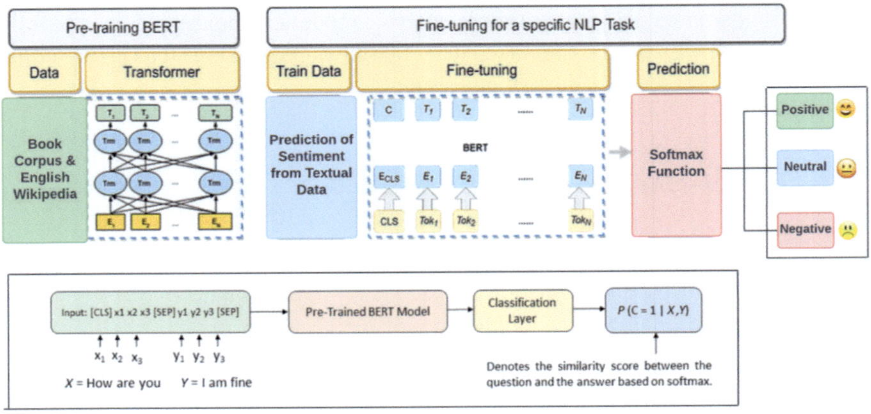

Fig. 7.2 Architectural diagram of BERT

7.5.4.2 Pseudocode for BERT Model for Sentiment Analysis

The pseudocode below is given for the BERT model for sentiment analysis:

Pseudocode 1: BERT Model Training and Evaluation Pipeline
• A pre-trained BERT model should be loaded. • First, clean and tokenize the text to get ready the input data. a. Convert tokens into the corresponding IDs. b. Create attention masks in order to discern between padding and real tokens. • Adjust the BERT model as follows: a. Configure the input, BERT, and output layers in the model architecture. b. Set up the model with an appropriate optimizer and loss function. • Educate the model: a. Divide the dataset into subsets for training and validation. b. While monitoring performance on the validation set, train the model on the training set. • Analyze the model: a. Use the validation data to gauge the model's performance and accuracy. b. Store the learned model for later application.

This pseudocode provides a high-level overview of the steps required to implement and train a BERT model for sentiment analysis.

7.5.4.3 How Bert Functions

The basis of BERT's operation is its bidirectional processing and attention mechanisms. Here's an explanation of how BERT works.

Processing of Text Input: BERT adds special tokens ([SEP] for sentence separation and [CLS] for classification) to the text after tokenizing the input. It then generates token IDs and attention masks in order to prepare the input for the model.

Bidirectional Contextualization: BERT uses an MLM (masked language model) objective for training, where words that are masked in the input are predicted by the model according to their context. This bidirectional training allows BERT to acquire a deep understanding of language context.

Self-attention Mechanism: The self-attention mechanism allows BERT to take into account the relative importance of different phrases in a sentence when making predictions. It allows the model to focus on relevant words regardless of where they are in the sequence by computing attention scores.

Fine-Tuning: Following pre-training, BERT can be made better for certain tasks like sentiment analysis. Using labeled data to train the model and adjusting the BERT layer weights to optimize performance for the task at hand are the steps involved in fine-tuning.

Prediction: After text input has been fine-tuned for sentiment analysis, BERT may classify it into distinct sentiment categories (positive, negative, and neutral) based on the learned contextual representations. The output layer uses a softmax algorithm to produce probability ratings for each sentiment class. In summary, BERT's innovative design and training methodology make it a powerful tool in the artificial intelligence space, allowing it to achieve cutting-edge results in sentiment analysis and other natural language processing applications.

7.5.5 Implementation of the LSTM Model

In natural language processing (NLP), the application of the LSTM (Long Short-Term Memory) model is a notable development. This section describes the architecture of LSTM, offers implementation-ready pseudocode, and elucidates its operation.

7.5.5.1 Architecture of LSTM

Recurring neural networks (RNNs) with Long Short-Term Memory (LSTM) are designed to recall long-term dependencies and resolve the vanishing gradient issue that ordinary RNNs encounter. LSTMs have memory cells, which store information over time. Three gates are present in every cell: input, forget, and output gates.

(i) Input Gate: Regulates the amount of fresh data that should be added to the cell.
(ii) Forget Gate: Chooses what part of the cell's present state should be forgotten.
(iii) Output Gate: It will regulate the output from memory cell.

LSTMs are apt for sequential data like text, where comprehending the context over time (i.e., words in a sentence) is important. Long-term dependencies can be captured by LSTMs without being caught off guard by short-term memory problems by managing the information flow (Fig. 7.3).

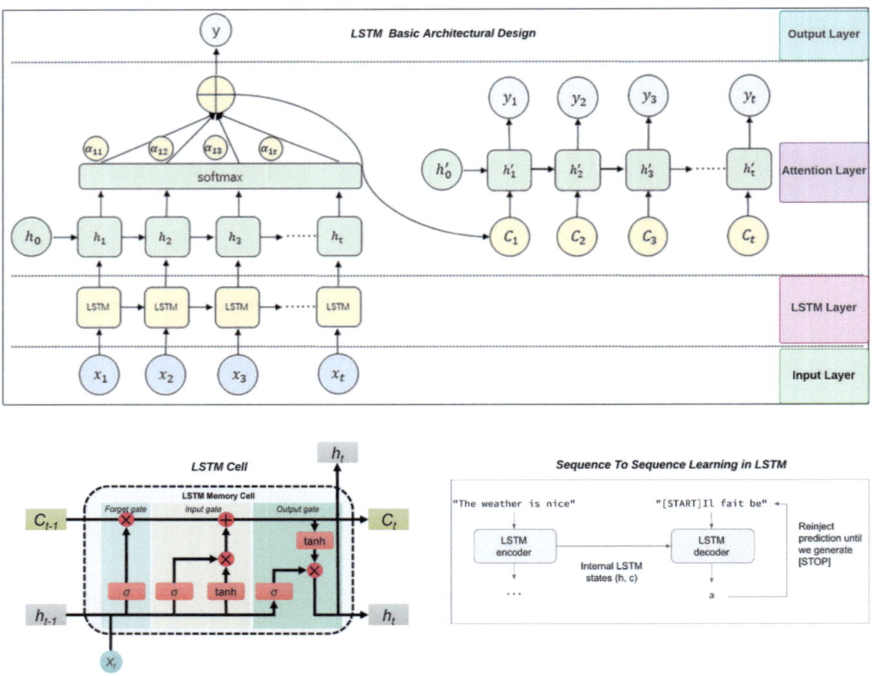

Fig. 7.3 Architectural diagram of LSTM

7.5.5.2 Pseudocode for LSTM Model for Sentiment Analysis

Pseudocode 2: LSTM Model Training and Evaluation Pipeline

- Preprocess the Data:
  a. Load the dataset and clean the text data (remove punctuation, convert to lowercase, etc.).
  b. Tokenize the text data and convert it into sequences of integers. c. Pad the sequences to ensure uniform input length. d. Split the dataset into training and validation sets.
- Prepare the Embedding Layer:
  a. Initialize an embedding layer to convert input tokens into dense vectors.
  b. If using pre-trained word embeddings (e.g., GloVe or Word2Vec), load and map the embeddings to the vocabulary.
- Build the LSTM Model Architecture:
  Define the input layer with the shape of padded sequences. Add an embedding layer to the model. Add one or more LSTM layers:
  a. Set the number of units (hidden state size) for each LSTM layer. Optionally, add dropout for regularization.
  b. Add a dense layer with an activation function (e.g., ReLU or Sigmoid).
  c. Add the output layer:
- Use a Softmax activation for multi-class classification (positive, negative, neutral). Use a Sigmoid activation for binary classification.
- Compile the Model: a. Set the optimizer (e.g., Adam) and loss function (e.g., categorical crossentropy for multi-class classification). b. Define the evaluation metrics (e.g., accuracy).
- Train the Model: a. Use the training set to fit the model. b. Specify the number of epochs and batch size. c. Monitor performance using the validation set.
- Evaluate the Model: a. Assess the model's performance using the validation data. b. Calculate evaluation metrics (accuracy, precision, recall, F1-score). c. Save the trained model for future use.
- Make Predictions: a. Load the saved model. b. Preprocess new text input. c. Use the model to predict sentiment labels for the input text.

7.5.5.3 How LSTM Functions

The LSTM network [22–24] works by passing the input sequence through a series of gates and cells. It processes the data as explained below.

Input Processing: Firstly, tokenization of text takes place which is followed by conversion of these tokens into numerical format. Lastly, padding is added to these tokens to ensure they are of same length.

LSTM Layer: The input sequence is fed to the LSTM Layer which decides if the information is to be retained or discarded based on the three gates.

State Updates: The LSTM updates its cell state using input from earlier steps and new data, making it retain important information over long sequences.

Output Generation: After processing the sequence, the LSTM produces an output, passed to a dense layer for classification, using softmax for sentiment analysis.

Training & Fine-Tuning: Throughout training, the LSTM changes weights using backpropagation through time (BPTT), determining which part has higher priority. LSTMs outshines in sentiment analysis by apprehending both short- and long-term word relationships, enhancing context understanding.

7.6 Test and Validation

The dataset was separated into training and test sets in order to evaluate the effectiveness and generalization potential of the LSTM and BERT models. The divide guarantees a strong assessment of the models on unseen data.

7.6.1 Dataset Division

Two subsets of the dataset were created: 80% of the training set is used to teach the models and identify trends in the data. Test Set (20%) used solely after training is finished to assess how well the model performs in the end on data that has never been seen before.

In order to guarantee that the models have enough data for learning while still having adequate data for validation and testing, the split ratio of 80–20 was selected.

7.6.2 Evaluation Metrics

The following metrics were used to evaluate the performance of the models:

- **Accuracy**: Accuracy is defined as the proportion of correct predictions out of all predictions made. It is calculated as:

  $$\text{Accuracy} = \text{Number of Correct Predictions/Total Number of Predictions}$$

  In cases where the dataset is imbalanced, accuracy alone cannot provide a complete picture of the model's performance. That is why additional metrics are often required.

  Precision: The proportion of true positive predictions out of all positive predictions made by the model. It helps measure the quality of positive predictions.
- **Precision**: Precision is a metric used to measure the accuracy of the positive predictions made by a model. It tells us how many of the instances that the model predicted as positive are actually correctly positive.

  The formula for Precision is:

  $$\text{Precision} = \text{True Positives/True Positives} + \text{False Positives}$$

- **Recall (Sensitivity)**: Recall, also known as Sensitivity, measures how well the model identifies all the actual positive instances in the dataset. In other words, it tells us how many of the positive cases the model was able to identify correctly.

  The formula for Recall (or Sensitivity) is:

  $$\text{Recall} = \text{True Positives/True Positives} + \text{False Negatives}$$

- **F1-Score**: The F1-score gives us a single measure of the ability of a model to predict both correct positives (precision) and catch all the actual positives (recall), while avoiding both false positives and false negatives.

  The F1-score is calculated as the harmonic mean of precision and recall. It is defined as:

  $$\text{F1} = 2 \times \text{Precision} \times \text{Recall/Precision} + \text{Recall}$$

  If both precision and recall are high, the F1-score will be high, indicating good performance. If either precision or recall is low, the F1-score will be lower, reflecting poor performance.
- **Confusion Matrix**: A confusion matrix is a powerful tool used to evaluate the performance of a classification model. It provides a detailed breakdown of how well the model is performing by showing how its predictions match the actual labels in the dataset.

  The confusion matrix is essential for calculating not just accuracy, but also other important metrics like Sensitivity, Precision, and Recall.

For a 3-class classification task, the confusion matrix structure is as follows:

	Predicted Class 0 (Positive)	**Predicted Class 1 (Neutral)**	**Predicted Class 2 (Negative)**
Actual Class 0 (Positive)	True Positives (TP0)	False Negatives (FN0,1)	False Negatives (FN0,2)
Actual Class 1 (Neutral)	False Negatives (FN1,0)	True Positives (TP1)	False Negatives (FN1,2)
Actual Class 2 (Negative)	False Negatives (FN2,0)	False Negatives (FN2,1)	True Positives (TP2)

7.6.3 Testing and Final Evaluation

After selecting the best model based on validation performance, the test set was used for final evaluation. The test set contains samples that were not seen during training, providing an unbiased assessment of model performance. The results, including accuracy, precision, recall, and F1-score, were recorded and compared for both LSTM and BERT models.

7.7 Results and Discussion

7.7.1 Performance Overview

The performance of LSTM and BERT models was evaluated using key metrics: precision, recall, F1-scorefor each of the 3 sentiment classes: negative (0), neutral (1), and positive (2). The detailed results for both models are as follows (Tables 7.3 and 7.4).

Table 7.3 LSTM performance table

Model	Class	Precision	Recall	F1-Score
LSTM	Negative (0)	0.88	0.55	0.68
	Neutral (1)	0.40	0.60	0.48
	Positive (2)	0.88	0.91	0.89
	Accuracy			0.79

Table 7.4 BERT performance table

Model	Class	Precision	Recall	F1-Score
BERT	Negative (0)	0.84	0.90	0.87
	Neutral (1)	0.45	0.60	0.51
	Positive (2)	0.94	0.86	0.90
	Accuracy			0.83

7.7.2 LSTM Model Results

- Precision: The LSTM model had high precision for negative (0) and positive (2) sentiments (0.88), but lower precision for neutral (1) sentiment (0.40).
- Recall: The recall for negative sentiment (0) was 0.55, neutral sentiment (1) was 0.60, and positive sentiment (2) was 0.91, indicating strong performance for positive sentiment.
- F1-Score: The F1-score was 0.68 for negative, 0.48 for neutral, and 0.89 for positive sentiment, with the model performing best on positive sentiment.
- Accuracy: The overall accuracy of the LSTM model was 0.79, showing decent performance but weaker results for neutral sentiment (Figs. 7.4 and 7.5).

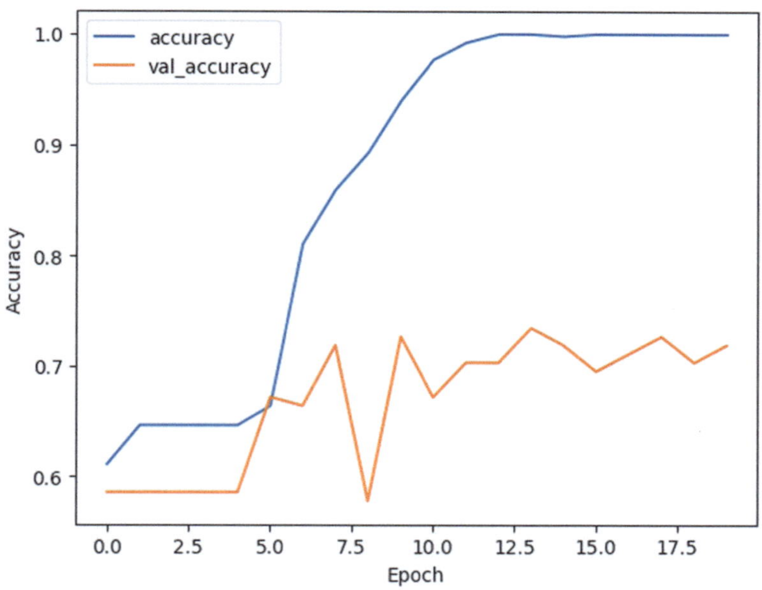

Fig. 7.4 LSTM sentiment analysis accuracy

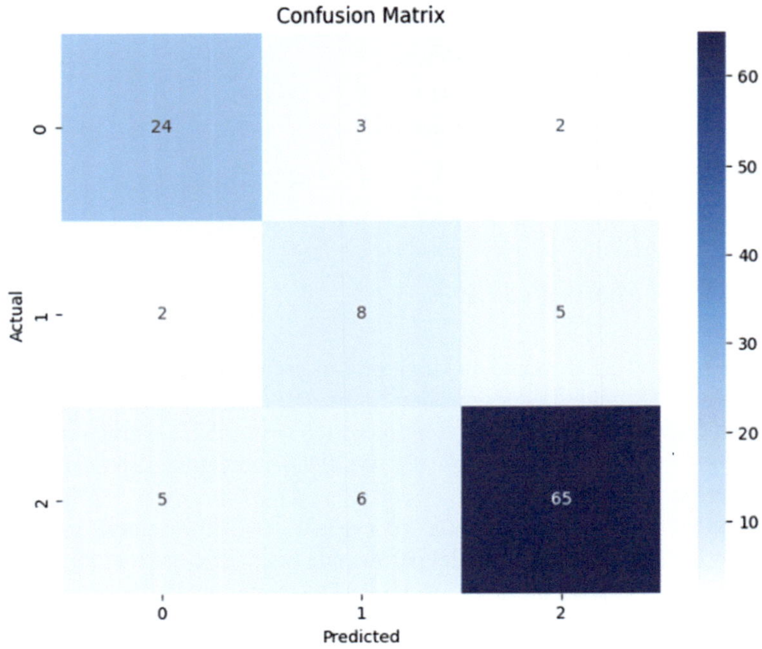

Fig. 7.5 LSTM model confusion matrix

7.7.3 *BERT Model Results*

- Precision: BERT had high precision for negative (0) and positive (2) sentiments (0.84 and 0.94), with moderate precision for neutral (1) sentiment (0.45).
- Recall: The recall for negative sentiment (0) was 0.90, neutral (1) was 0.60, and positive (2) was 0.86, indicating strong identification of negative and positive sentiments.
- F1-Score: The F1-score was 0.87 for negative, 0.51 for neutral, and 0.90 for positive sentiment, with the model excelling in negative and positive categories.
- Accuracy: The overall accuracy of BERT was 0.83, outperforming LSTM, particularly in positive sentiment classification (Figs. 7.6 and 7.7).

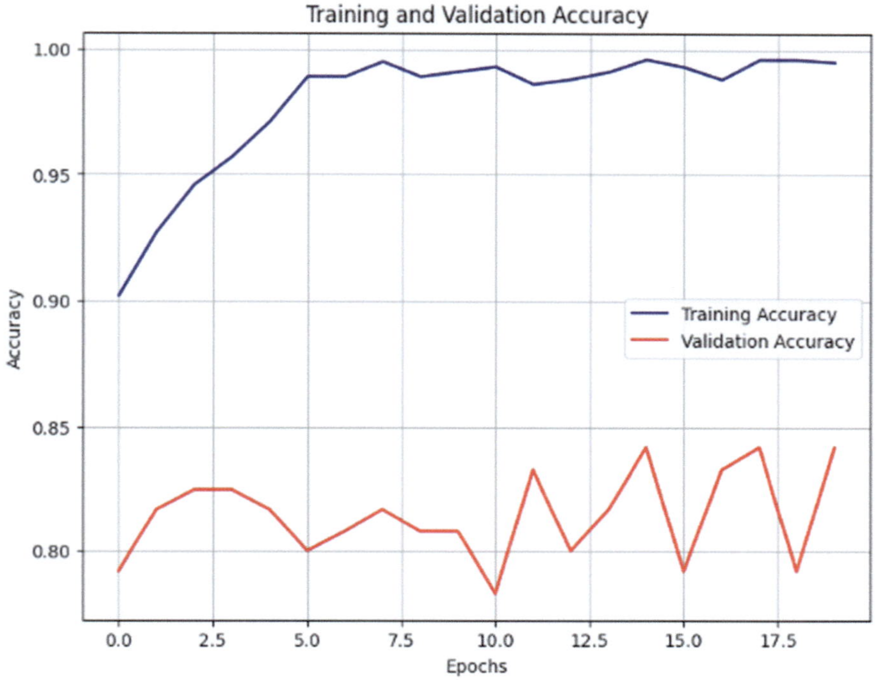

Fig. 7.6 BERT sentiment analysis accuracy

7.7.4 Comparing the Models: BERT and LSTM

The BERT model outperformed the LSTM model in terms of overall accuracy (0.83 vs. 0.79). BERT exhibited higher precision and recall for both negative and positive sentiments, while the LSTM struggled, particularly with neutral sentiment. This highlights BERT's strength in handling complex language understanding and a wider range of sentiments. Both models had difficulty with neutral sentiment (class 1), with significantly lower precision, recall, and F1-score for neutral instances, especially in the case of the LSTM. This could be attributed to the challenge of classifying neutral sentiments, which are subtler and harder to distinguish. Overall, while both models performed well, BERT's pre-trained transformer architecture allows for better contextual understanding, making it more effective for sentiment analysis tasks compared to the simpler LSTM model (Table 7.5).

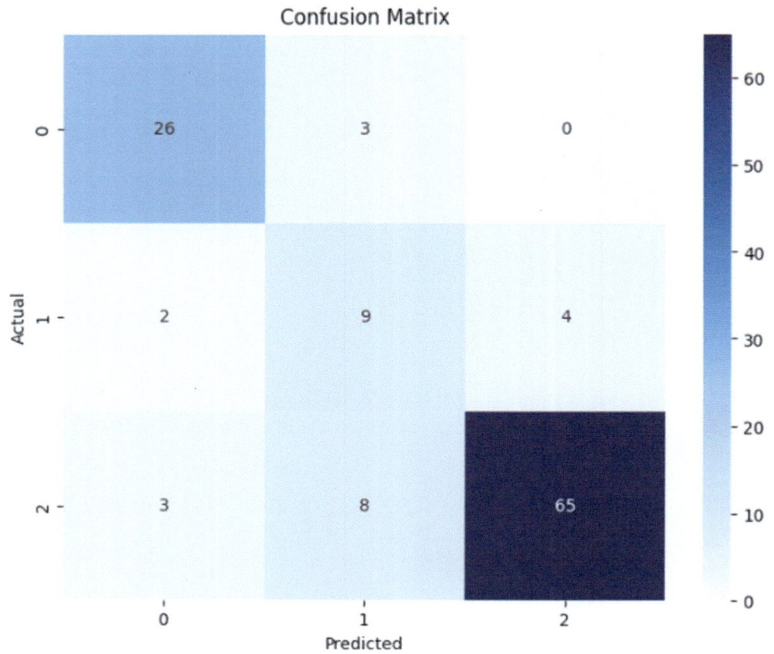

Fig. 7.7 BERT model confusion matrix

Table 7.5 Comparison of BERT & LSTM

Metric	LSTM model	BERT model
Accuracy	0.79	0.83
Average Precision	0.72	0.74
Average Recall	0.69	0.78
Average F1-Score	0.68	0.76

7.8 Conclusion

The study presents a comparative analysis of two models, Long Short-Term Memory (LSTM) networks and Bidirectional Encoder Representations from Transformers (BERT). The comparison is made with respect to sentiment analysis and also illustrates how sentiment analysis has improved over the years, from traditional to advanced models. LSTM is good at understanding and recognizing the order of words and long-term patterns, which makes it useful for analyzing social media content. On the other hand, BERT has the capability of reading text in both directions, allowing it to capture more complex sentiments or emotions like sarcasm or irony. Because of its transformer architecture, which scrutinizes input in parallel, BERT usually needs more time and resources than LSTM from a computational viewpoint. But more frequently than not, the trade-off leads to better performance, especially in challenging language tasks. Important performance metrics that are critical for assessing the effectiveness of both models were also discovered by the study, including accuracy, F1-score, sensitivity and precision. LSTM and BERT models have gained a lot of popularity, proving to be very useful tools. Still, improving their effectiveness in real-life applications requires overcoming dataset restrictions and various technical challenges. The study also highlights research gaps, such as the limited variety of data sources and difficulties with words that have multiple meanings. There is also room for improvement in how emotions are classified, beyond just positive or negative. Overall, the comparison between LSTM and BERT models for sentiment analysis shows their strengths and weaknesses. BERT performed better overall, with higher accuracy (0.83 vs. 0.79) and better precision and recall for both positive and negative sentiments. This shows that BERT is better at understanding complex language and capturing a wider range of emotions. Both models had trouble with neutral sentiments and showed much lower accuracy for this category. Neutral sentiments are harder to identify because they do not have clear emotional clues. LSTM had an even harder time with neutral sentiments compared to BERT, showing that it struggles more with understanding context. Future research can focus on solving current problems, like the lack of diverse datasets, the confusion caused by words with multiple meanings, and the need to better classify emotions beyond just positive or negative. Improving datasets and model designs can make sentiment analysis more accurate and useful in real-life situations.

References

1. L. Zhang, S. Wang, B. Liu, Deep learning for sentiment analysis: a survey. Wiley Interdisc. Rev.: Data Min. Knowl. Discov. **13**(1), e1471 (2023)
2. A. Yadav, D.K. Vishwakarma, Sentiment analysis using deep learning architectures: a review. Artif. Intell. Rev. **53**(4), 4335–4385 (2023)
3. J. Singh, V. Gupta, A resource-efficient LSTM model for sentiment analysis in resource-constrained environments. Neural Comput. Appl. **35**(8), 6123–6138 (2023)
4. J. Devlin, M.W. Chang, K. Lee, K. Toutanova, BERT: pre-training of deep bidirectional transformers for language understanding. Proc. NAACL-HLT **2019**, 4171–4186 (2019)
5. X. Li, L. Bing, W. Zhang, W. Lam, Exploiting BERT for end-to-end aspect-based sentiment analysis. Proc. EMNLP **2023**, 34–43 (2023)
6. Y. Liu, M. Ott, N. Goyal, et al., RoBERTa: a robustly optimized BERT pretraining approach. computer science, computation and language
7. C. Sun, L. Huang, X. Qiu, Utilizing BERT for aspect-based sentiment analysis via constructing auxiliary sentences. Proc. NAACL **2023**, 380–385 (2023)
8. A. Sharma, et al., *Fundamentals of Recurrent Neural Network (RNN) and Long Short-Term Memory (LSTM) Network* (ScienceDirect, 2020)
9. A. Alharthi, et al., *Sentiment Analysis Based on Deep Learning: A Comparative Study* (MDPI, 2020)
10. Hua, Y, *Understanding BERT Performance in Propaganda Analysis* (2019). https://doi.org/10.48550/arXiv.1911.04525
11. J. Liu, et al., *Improving BERT with Self-supervised Attention* (IEEE, 2021)

12. C. Rojas, et al., *Sentiment Analysis in Social Media and Its Application: Systematic Literature Review* (ScienceDirect, 2020)
13. A. Zadeh, et al., *Hybrid Approaches for Sentiment Analysis* (CAI, 2018)
14. M.Z. Islam, et al., A review on recent advances in deep learning for sentiment analysis: performances, challenges, and limitations (2020)
15. A. Bhowmick, et al., Recent improvements in sentiment analysis of tweets. ACL Anthology (2023)
16. A. Zadeh, et al., *Hybrid Approaches for Sentiment Analysis* (CAI, 2017)
17. P. Tiwari, et al., *Increasing Text Filtering Accuracy with Improved LSTM* (IEEE, 2023)
18. S. D'Andrea, et al., *Sentiment Analysis on Facebook* (Stanford NLP, 2010)
19. A. Noorian, A. Harounabadi, M. Hazratifard, A sequential neural recommendation system exploiting BERT and LSTM on social media posts. Complex Intell. Syst. **10**(1), 721–744 (2024)
20. S. Jamshidi, M. Mohammadi, S. Bagheri, H.E. Najafabadi, A. Rezvanian, M. Gheisari, … Z. Wu, Effective text classification using BERT, MTM LSTM, and DT. Data Knowl. Eng. **151**, 102306 (2024)
21. H. Murfi, T. Gowandi, G. Ardaneswari, S. Nurrohmah, BERT-based combination of convolutional and recurrent neural network for Indonesian sentiment analysis. Appl. Soft Comput. **151**, 111112 (2024)
22. K.K. Mohbey, G. Meena, S. Kumar, K. Lokesh, A CNN-LSTM-based hybrid deep learning approach for sentiment analysis on Monkeypox tweets. New Gen. Comput. **42**(1), 89–107 (2024)
23. F. Khanam, A. Chakraborty, M.A. Habib, M.S. Iqbal, Bangla sentiment analysis on highly imbalanced data using hybrid CNN-LSTM & Bangla BERT, in *2024 3rd International Conference on Advancement in Electrical and Electronic Engineering (ICAEEE)* (IEEE, 2024), pp. 1–5
24. R. Patel, J. Chauhan, N.K. Tiwari, V. Upaddhyay, A. Bajpai, A deep learning framework for hourly bitcoin price prediction using bi-LSTM and sentiment analysis of Twitter data. SN Comput. Sci. **5**(6), 767 (2024)

Part III
Smart Mobility Systems

Chapter 8
Adaptive User Interface for Electric Vehicle Route Information in Urban Mobility Services

Ana Vigário, Joana Oliveira, Rodrigo Fernandes, Tiago Pinto, Arsénio Reis, Tânia Rocha, and João Barroso

Abstract Growing urbanisation, the development of smart cities, and environmental concerns have driven the implementation of advanced technologies and the modernisation of transport systems. Electric motorcycles have emerged as an effective solution for mobility, but they also present specific challenges, particularly related to the mode of riding, which is more complex than that of other vehicles and requires greater attention, skill, and preparation. Therefore, the interaction between the rider and the support system must be carefully designed, with particular emphasis on the interface and the adaptation of route information. This interface should be intuitive, accessible, and capable of presenting relevant information in a clear and objective manner, minimising distractions while riding. In addition, it must be adaptable to user preferences, allowing for customisation such as colour themes, levels of detail in the information displayed, or specific notifications regarding adverse weather and road conditions. Adapting route information provides a more efficient, safe, and satisfactory user experience. It enables riders to access personalised information, continuously updated in real time and tailored to the situation and their specific needs, including traffic conditions, road surface state, and weather conditions. This optimisation leads to better time management, energy consumption, and overall ride quality, enhancing urgent and non-urgent services. Moreover, integrating clearly and objectively features such as voice commands and compatibility with mobile or wearable devices (e.g., smartwatches) can facilitate real-time interaction without compromising safety. The interface should also offer advanced functionalities in line with technological developments and user needs, adapting to each rider's specific requirements. This not only improves the individual experience but also promotes efficiency and sustainability, contributing to the advancement of smart cities and innovative mobility solutions.

A. Vigário · J. Oliveira · R. Fernandes · T. Pinto · A. Reis · T. Rocha · J. Barroso
University of Trás-os-Montes and Alto Douro, Vila Real, Portugal

T. Pinto (✉) · A. Reis · T. Rocha · J. Barroso
INESC-TEC, Vila Real, Portugal
e-mail: tiagopinto@utad.pt

G. A. Tsihrintzis et al. (eds.), *Human-Smart City Interactions and User-Citizen Experiences*, Learning and Analytics in Intelligent Systems 58,
https://doi.org/10.1007/978-3-032-06364-9_8

Keywords Electric motorcycles · User interface · Intelligent transportation systems · Real-time route adaptation · Smart mobility · Human–machine interaction · User-centered design

8.1 Introduction

The concept of smart cities has been taking shape in recent times as a response to the urban challenges of our century, due to population growth and pollution. The integration of digital technologies and information systems is one of the solutions to optimize urban resources and improve the quality of life of the population.n Albuquerque et al. [1], which analyses traffic and accident patterns in the city of Lisbon, it was found that in 2019, 96.9% of traffic congestion occurred on primary roads and that the most common type of accident was collisions (56.8%), followed by pedestrian run-overs (24.3%), influenced by temporal, geographical, and environmental factors.

Analyzing the lifestyle of citizens in large cities, we can conclude that one of the critical points is transport. Urban mobility is clearly at the heart of this modernization, due to the inefficiency of transport systems, which negatively affect the environment, with high carbon emissions and drastically reduce the well-being of citizens [2]. The management of smart cities is therefore crucial in this process, in order to redesign the way people move around.

This modernization of transport systems involves the creation of integrated mobility ecosystems, in which various modes of transport–including electric scooters, bicycles, shared cars, and public transport–coexist efficiently and sustainably [3]. The objective is to promote a multi-modal model, personalised and adapted to the real needs of users.

In this context, the interface between the citizen and mobility systems plays a crucial role, as the user makes informed decisions regarding routes and alerts, thereby interacting with the urban environment. The existence of effective, intuitive, and adaptive interfaces is essential to ensure that smart city technology becomes a facilitator of mobility and sustainability.

It is at this point that the integration of electric scooters, in urban context, emerges as a practical response to the demands of the environment and to the need for effective citizen mobility [4]. Supported by the appropriate technology and physical infrastructure, electric scooters become a fundamental pillar in the transformation of these cities, requiring interfaces that communicate with urban systems, provide optimised navigation, and adapt to users in real time.

This chapter is organised as follows: Sect. 8.2 addresses the growth of electric mobility and its benefits in smart cities. Section 8.3 explores the need for adaptive interfaces for scooters, with an emphasis on the importance of real-time information and route personalisation. Section 8.4 discusses user-centred design applied to urban mobility. In Sect. 8.5, the relationship between route optimisation, sustainability, and two-wheeled mobility is analysed. Section 8.6 presents the A-MoVeR case study, detailing the developed solution and the algorithms applied. Finally, Sect. 8.7 presents the conclusions and future perspectives.

8.2 Sustainable Urban Mobility and Electric Vehicles

The growing role of electric vehicles, particularly scooters, in the context of sustainable urban mobility plays a central part in the transformation of smart cities. The widespread adoption of these vehicles, along with the crucial role of route optimisation, ensures a safer, more personalised, and more sustainable driving experience.

8.2.1 The Growth of Electric Mobility in Smart Cities

One of the fundamental pillars of sustainable mobility in smart cities is the transition from combustion vehicles to electric solutions. Electric vehicles markedly reduce emissions of polluting gases, which have a direct impact on the health and well-being of citizens, as well as on air quality. They are also more energy-efficient, wasting less energy in the from of heat and noise [5].In addition to the environmental aspect, there are also benefits such as reduced noise pollution, lower dependence on fossil fuels, and decreased vehicle maintenance costs [6].

All these factors reinforce the attractiveness of electric mobility in urban centres, where population density demands increasingly clean, quiet, and efficient solutions [4]. More specifically, electric scooters occupy less physical space compared to cars, making parking easier and offering excellent alternatives for short- to medium-distance journeys, helping to relieve congestion and improve urban flow [4]. When integrated into intelligent mobility systems, electric scooters become essential components of urban ecosystems, making them more efficient, sustainable, and user-centred.

Despite all the benefits mentioned above, the integration of these vehicles into smart cities still faces several structural, technological, and behavioural obstacles. One of the main barriers is the charging infrastructure for this type of vehicle. The scarcity of dedicated charging stations limits their autonomy and results in a lack of interest among citizens in adopting electric scooters as a means of transport [7].

Another challenge lies in the integration of technology with urban information systems. For electric scooters to fully benefit from smart solutions, such as optimised routing, traffic forecasts, or even environmental conditions, it is essential that they are connected to a real-time urban data network, which is still limited.

Route optimisation also plays a key role in urban electric mobility, particularly in the case of electric scooters. When users are able to choose the shortest, fastest, or most energy-efficient routes, this feature enhances battery efficiency and reduces time spent in traffic, two critical factors for users of this type of electric vehicle. From a sustainability perspective, reducing travel time and avoiding traffic jams leads to lower emissions and improved traffic flow [4]. A city equipped with intelligent navigation systems can prevent overloading in specific areas, thereby contributing to a more balanced management of urban mobility.

From the user's perspective, personalising the interface of electric scooters according to their preferences significantly improves the driving experience, making it more comfortable, safer, and more efficient. Moreover, adaptive interfaces that visually and intuitively integrate such optimisation can provide real-time adjustments based on traffic, weather, vehicle status, and user preferences [6].

8.2.2 The Benefits of Electric Scooters

Electric scooters are gaining ground among urban citizens as a practical and intelligent solution to the mobility challenges faced in cities. Unlike combustion engine vehicles, these types of vehicles do not emit polluting gases during use, which directly contributes to the improvement of urban air quality, one of the main problems in metropolitan areas [5].

Exposure to environmental pollution results in millions of deaths annually, with the transport sector being the largest contributor to pollution levels in major cities. The noise produced by combustion vehicles is also a form of pollution that affects individuals' health. In the European Union, 65% of the urban population is exposed to noise levels above 55 dB [4].

In terms of energy efficiency, electric motors are notably more efficient than combustion engines, consuming less energy to perform the same task, an important point for drivers. Moreover, the accessibility of this type of electric vehicle is an evident advantage. Electric scooters are smaller and more agile, allowing them to navigate through traffic, park in tighter spaces [4], and even access low-emission zones—areas that are increasingly common in smart and sustainable cities. In short, by combining environmental sustainability, energy efficiency, and agility in urban traffic, electric scooters offer a complete package for cleaner, faster, and more practical mobility in the cities of today and the future [8].

8.2.3 The Benefits of Route Optimisation

Route optimisation is essential in electric mobility, particularly when it comes to scooters, where battery autonomy is still a limiting factor. With the availability of intelligent algorithms and real-time data, such as traffic, roadworks, weather, or charging point availability, it is possible to map out shorter, faster, and more efficient routes. For electric scooter riders, these factors are particularly impactful, as they result in less time on the road, lower energy consumption, and reduced stress related to route uncertainty.

In urban contexts, where driving can be chaotic, route optimisation helps to avoid congestion and navigate around restricted areas, making the most of the vehicle's flexibility. Moreover, the automatic selection of routes can help extend battery life, a critical factor for those who rely on the vehicle for daily travel.

Finally, this optimisation also enables a more personalised and safer riding experience, adapted to the rider's driving style, current conditions, and even the battery level of the scooter: better planning, greater sustainability, and a smarter overall driving experience [6].

8.3 The Need for Adaptive Interfaces for Electric Scooters

Adaptive interfaces are emerging as fundamental elements in making riding safer, more efficient, and more personalised. These interfaces must respond in real time to the variables of the urban environment and to the individual needs of the rider, providing only the essential information at the right moment, in an intuitive and distraction-free manner.

8.3.1 Riding Challenges and the Importance of Real-Time Information

Riding an electric scooter in an urban environment demands higher levels of attention, coordination, and anticipation compared to other means of transport. The lack of physical protection, the constant need for balance, and direct exposure to weather conditions make riders more vulnerable and dependent on quick and accurate decisions.

With electric scooters, new challenges arise: the absence of engine noise reduces auditory awareness for other road users, increasing the risk of accidents, and the vehicle's limited range requires greater route planning. Additionally, the behaviour of electric scooters differs significantly from combustion models, requiring a specific learning curve.

In this context, real-time information becomes vital. Data on traffic, the presence of low-emission zones and accidents, weather changes, or even the availability of charging points can make the difference between a smooth journey and an avoidable setback [6]. Intelligent interfaces that provide this information quickly, intuitively, and non-intrusively are therefore key elements in enhancing user safety, efficiency, and confidence[9]. In a scenario where an electric motorbike rider is already facing the inherent challenges of riding, any additional element may compromise their safety. This is where the interface plays a fundamental role [10].

A well-designed interface acts as a facilitator, not a complication. By presenting only relevant information, in a clear, organised way and adapted to real-time context, the interface reduces cognitive overload. For instance, displaying visual alerts only when urgent action is needed (such as rerouting due to heavy traffic) or enabling voice commands are effective ways to help the rider stay focused on what truly matters: the road [6].

Moreover, adaptive interfaces can automatically adjust the level of detail or complexity of the information displayed based on speed, location, or even the user's profile [11]. An experienced rider may prefer more technical data, while a novice might benefit from simplified notifications: the interface is not merely an information panel, but an active safety tool. By minimising distractions and supporting quick and accurate decision-making, it helps protect both the rider and those sharing the road [6].

8.3.2 Route Personalisation

Route optimisation plays a crucial role in electric mobility, particularly in the case of electric scooters, where battery autonomy remains a limiting factor. By using intelligent algorithms and real-time data, such as traffic, roadworks, weather conditions, or charging station availability, it is possible to map out shorter, faster, and more efficient routes [12].

For electric motorbike riders, this translates into less time on the road, lower energy consumption, and reduced stress associated with route uncertainty [13]. In urban environments, where navigation can be chaotic, route optimisation enables riders to bypass congestion and restricted areas with agility, taking full advantage of the vehicle's flexibility. Moreover, the automatic selection of routes that favour regenerative braking, for example, can help extend battery range–an essential benefit for those who rely on the vehicle for daily commuting.

Finally, this optimisation also paves the way for a more personalised and safer riding experience, tailored to the rider's style, current conditions, and even the motorbike's battery level. In short: better planning, greater sustainability, and a smarter riding experience [6].

8.4 User-Centred Design and Interfaces for Urban Mobility

User-Centred Design (UCD) is a design methodology that prioritises the needs, preferences, and limitations of end users throughout the entire process of designing and developing digital systems [14]. By adopting UCD, the main goal is to ensure that the final interface or product is intuitive, easy to use, and adaptable to the specific characteristics of its users, providing a positive user experience.

When applied to the development of interfaces for electric scooters, the objective is to create a navigation and control system that not only delivers the necessary information, but does so in an intuitive and efficient manner, taking into account the specific conditions of riding in urban environments [10] . UCD facilitates interaction with interfaces. These must be simple and straightforward, allowing users to access and interact with information while remaining focused on riding. The use of gestures, voice commands, or easily accessible buttons is crucial to minimise the need for screen interaction while on the move [14].

The interface should be accessible to all types of users, including those with visual, auditory, or motor impairments[12] . This can be achieved through the implementation of audio feedback, real-time notifications, and adjustable text sizes or colour contrasts to ensure readability under varying lighting conditions.

One of the pillars of UCD is the ability to customise the interface according to user preferences [11]. This may include personalisation of route information, map views, sound alerts, and vibrations, tailoring the interface to suit each rider's style and preferences.

The interface should also provide real-time information about battery status, traffic on suggested routes, and weather conditions, helping users to make informed decisions and optimise their riding experience [14]. Feedback should also be immediate and discreet, preventing the rider from becoming distracted or overwhelmed by excessive information.

These principles ensure that electric scooter interfaces are not only functional but also enhance the overall user experience, making riding safer, more personalised, and more efficient within a dynamic urban context [11]. In fast-changing city environments, it is essential that users have easy interaction, access to critical information, and full control over the personalisation of their experience, providing them with the necessary confidence and safety to enjoy their ride.

Interface intuitiveness is an essential tool for rider comprehension [14]. The interface should be simple and direct, allowing quick access to information while keeping the rider fully focused on the road. Visual elements such as intuitive icons, simplified menus, and minimal interactions are crucial to avoid distractions and ensure that the rider can operate the scooter while maintaining full attention on driving. The information presented should be logically organised, with routes, alerts, and performance metrics easy to understand without overloading the user with unnecessary details [14]. Interfaces must prioritise clarity, with a clear information hierarchy that enables the rider to make quick decisions based on traffic or route conditions.

Accessible interfaces are designed to meet the needs of a wide range of users, including those with visual, auditory, or motor impairments [12]. This may involve the use of large fonts, high colour contrast, audio and haptic (vibratory) feedback, and support for voice commands and other interaction methods–enabling riders to stay focused on the road while receiving important information. Interface design must also take into account different lighting conditions, such as daytime or nighttime, when visibility is reduced. The use of high-contrast color schemes and readable displays in various lighting scenarios improves accessibility in any context [14].

Each rider has a unique style and individual needs. Customisable interfaces allow adjustment of variables such as route information, sound alerts, colour themes, and levels of detail to match the rider's driving preferences. One rider may prefer a simplified display with discreet sound alerts, while another might opt for a more detailed interface with vibrating alert systems. Personalisation also extends to battery management, visualisation of traffic and weather conditions, and navigation preferences. Allowing users to choose how they wish to interact with these variables creates a smoother and more pleasant experience tailored to their specific needs [11].

In the context of electric scooters, designing interfaces that combine these three pillars—intuitiveness, accessibility, and personalisation—is key to maximising safety, comfort, and efficiency for the rider. Effective design not only enhances the user experience, but also minimises distractions while riding, ensuring that the interaction with the interface is natural, straightforward, and complication-free [15].

8.5 Efficiency and Sustainability: The Intersection between Route Optimization, Motorcycle, and the Environment

Last Mile Delivery, in urban environments, faces significant challenges due to the growth of cities, chaotic vehicle traffic, and the increasing demand for fast deliveries [16]. These factors, combined with the pressure to provide services with greater speed and efficiency, make urban logistics a complex and constantly evolving sector. The focus on improving the quality of life and ensuring the satisfaction of inhabitants is closely linked to the development of smart and sustainable cities. In this context, electric motorcycles emerge as an effective solution for logistics services, although they also present specific challenges, particularly related to driving behavior. Aligned with the development and implementation of route optimization algorithms and the creation of a front-end adaptable to driving modes, this approach will enable a more effective response to the demands of urban environments, contributing to the transformation of smart cities.

8.5.1 The Use of Electric Motorcycles in Logistics and Urban Mobility

The growing concern about the environmental impact of current road transport systems has accelerated the option of clean and energy-efficient alternatives that aim to reduce the carbon footprint. The electric vehicles have gradually been recognised as promising alternatives to combustion engine vehicles, and so promoting their use could potentially have a significant positive impact on the environment and energy systems both globally and locally [7]. In addition, the shift from internal combustion engine vehicles to light electric vehicles (such as electric motorcycles and mopeds) is considered a promising solution for improving transport sustainability [5].

Electric motorcycles have emerged as a promising alternative, offering advantages in terms of sustainability and operational efficiency. They stand out as a solution that aligns the reduction of environmental impacts with the need for greater dynamism in urban logistics operations. From an environmental sustainability perspective, the large-scale adoption of electric two-wheelers represents an important contribution to the reduction of greenhouse gas emissions and air pollution in urban areas [6]. As is growing number of people choose to replace combustion vehicles with electric two-wheeled alternatives, there is an effect on reducing polluting gas emissions and improving air quality, especially in urban areas, where transport-related pollution is a major concern [8]. Another relevant issue is the significant reduction in noise levels [17], as conventional motorcycles and cars are among the main sources of urban traffic noise.

Two-wheeled vehicles are able to access congested areas, restricted spaces, and difficult terrain much more quickly and easily than four-wheeled or larger vehicles [18]. However, they are still more complex to drive than other vehicles, requiring greater attention, skill, and preparation from the part of the driver. Drive assistance platforms, designed to support navigation or provide real-time information, can also become a source of distraction. Studies indicate that periods as brief two seconds, during which the rider's eyes are diverted from the road, are directly associated with an increased risk of near-crashes and collisions. Moreover, operating a motorcycle is inherently more demanding than driving a car, and consequently, the absolute effects of distraction may be greater when riding a motorcycle [15].

8.5.2 Mobility Interface Designed for Motorcyclists

Motorcycles, which offer clear advantages in terms of accessibility, are often overlooked or negatively perceived due to concerns related to pollution, noise, road safety and rider behaviour [5]. In a study conducted by Maestracci et al. [19], in fact, the most frequent causes of accidents in Paris involving at least one "motorised two-wheeler" were identified as situations such as vehicles changing direction (right or left turns), falls or skidding (often resulting from the need to avoid another driver) and pedestrians crossing outside designated areas or against traffic signals. These findings highlight the constant need for attention to the road and its surrounding environment.

Klauer et al. [15] argued that riding a motorcycle is more demanding than driving a car and, as a result, the absolute effects of distraction can be even greater for motorcyclists. The overall impact of distraction across a population of drivers may be reduced through the use of safer and more user-friendly navigation devices. A rider assistance interface should therefore provide support without neglecting the limitations and challenges of urban riding.

8.5.3 The Route Optimization as a Strategy

One of the biggest problems facing the logistics sector is the heavy traffic flow in urban areas, which causes delays in travel times, increased fuel consumption, and higher transport-related costs. In addition to the burdens inherent in the service, these journeys also contribute to worsening air and noise pollution, negatively affecting the quality of life of local inhabitants. These impacts can be mitigated through the implementation of route optimization techniques [6].

It is important to note that route optimization is not limited to determining the shortest path, but can also be guided by additional criteria, such as energy consumption and associated emissions. In this way, it becomes possible to identify routes that simultaneously minimize fuel consumption and CO_2 emissions, thereby contributing to more efficient and sustainable mobility [20].

The electric vehicle routing problem has been attracting growing interest among researchers and industry professionals in the context of the transition from combustion engine vehicles to more sustainable and efficient solutions such as electric vehicles. Although at first glance, route optimisation for electric vehicles may appear similar to classical vehicle routing problems, it presents its unique challenges that distinguish it significantly [21]. In the case of electric scooters, beyond these general challenges, additional factors further increase the complexity of the problem, such as limited capacity for transporting passengers or goods, long charging times, and the scarce availability of charging infrastructure.

8.6 Case Study A-MoVeR

The A-Mover Agenda [22] aims to promote a substantial transformation in the industrial development model of the Trás-os-Montes and Alto Douro region. In line with sustainability and development goals, and driven by advances in technology and energy, along with a growing sense of environmental responsibility, the initiative sought to develop a solution capable of addressing navigation challenges when riding two-wheeled vehicles for logistics services. The use of electric scooters has emerged as a reliable alternative for both urgent and non-urgent services, highlighting their viability for such operations.

The creation of an interface for this type of vehicle, although showing great potential for improving safety and efficiency in logistics and last-mile transport operations, faced a critical challenge: the need for riders to maintain full focus on the road while riding. Riding a motorcycle demands heightened attention, and any distraction can compromise the rider's safety. This characteristic represented a significant barrier to the use of informational devices, such as small handlebar-mounted screens, which could support navigation and provide real-time data, but could also serve as a source of distraction [23].

Urban riding presented numerous logistical challenges, particularly in the context of sustainable mobility using electric scooters. In this scenario, the development of tools to support route planning became essential, especially when considering the specificities of riding electric scooters. Optimisation algorithms contributed to the creation of efficient routes, reducing costs and time, while accounting for vehicle-specific constraints such as cargo capacity and the limited battery range. Route planning is practically and safely needed not only to determine the ideal path but also to identify the location of charging stations and assess the feasibility of completing the entire journey—departure, service delivery, and return to base—within the available energy limits. Additionally, these tools had to be robust enough to handle adverse conditions, such as poor road surfaces or extreme weather, enabling dynamic reconfiguration of routes.

In light of these challenges, the development of a compact, useful, and efficient mobility frontend was proposed, specifically designed for electric scooters with an optimised interface. This would allow users to access necessary services in a practical and safe manner while on the move, taking into account the specific characteristics and limitations of this type of vehicle. To cope with the dynamic nature of the information, the interface needed to be flexible and intuitive, capable of adapting to updates and responding effectively to unexpected changes, such as route deviations or the altered availability of charging points.

The solution involved developing a services frontend to support professional scooter use, integrated with a robust backend that would feed the interface with a complete and regularly updated database. The objective was to create a user-oriented, adaptable interface to be implemented on display devices, capable of presenting the outputs of route optimisation models clearly, transparently, and intuitively—based on rider interaction and usage patterns Fig. 8.1.

Two interface models were designed, offering flexibility to the user. One of the models was intended to be integrated directly into the electric scooter's display, with a dedicated area for presenting task-related and navigation information. The other model was developed for use on a portable device, allowing it to be carried outside the vehicle and, when needed, connected to the scooter's system via a wireless or wired connection.

8.6.1 Optimisation Algorithms

In both models, the route display was expected to show relevant information and ensure that the results from the route optimisation models were presented clearly and transparently, minimising distractions and prioritising safety. The main objective of studying optimisation algorithms was to compare solutions that maximise operational efficiency, with each technique adapted to consider the particularities of electric vehicles, namely their limited range and the need for recharge planning along the route [24].

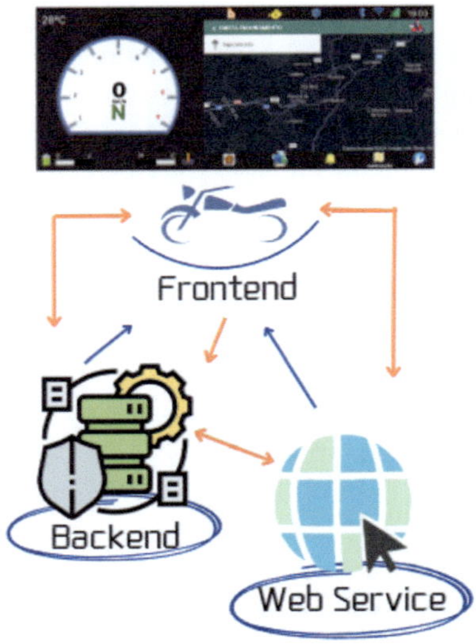

Fig. 8.1 Platform overview

Several algorithms were explored for solving the Vehicle Routing Problem (VRP) [25], specifically adapted to the context of electric vehicles [26]. Various relevant scenarios were considered, such as Capacitated Vehicle Routing (CVRP) [27], Vehicle Routing Problem with Time-Windows (VRPTW) [28], Multi Depot Vehicle Routing Problem (MDVRP) [29], and Dynamic Vehicle Routing Problem (DVRP) [30], where new requests may arise during ongoing deliveries. To effectively address these different variants, exact, heuristic, and metaheuristic approaches were applied, including techniques such as Branch and Bound.

The route optimisation models were tested using custom-created scenarios for evaluation purposes, such as:

- *Simple Delivery Scenario*: A fleet of electric vehicles was required to serve six customers from a single depot, with vehicles having limited cargo and battery capacity, aiming to minimise the total distance travelled;
- *Scenario with Customer Time Constraints*: Vehicles with limited cargo capacity and customers with time windows for service, which influence the travel time between locations;
- *Real-Time Routing Scenario with Battery Recharging*: Vehicles with limited cargo capacity, where new orders emerged during ongoing deliveries, and battery recharging was required, making it necessary to include stops at designated charging stations;
- *Multi-Depot Vehicle Routing Scenario*: Two depots serve a city, each with its own fleet. Customers are assigned to the nearest depot, aiming to minimise the total distance travelled .

These, among other scenarios explored, allowed for the evaluation of the effectiveness of the proposed algorithms under different operational conditions. To this end, tests were conducted using the scenarios described above, applying the Dijkstra algorithm [31], as well as Nearest-Neighbor [32] and Savings [33], which are widely recognized in solving routing problems. The results obtained made it possible to identify the advantages and limitations of each approach, as illustrated in the examples below:

– ***Scenario with Time Windows***

- *Description*: Electric vehicles are required to make deliveries within specific time intervals set by the customers.
- *Results*:
  - Dijkstra:
    - Total Distance: 7.50 km;
    - Total Travel Time: 0h 11m;
    - Number of Depot Visits: 2;
    - Execution Time: 1.39 seconds .
  - Nearest Neighbor:
    - Total Distance: 7.50 km;
    - Total Travel Time: 0h 11m;
    - Number of Depot Visits: 2;
    - Execution Time: 0.30 seconds.
  - Savings:
    - Total Distance: 10.50 km;
    - Total Travel Time: 0h 15m;
    - Number of Depot Visits: 3;
    - Execution Time: 0.26 seconds.
- *Conclusion*: The Dijkstra and Nearest Neighbour algorithms were the most efficient in terms of total distance and travel time, although Dijkstra had a longer execution time. The Savings algorithm showed intermediate performance but required more depot visits.

- ***Dynamic Recharge Scenario***
  - *Description*: During the delivery process, new requests arise that need to be incorporated into the routes, also taking into account stops for recharging the vehicle batteries.
  - *Results*:
    - Dijkstra:
      - Total Distance: 28.62 km;
      - Total Travel Time: 10h 11m;
      - Number of Depot Visits: 2;
      - Execution Time: 0.55 seconds.
    - Nearest Neighbor:
      - Total Distance: 28.62km;
      - Total Travel Time: 10h 03m;
      - Number of Depot Visits: 2;
      - Execution Time: 0.34 seconds.
    - Savings:
      - Total Distance: 28.62 km;
      - Total Travel Time: 10h 03m;
      - Number of Depot Visits: 2;
      - Execution Time: 0.26 seconds.
  - *Conclusion*: All three algorithms produced very similar results in terms of efficiency and time, with a slight advantage for the Savings algorithm, which combined the shortest total travel time with the lowest execution time.

8.6.2 Interfaces

The objective is to create a user-oriented, adaptable, and customisable interface that allows the rider to adjust the settings according to their needs and preferences, regardless of the type of device used. A well-designed user interface minimises distraction and confusion; with clear and intuitive controls and displays, the user will spend less time interacting with commands and more time focused on the road. Customisable settings contribute to an improved experience by striking a balance between functionality, usability, and safety—a significant challenge in the automotive design of graphical user interfaces (UI) and user experience (UX) systems.

Display-embedded interface solution for the scooter

The layout design of this solution presents all information within a dedicated area, so as not to interfere with the scooter's operational data. This ensures a clear separation between navigation elements and the scooter's functional indicators, minimising the rider's cognitive and visual load Fig. 8.2.

Priority was given to the simple and clear presentation of the available options, such as tasks and history, in order to facilitate navigation and enhance the usability of the interface. The elements were sized according to user perception, bearing in mind that a scooter display is typically non-touchscreen. Therefore, components such as buttons, menus, and information displays must be easily accessible and quickly understandable, allowing for option selection via the scooter's handlebar controls.

The points to be visited are presented in the order defined by the optimisation algorithms. This ordering takes into account the constraints imposed by the user, such as priority locations, route preferences, time limits, and charging points, among other factors. The main objective of the application is to display only the crucial details needed for navigation; for this reason, only the most relevant information, such as the next destination, will be shown in Fig. 8.3.

When movement begins, the navigation layout should display only the map with directional guidance, speed limits, and other driving support information. To handle the dynamic nature of information, the interface must be flexible and intuitive, capable

Fig. 8.2 Display

Fig. 8.3 Display–Task in progress

of adapting to updates and efficiently responding to unexpected changes, such as route deviations or changes in the availability of charging points.

The navigation interface should present directional information clearly and in advance, allowing the rider to prepare in good time for manoeuvres or route changes. This approach is essential to ensure safe riding, particularly in complex urban environments or on unfamiliar routes.

Instructions must be visible and easily understandable, integrating visual elements and, when necessary, audio cues, in order to minimise distractions and reduce the user's cognitive load. Early presentation of guidance promotes quicker reaction times from the rider, contributing to a smoother and safer navigation experience Fig. 8.4.

Notifications, whether related to the vehicle's status, driving alerts, environmental information, or general messages, should be displayed using the minimum space necessary without compromising readability. A distinct warning sound may be emitted to indicate an alert, and the notification should remain on screen only for the time required for the rider to view it safely.

The history of vehicle status information, as well as notifications sent by third parties, will be presented in a dedicated layout, accessible at a later time, only when the scooter is stationary Fig. 8.5.

The display of the history of visited locations will only be available when the application is not in navigation mode and the vehicle is detected to be stationary. The history will be organised chronologically, based on the time of visit, providing a clearer overview of the points visited throughout the day. Additionally, it will be possible to perform searches using criteria such as destination name and alternative dates Fig. 8.6.

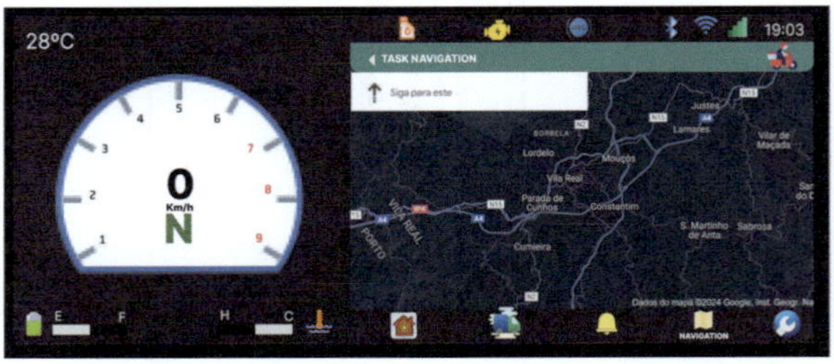

Fig. 8.4 Display–Task navigation

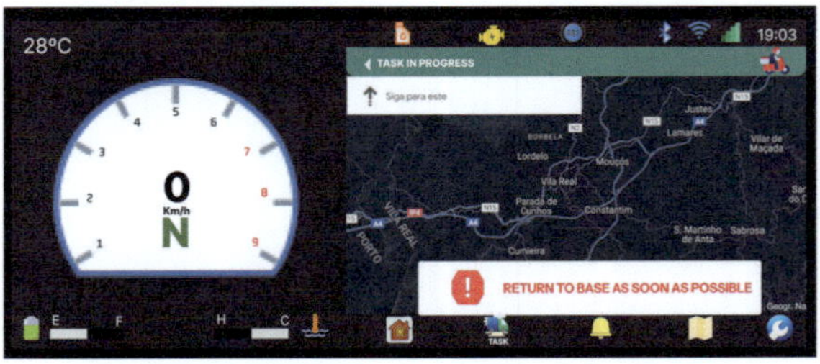

Fig. 8.5 Display–Task notification

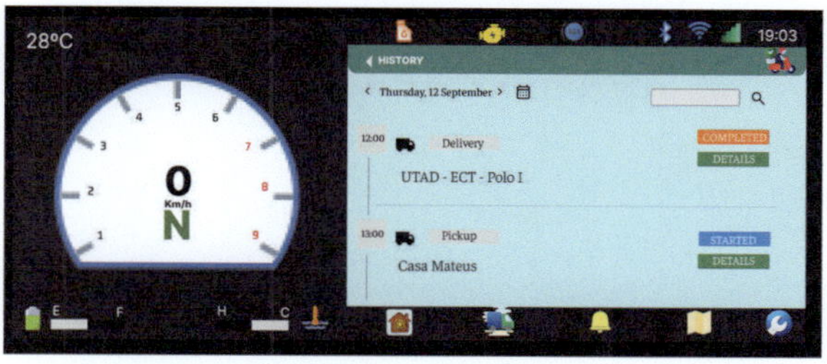

Fig. 8.6 Display–History

Fig. 8.7 App smartphone example

Portable Application Solution

Despite technological advances, many drivers still prefer to use their own smartphones instead of the vehicle's built-in systems, a warning sign for automotive UX designers. The future of vehicles is increasingly likely to resemble that of connected mobile devices. The digital experience (before, during, and after driving) will be a key differentiating factor. Automotive UX is still in its early stages, but its potential is immense [34] Fig. 8.7.

For this solution, the interface will be developed on a removable device, such as a smartphone, which can be connected to the scooter's system. A specific area will be reserved for displaying driving-related information, such as the navigation map, route guidance, notifications, and other relevant data, without compromising the rider's visibility or safety Fig. 8.8.

The application will be developed to ensure its functionality even when disconnected from the scooter's device, operating as a support system for delivery and collection logistics. In this standalone mode, the user will be able to access essential information related to assigned tasks, consult routes, manage delivery status, and confirm receipts—thereby ensuring service continuity regardless of connection to the vehicle. This approach enhances the versatility of the application, allowing it to

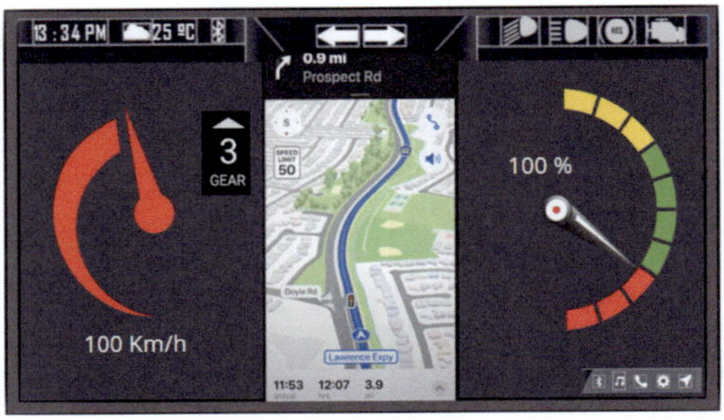

Fig. 8.8 Display–Navigation

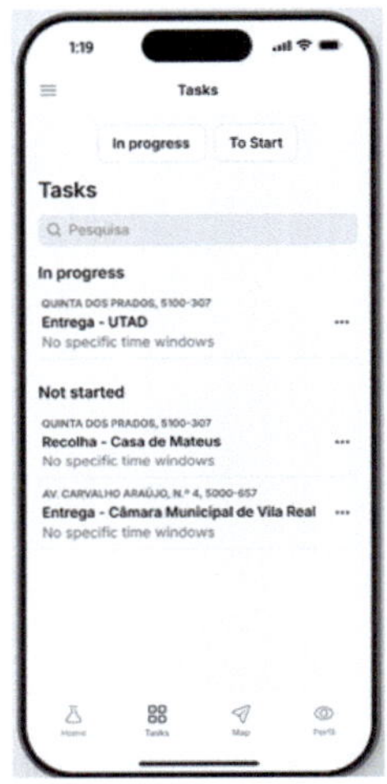

Fig. 8.9 Tasks

be used in different operational contexts and reinforcing user autonomy in managing their logistical tasks Fig. 8.9.

The user will have the freedom to use the application independently on their smartphone, with access to navigation functionalities Fig. 8.10.

However, when the device is connected to the scooter, it will be possible to project the navigation directly onto the vehicle's display. This integration provides a safer and more optimised riding experience, allowing the rider to view route instructions, notifications, and relevant information without needing to interact directly with the smartphone. In this way, greater ergonomics and safety are ensured, promoting a seamless digital experience between mobile and vehicle modes Fig. 8.11.

All notifications, alerts, and messages must be displayed ergonomically, with strategic positioning within the rider's field of vision and for an appropriate duration, so as not to compromise attention to the road. The interface should ensure that critical information is visible and easily understood without effort, adhering to road safety principles and minimising cognitive load. This careful presentation of information contributes to a smoother, more comfortable, and safer riding experience Fig. 8.12.

Fig. 8.10 Smartphone navigation

Fig. 8.11 Smartphone-connected navigation

Fig. 8.12 Smartphone-connected navigation-Notification

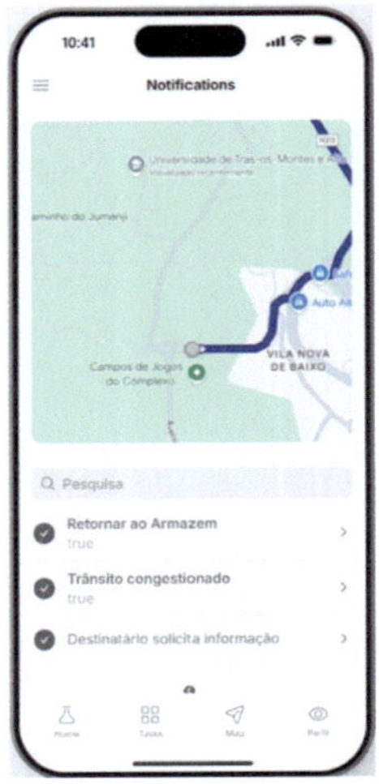

Fig. 8.13 Smartphone-Notifications

Fig. 8.14 Interface login

All notifications and messages will be available in a dedicated layout, accessible when the scooter is stationary Fig. 8.13.

The interface will be developed for use on a smartphone and, when connected to the scooter's display during navigation, will operate in an integrated manner. Outside the riding context, it can be used to view travelled routes, visited locations, and other additional features. This approach prioritises the reduction of distractions while riding, allowing the rider to stay focused on the road, with brief and occasional interactions for checking navigation and notifications Fig. 8.14.

The proposed interface will include the possibility for user customisation, allowing adjustments according to individual preferences, such as visual themes, element layout, and types of interaction Fig. 8.15.

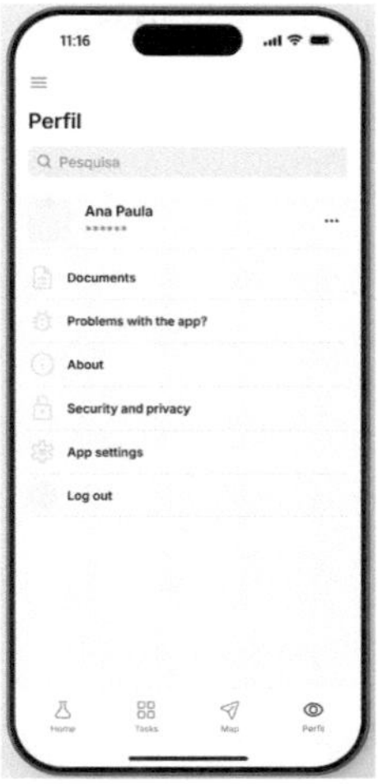

Fig. 8.15 Profile

Updates will be carried out remotely via OTA (Over-the-Air) technology, which enables the updating of software, firmware, or device settings through wireless connections such as Wi-Fi, mobile data, or Bluetooth. This will allow for UI modifications, the addition of new menus, buttons, or colour schemes, and an improved user experience through adjustments based on real user feedback. It will also enable the correction of navigation, connectivity, or display issues, and adaptation of the interface to support new features. In automotive UX, the ability to update the interface via OTA is a key trend, allowing for continuous interface improvement without the need for a technician [35].

8.7 Conclusion

In summary, the implementation of adaptive interfaces for electric scooters within the context of smart urban mobility represents a significant advancement in the convergence of technology, sustainability, and user experience. By integrating route optimisation algorithms, user-centred interfaces, and continuous personalisation and update capabilities, it is possible to offer a safer, more efficient riding experience aligned with the contemporary urban challenges faced by smart cities.

The design of these interface solutions will follow UI/UX best practices, focusing on the needs and preferences of the user. Controls will be intuitive, with easy access to information and personalisation options, including adjustments to colours and font size, ensuring accessible and adaptable use across various contexts, such as weather conditions, lighting, and the rider's level of technical experience. The interface should align with the general principles of user experience, promoting a coherent, satisfying, and rider-focused interaction. Additionally, with an international audience in mind, the design will consider language localisation, cultural sensitivities, and global accessibility, ensuring the interface is understandable and usable by individuals from diverse backgrounds.

The design must be future-proof, capable of adapting to technological evolution, including artificial intelligence, while consistently maintaining ease of use. UX refers to the overall user experience, while UI (User Interface) relates to the visual and functional aspects of the system. Together, they are essential for ensuring a safe, efficient, and satisfying interaction with the vehicle, highlighting the importance of minimising distractions and effectively managing mental workload.

The flexibility of the proposed solutions—whether integrated displays or mobile devices—reinforces the importance of an inclusive, accessible approach tailored to real usage contexts. These solutions contribute not only to improved logistics operations and personal mobility but also to the development of smarter and more environmentally responsible cities.

The future of urban mobility will inevitably rely on increasingly sophisticated interactive systems, capable of responding in real time to user needs and the dynamics of the urban environment. In this scenario, adaptive interfaces will play a central role in transforming two-wheeled travel into a more connected, safe, and sustainable digital experience.

Acknowledgements The study was developed under the project A-MoVeR—"Mobilizing Agenda for the Development of Products and Systems towards an Intelligent and Green Mobility", operation n.o 02/C05-i01.01/2022.PC646908627-00000069, approved under the terms of the call n.o 02/C05-i01/2022—Mobilizing Agendas for Business Innovation, financed by European funds provided to Portugal by the Recovery and Resilience Plan (RRP), in the scope of the European Recovery and Resilience Facility (RRF), framed in the Next Generation UE, for the period from 2021–2026

References

1. Albuquerque, V.,et al.: Smart Cities: Data-Driven Solutions to Understand Disruptive Problems in Transportation–The Lisbon Case Study. Energies 2021, 14, 3044.(2021) https://doi.org/10.3390/en14113044
2. Xu, Jiayuan et al.: Effects of urban living environments on mental health in adults. Nature medicine vol. 29,6 (2023): 1456-1467. https://doi.org/10.1038/s41591-023-02365-w
3. A. Machado, J. Santos, A. Sacavém, M.J. Sousa, Digital Transformation: Management of Smart. Cities (2023). https://doi.org/10.1108/978-1-80455-994-920231004
4. Bakker, S.: Electric Two-Wheelers, Sustainable Mobility and the City. In: *InTechOpen* (2018). https://doi.org/10.5772/intechopen.81460
5. S. Will, C. Luger-Bazinger, M. Schmitt, C. Zankl, Towards the Future of Sustainable Mobility: Results from a European Survey on (Electric) Powered-Two Wheelers. Sustainability **13**, 7151 (2021). https://doi.org/10.3390/su13137151
6. Huang, Fei-Hui. *Understanding User's Experiences of Riding a Two-Wheeler Vehicle and Their Intentions of Purchasing Eletric Two-Wheelers*, Oriental Institute of Technology (2019). Available at: https://hrcak.srce.hr/file/331737. https://doi.org/10.7307/ptt.v31i5.3014
7. C. Zhuge, C. Shao, Agent-based modelling of locating public transport facilities for conventional and electric vehicles. Netw. Spat. Econ. **18**(4), 875–908 (2018). https://doi.org/10.1007/s11067-018-9412-3

8. Karpurapu, S., Saikumar, K., Kocharla, L., Rao, P. S., Govathoti, S.: Scalable Innovative Factors for Shaping Consumer Intentions on Electric Two-Wheelers Adoptions. *Scalable Computing: Practice and Experience*, **25**(6), 4507-4517 (2024). https://doi.org/10.12694/scpe.v25i6.3212
9. Vermeeren, A.: *User Experience White Paper: Bringing clarity to the concept of user experience*. Available at: https://www.academia.edu/853674/User_Experience_White_Paper_Bringing_clarity_to_the_concept_of_user_experience
10. B. Shackel, Usability - Context, framework, definition, design and evaluation. Interact. Comput. **21**(5), 339–346 (2009) https://doi.org/10.1016/j.intcom.2009.04.007.https://www.sciencedirect.com/science/article/pii/S0953543809000241
11. R. Juárez-Ramírez, User-centered design and adaptive systems: toward improving usability and accessibility. Univ. Access Inf. Soc. **16**, 361–363 (2017). https://doi.org/10.1007/s10209-016-0480-1
12. Hassenzahl, M.: *User experience (UX): Towards an experiential perspective on product quality*. In: *ACM International Conference Proceeding Series*, **339**,11–15. (2008) https://doi.org/10.1145/1512714.1512717
13. S. Kujala, V. Roto, K. Väänänen-Vainio-Mattila, E. Karapanos, A. Sinnelä, UX Curve: A method for evaluating long-term user experience. Interact. Comput. **23**(5), 473–483 (2011). https://doi.org/10.1016/j.intcom.2011.06.005
14. Emma, L.: User-centered design to enhance accessibility and usability in digital systems. (2024). Available at: https://www.researchgate.net/profile/Lawrence-Emma/publication/386339454_User-centered_design_to_enhance_accessibility_and_usability_in_digital_systems/links/674ded153d17281c7df33009/User-centered-design-to-enhance-accessibility-and-usability-in-digital-systems.pdf
15. Klauer, S.G., Dingus, T.A., Neale, V.L., Sudweeks, J.D., Ramsey, D.J.: The Impact of Driver Inattention on Near-Crash/Crash Risk: An Analysis Using the 100-Car Naturalistic Driving Study Data. National Highway Traffic Safety Administration (NHTSA) (2006)
16. W.A. Mohammad, Y. Nazih Diab, A. Elomri, C. Triki, Innovative solutions in last mile delivery: concepts, practices, challenges, and future directions. Supply Chain Forum: An International Journal **24**(2), 151–169 (2023). https://doi.org/10.1080/16258312.2023.2173488
17. Sheng N, Zhou X, Zhou Y. Environmental impact of electric motorcycles: Evidence from traffic noise assessment by a building-based data mining technique. Sci Total Environ. 2016 Jun 1;554-555:73-82. (2016) https://doi.org/10.1016/j.scitotenv.2016.02.148. Epub Mar 4. PMID: 26950621
18. K. Apiratwarakul, S. Tiamkao, L.W. Cheung, I. Celebi, T. Suzuki, K. Ienghong, Application of Automated External Defibrillators in Motorcycle Ambulances in Thailand's Emergency Medical Services. Open Access Emergency Medicine **14**, 141–146 (2022). https://doi.org/10.2147/OAEM.S361335
19. M. Maestracci, F. Prochasson, A. Geffroy, F. Peccoud, Powered two-wheelers road accidents and their risk perception in dense urban areas: Case of Paris. Accid. Anal. Prev. **49**, 114–123 (2012). https://doi.org/10.1016/j.aap.2011.05.006
20. Kumar,A., Singh, S., Sharma, S.: *Route Optimization for Fuel Consumption and CO_2 Emission Considering Two-Wheelers*, in *Proceedings of the 2023 3rd International Conference on Technological Advancements in Computational Sciences (ICTACS)*, 1170–1175, ISBN: 979-8-3503-4233-8 (2023). https://doi.org/10.1109/ICTACS59847.2023.10390305
21. Rajesh, K., Jain, E., Kotecha, P.: *A Multi-Objective Approach to the Electric Vehicle Routing Problem*(2022). https://doi.org/10.48550/arXiv.2208.12440
22. A-MoVeR. *Agenda Mobilizadora para o Desenvolvimento de Produtos e Sistemas Inteligentes de Mobilidade Verde*. (2022) Available at: https://www.utad.pt/gpfe/a-mover-agenda-mobilizadora-para-o-desenvolvimento-de-produtos-e-sistemas-inteligentes-de-mobilidade-verde/. Accessed: 14 Apr. 2025
23. M.A. Regan, G. Lintern, R. Hutchinson, C. Turetschek, Use of Cognitive Work Analysis for exploration of safety management in the operation of motorcycles and scooters. Accident Analysis & Prevention **74**, 279–289 (2015). https://doi.org/10.1016/j.aap.2014.07.009

24. Routific. *Route Optimization Software for Delivery Businesses*. Available at: https://www.routific.com/route-optimization. Accessed: 14 Apr. 2025
25. Chen, Z.: Algorithm and Application in Vehicle Routing Problem: A Review. *J. Electron. Res. Appl.***9**(2), 166-174 (2025). https://doi.org/10.26689/jera.v9i2.9939
26. Upper. *The Green Vehicle Routing Problem (GVRP)*. Available at: https://www.upperinc.com/blog/vehicle-routing-problem/. Accessed: 14 Mar. 2025
27. Hexaly. *Capacitated Vehicle Routing (CVRP)*. Available at: https://www.hexaly.com/docs/last/exampletour/vrp.html. Accessed: 14 Mar. 2025
28. Google OR-Tools. *Vehicle Routing Problem with Time Windows*. Available at: https://developers.google.com/optimization/routing/vrptw. Accessed: 14 Mar. 2025
29. Hexaly. *Multi Depot Vehicle Routing Problem (MDVRP)*. Available at: https://www.hexaly.com/docs/last/exampletour/multi-depot-vehicle-routing-problem-mdvrp.html. Accessed: 14 Mar. 2025
30. E. Kucharska, Dynamic Vehicle Routing Problem-Predictive and Unexpected Customer Availability. Symmetry **11**(4), 546 (2019). https://doi.org/10.3390/sym11040546
31. Upper. *Understanding Dijkstra's Algorithm – Comprehensive Guide*. Available at: https://www.upperinc.com/glossary/route-optimization/dijkstras-algorithm/. Accessed: 14 Mar. 2025
32. DATASTAX.*What is the Nearest Neighbor Algorithm? Method & Examples*.Available at: https://www.datastax.com/guides/what-is-nearest-neighbor. Accessed: 14 Mar. 2025
33. Wang, X., Zhao, S.-Z., Chu, H., Li, Y.: An Improved Savings Method for Vehicle Routing Problem. In: Proceedings of the CCSSE 2016, 1–4. (2016). https://doi.org/10.1109/CCSSE.2016.7784340
34. Khanna, J.: *A Breakdown on Automotive UX: Taking User Experiences to The Next Level*. Automotive Design Planet, 2023. Available at: https://automotivedesignplanet.com/stories/1698845374-a-breakdown-on-automotive-ux. Accessed: 24 Mar. 2025
35. UserTesting,*Automotive UX/UI – User Research – User Interface for Autos*, UserTesting Blog, 18 de agosto de 2023. Available at: https://www.usertesting.com/blog

Part IV
IoT-Enabled Infrastructure

Chapter 9
Design and Implementation of a Smart Recycling Bin Using Multi-sensor Integration and Machine Learning

Ioannis Gkini, Griseld Gkini, and Dimitrios Tsipianitis

Abstract As the world's population expands and consumption rises, urban waste continues to pile up, putting pressure on conventional waste management systems. This underscores the critical necessity for efficient recycling methods. Traditional approaches, which typically rely on extensive manual sorting, are not only ineffective but also neglect the mounting environmental challenges such as pollution and the depletion of resources. This paper presents a novel Smart Recycling Bin that integrates cameras, audio detection, and various sensors with machine learning to optimize waste sorting. Designed for both home and public use, this technology adapts to different environments, ensuring versatility. The bin's sophisticated sensor technology combined with machine learning algorithms enables it to accurately identify and categorize various waste types, offering a scalable solution for enhancing recycling rates and reducing contamination. This advancement represents a significant step forward in making recycling more efficient and adaptable to varying waste challenges.

Keywords Recycling · Smart bin · Machine learning · ML · AI · Audio · Vision · Sensors

I. Gkini · G. Gkini
Post Graduate Student, University of Patras, Patras, Greece
e-mail: up1066689@ac.upatras.gr

G. Gkini
e-mail: 1080555@ac.upatras.gr

D. Tsipianitis (✉)
Professor, University of Patras, Patras, Greece
e-mail: dtsipianitis@ece.upatras.gr

G. A. Tsihrintzis et al. (eds.), *Human-Smart City Interactions and User-Citizen Experiences*, Learning and Analytics in Intelligent Systems 58,
https://doi.org/10.1007/978-3-032-06364-9_9

9.1 Introduction

Recycling challenges stem from sorting inaccuracies and limited public awareness about proper disposal methods, leading to contamination that degrades recyclable materials and compromises the overall effectiveness of the recycling process [1, 2]. To combat these issues, our Smart Recycling Bin incorporates a comprehensive system utilizing weight, material properties, sound, and visual sensors, along with machine learning. This sophisticated setup not only allows for accurate waste classification under realistic conditions but also identifies specific characteristics of the waste item, enabling it to provide immediate and detailed feedback to the user. Such feedback can help correct potential disposal errors, ensuring more effective recycling practices. To showcase the feature extraction capability, we trained two machine learning models: one that can distinguish between five categories—Aluminum, Plastic, Paper, Glass, and Containers with Content—and another that detects lids on containers. Lid detection is crucial for improving recycling efficiency by ensuring materials like plastic lids on glass bottles are sorted separately. This precise sorting helps enhance the overall effectiveness of the recycling process [3]. The two models we trained with our custom dataset are lightweight enough to run efficiently on a Raspberry Pi. The device we implemented can be used in all spaces and would be particularly useful for home use, functioning as a personal recycling assistant in every household. It would also be highly valuable to place such a bin in educational spaces, where students could be sensitized from a young age to good recycling practices.

9.2 State of the Art

In this field, a number of research articles were already published. Attention is being paid to this area increasingly due to its significance in maintaining ecological balance and promoting sustainable development. Recent advancements in deep learning have significantly enhanced capabilities in computer vision. The convolutional neural network (CNN), a prominent deep-learning algorithm, is extensively recognized for its effectiveness in image classification, segmentation, and detection [4–6], and is thus frequently employed in waste classification as well [7–9].

One study introduced a smart recycling bin utilizing waste image classification to boost recycling efficiency, incorporating advanced machine learning models on embedded systems like Jetson Nano and K210. The bin uses the EfficientNet B0 model, a compact yet powerful CNN architecture known for its efficiency in image recognition tasks, achieving an accuracy of 95.38% with rapid inference times suitable for real-time applications. This system innovatively minimizes power consumption while maintaining high performance, integrating lighter models like MobileNet and the energy-efficient K210. The bin classifies waste into categories such as paper, metal, and plastic, enhancing urban waste management by tackling practical application challenges effectively [7].

WasteNet classifies waste by using a deep neural network-based model that processes images captured by cameras fitted to the bins. The system employs transfer learning, leveraging pre-trained models on the ImageNet dataset, which allows it to adapt to the specific task of waste classification more effectively. These models are then fine-tuned or adjusted through a hybrid approach, combining both fixed feature extraction from earlier network layers and fine-tuning of the later layers. This method helps the model learn and differentiate between various types of waste such as plastic, paper, glass, metal, and cardboard. The neural network analyzes the images to determine the correct category of waste and directs it to the appropriate compartment within the smart bins, ensuring accurate disposal and recycling [8].

A research has focused on developing ML-CNN models for waste classification, which combine the robust feature-detection capabilities of CNNs with the decision-making prowess of multilayer perceptrons. Specifically, the system utilizes a two-stage classification approach: a multilayer perceptron first determines if waste is metallic or not, while the CNN classifies non-metallic materials. This setup aids in effectively categorizing various types of waste, supporting efficient recycling practices in urban environments [9].

The GREENBIN system is an innovative solution for automated waste segregation at the household level, specifically designed for environments like apartments and educational institutions where waste quantity is manageable. This system incorporates advanced sensor technology, including capacitive moisture sensors for determining waste moisture content, inductive metal sensors for detecting metallic objects, methane sensors for identifying organic waste, and odor sensors for detecting biomedical waste [10].

Another paper discusses a Smart Bin system designed for effective waste segregation using capacitive sensing and IR spectroscopy, capable of distinguishing between dry, wet, and plastic wastes. The system includes automation features such as an automatic door and uses a mobile app to inform users of waste types and bin status. This innovation enhances waste management efficiency, aligning with sustainable practices and the smart cities initiative. The technology not only segregates waste accurately but also optimizes disposal routes through data analysis, significantly reducing human intervention and improving sanitation [11].

The GULP system utilizes Convolutional Neural Networks (CNNs) with ResNet-50 v2 architecture for waste classification, processing images captured by an integrated camera. By leveraging transfer learning from pre-trained models on the ImageNet dataset, the system effectively categorizes waste into biodegradable, non-biodegradable, and recyclable types. A Raspberry Pi 4 Model B powers the processing, while a solar energy setup ensures sustainable operation. The system also features SMS notifications to alert personnel when the bin is full, representing a practical and efficient integration of machine learning, IoT, and renewable energy for waste management [12].

Unlike existing studies, our bin offers versatile waste classification based on weight, material composition, sound and visual characteristics. Our advanced machine learning algorithms not only enhance real-time data processing but can also adapt better to realistic scenarios, providing more contextual information through the various features we can extract from the waste. This comprehensive approach revolutionizes recycling practices, paving the way for more sustainable urban environments. Additionally, by implementing the device, we demonstrated the feasibility of this idea, moving beyond mere ideological description.

9.3 Device Architecture and Functionality

In this section, we cover the essential components and operational mechanics of the device. We analyze the main features of the device in the following segments.

A. *Hardware Configuration*

The system is powered by a Raspberry Pi 4 Model B with 4 GB of RAM, which acts as the central processor for integrating sensors, actuators, and running the classification algorithm. The Raspberry Pi 4 was selected for its combination of strong computational performance and power efficiency, making it suitable for edge computing tasks in compact and energy-constrained environments. Its capability to run machine learning models has been validated in numerous studies, further establishing its effectiveness for this application [13–16].

To address challenges with the weight sensors, which did not deliver reliable performance when directly connected to the Raspberry Pi, an Arduino Mega 2560 was incorporated. The Arduino features an onboard Analog-to-Digital Converter (ADC), which provides a more precise and stable way to process the analog signals from the weight sensors. The Arduino then communicates the processed data to the Raspberry Pi via a serial connection (Fig. 9.1).

B. *Sensors*

1. Strain Gauge: We incorporated four 3 kg weight sensors (OEM), which effectively function as a scale. This setup allows for the precise measurement of the weight of the disposed items, providing critical data for identifying the type of waste based on its mass.
2. Inductive Sensor: Three LJ18A3-8-Z/BX inductive sensors were used, which are specifically used to detect metallic objects exclusively [17, 18]. This feature is crucial for separating metal from non-metal waste.
3. Capacitive Sensor: Three LJC18A3-B-Z/BX capacitive sensors, which are designed to detect both metallic and non-metallic objects. These sensors are

Fig. 9.1 **a** Raspberry Pi. **b** Arduino Mega 2560

equipped with a potentiometer on the back, allowing fine-tuning of their sensitivity threshold to suit specific applications. During testing, it was observed that these sensors are particularly effective at detecting metallic objects and items containing liquids, making them well-suited for distinguishing such materials in the waste stream [18, 19]. To optimize their performance for the intended use case, the sensitivity was carefully adjusted to a very low level. This configuration ensures that the sensors activate only in the presence of metallic objects or waste with liquid content, minimizing false detections caused by other materials.

4. Camera: The Raspberry Pi Camera Module V3—Standard 11.9MP with a 120-degree field of view was selected to optimize the spatial footprint of the bin. This compact camera setup allows for detailed visual analysis of the waste material as it is thrown into the bin.
5. Microphone: We observed that each type of material produces a characteristic sound when it hits a hard surface. So, we incorporated a usb microphone to record the sound made by the object upon impact, providing an additional layer of data to enhance the sorting accuracy and minimize uncertainty of the classification algorithm (Fig. 9.2).

C. *Device Structure*

The bin is divided in two main sections:

1. The Preliminary Chamber, serving as the initial processing area where waste is received and assessed. The floor of this chamber, where waste lands after disposal, features a unique design consisting of two metal plates, each 40 cm long and 15 cm wide, joined at an angle to form an inverted triangular shape. This configuration stabilizes the waste quickly, ensuring clear image capture and providing an ideal surface for sensor placement.

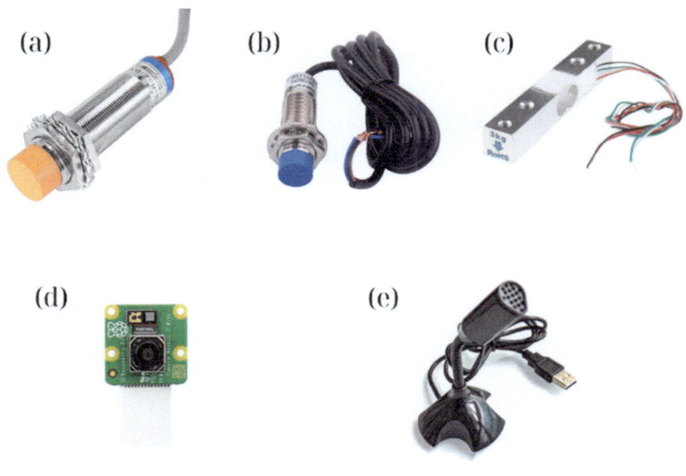

Fig. 9.2 Sensors: **a** OEM 3 kg load cell. **b** LJ18A3-8-Z/BX inductive sensor. **c** LJC18A3-B-Z/BX capacitive sensor. **d** Raspberry Pi camera module V3 with 120° FOV. **f** Mini USB Microphone

The capacitive sensors are located on the lower portion of the left plate, while inductive sensors are positioned on the corresponding area of the right plate. This arrangement ensures that any object longer than 10 cm contacts at least one sensor group, enabling accurate detection and categorization. The floor also acts as a scale, supported by four symmetrically placed weight sensors that provide precise weight measurements.

A rotational mechanism, powered by an MG996R servo motor and gears, allows the floor to tilt left or right, guiding waste into the next section. Additionally, a centrally mounted camera captures visual data, while a microphone near the landing point records the sound upon impact, adding another dimension to the waste analysis process.

2. Final Storage Chambers: This section serves as the final storage area for the waste after it has been categorized. The number of these chambers was chosen so that one side receives non-recyclable waste while the other collects recyclables, thus saving space. To utilize the precise sorting information obtained, a type of stamp could be placed on the recyclables to properly separate them at the final sorting line in recycling centers. This would ensure that the detailed categorization does not go to waste and facilitates further processing.

D. *Operational Process*

As the user approaches to dispose of an item, an IR sensor located at the lower part of the bin's entrance detects the presence of the waste. This triggers the weight sensors to calibrate and the microphone to begin recording for the next two seconds. Concurrently, two LEDs light up to provide sufficient illumination inside the bin, ensuring that the camera captures a clear image of the waste. At the same time, data from the weight, inductive, and capacitive sensors are collected. Following this, the classification algorithm processes the data to determine the type of waste. Based on the classification outcome, if the model determines with over 90% confidence that the item belongs to the categories of Aluminum, Glass, Plastic, or Paper, the floor of the bin rotates clockwise to direct the waste into the appropriate section of the final storage chamber for recyclables. Conversely, if the confidence level is below 90% or the item is categorized as a Container with Content, the floor rotates counterclockwise, directing the waste to the non-recyclable section. From the moment an object has entered the bin until it is categorized, approximately 5 s elapse and the prechamber is ready to accept another object. Finally, to display the results and observations, we could install a simple LCD screen (Figs. 9.3, 9.4, 9.5, 9.6, 9.7, 9.8 and 9.9).

9.4 Software Configuration

The classification system of the device uses a multimodal approach to sort waste into five categories: Aluminum, Plastic, Paper, Glass, and Containers with Content (regardless of material). To build the system, data was gathered by repeatedly

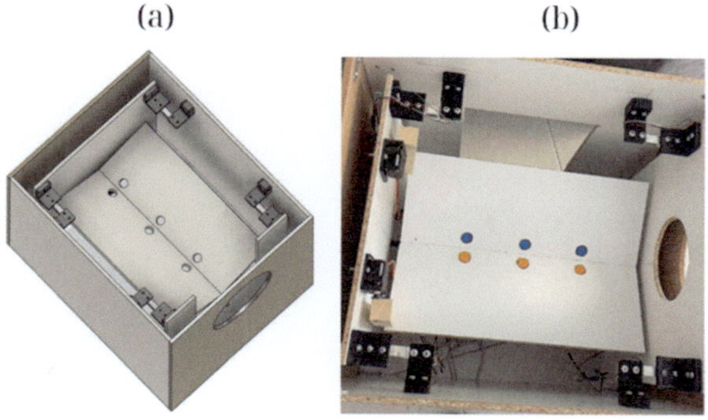

Fig. 9.3 Preliminary chamber seen from a top-down view **a** Fusion 360 design. **b** Actual implementation

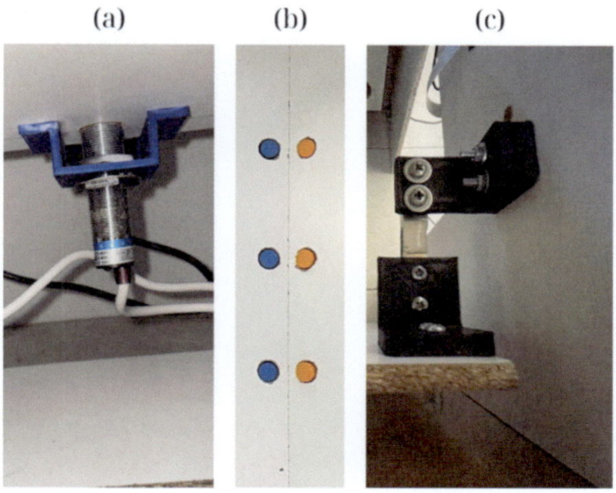

Fig. 9.4 **a** Inductive sensor under the floor of the pre-chamber. **b** Inductive-capacitive sensor groups. **c** Load cell holding the floor

Fig. 9.5 View from the entrance of the pre-chamber, featuring a microphone and camera mounted on the ceiling. Inductive and capacitive sensors are secured with protective tape to shield them from debris and contamination

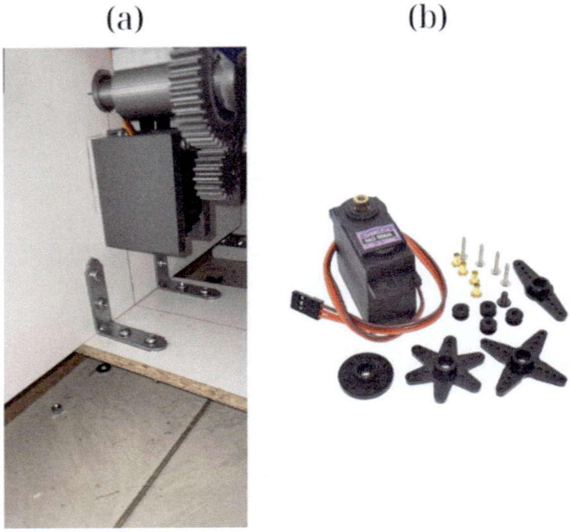

Fig. 9.6 **a** Detail of the operational axis beneath the pre-chamber floor, showcasing the servo mechanism with dual gears for rotational control. **b** MG996 Servo 360° high torque

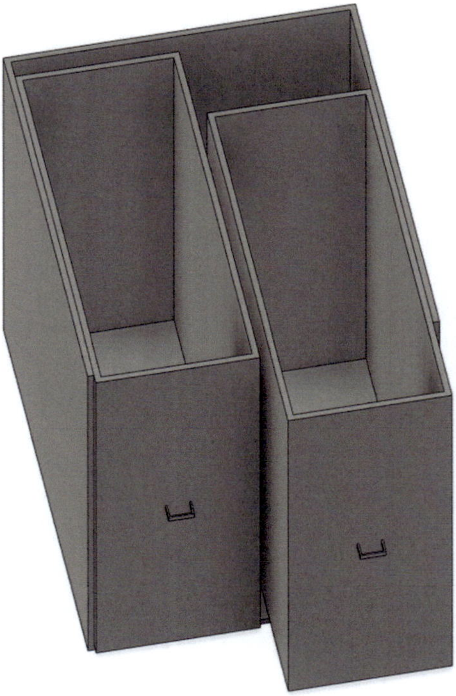

Fig. 9.7 Final storage chambers

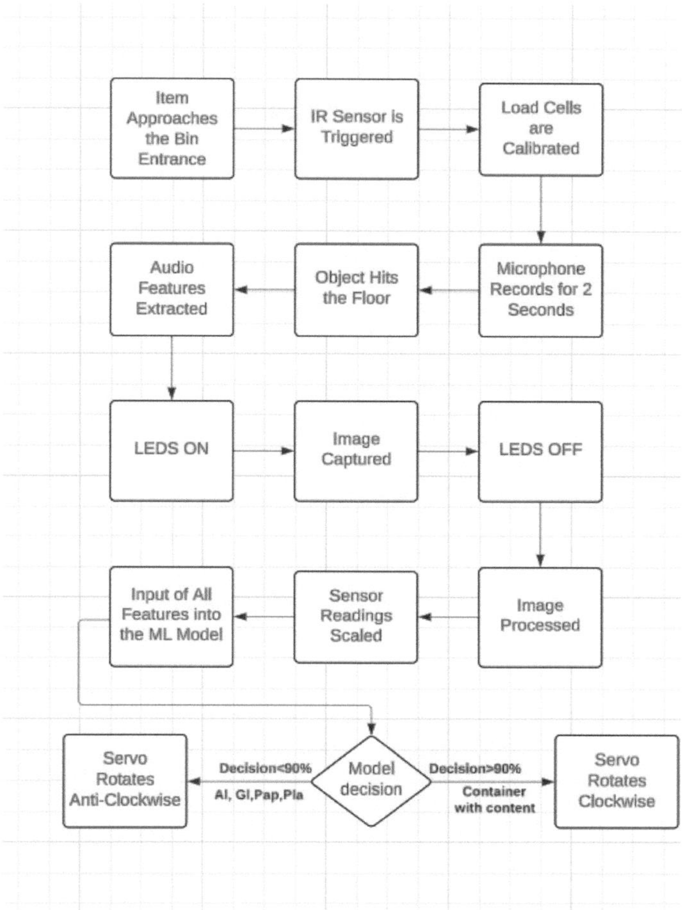

Fig. 9.8 Flowchart of the bin's operational process

disposing of various waste items into the bin. This process produced a dataset of 354 images and their corresponding data points, which were used to train a machine learning model implemented with TensorFlow. The dataset was carefully designed to include a wide variety of waste items, ensuring the model can generalize well to different types of trash encountered in real-world recycling situations.

To ensure robust performance, the dataset was divided into three parts: 60% for training, 20% for validation to fine-tune the model during development, and 20% for testing to evaluate its accuracy on unseen data. This structured approach helps the system perform reliably in classifying waste. Figure 9.10 provides examples of the waste items included in the dataset, highlighting the range of materials the system is trained to recognize.

Before integrating the camera and microphone data into a single structure, they were independently tested in separate models to verify the reliability of the data. This

Fig. 9.9 Image of object entering the bin's entrance, triggering the IR sensor (green light) on the entrance

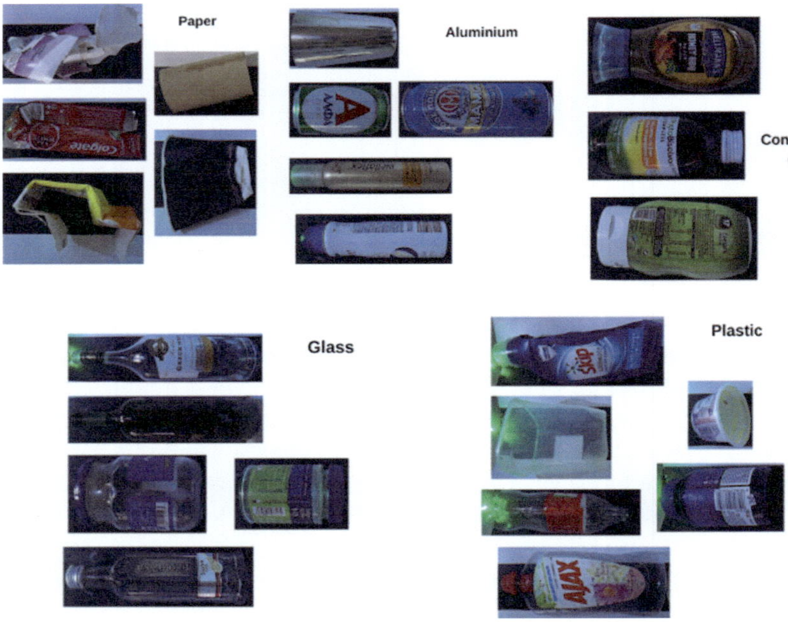

Fig. 9.10 Image dataset sample for each category

step was crucial to ensure that each type of sensor data could effectively contribute to the overall accuracy of the waste classification system without interference from the other modalities. Both individual models achieved an accuracy of over 92%.

In the following sections, we analyze the processing of all data collected from various sensors and their integration into the model.

A. *Image Processing*

The process begins with object detection, using the TensorFlow SSD MobileNet V2 FPNLite 320 × 320 model. This model was chosen for its speed and lightweight design [20], making it well-suited for deployment on resource-constrained devices like the Raspberry Pi. The model performs two key tasks. First, it detects whether a lid is attached to a container, allowing the system to prompt users to separate the lid from the container in future disposals if they are made of different materials. Second, it identifies the object inside the bin, ensuring accurate detection and classification. Once the object is detected, the region within the bounding box provided by the detection model is cropped from the image. The cropped RGB image is resized to 300 × 300 pixels and scaled so that its pixel values range between 0 and 1. This preprocessed image is then passed to the InceptionV3 network, which operates as a static feature extractor. Utilizing pre-trained ImageNet weights, the network processes the image while excluding the top classification layers. The result is a 3D tensor with dimensions (8, 8, 2048) produced by the final convolutional layers [21, 22]. To further refine the features, a GlobalAveragePooling2D layer is applied to the tensor. This layer condenses each feature map into a single value, simplifying the data while retaining meaningful information. The output is a 1D tensor containing 2048 extracted features, which serve as a compact yet descriptive representation of the object's visual characteristics. This streamlined feature set facilitates subsequent stages of analysis and classification (Fig. 9.11).

B. *Sensor Data Integration*

Simultaneously, sensor measurements are first scaled using the StandardScaler from the scikit-learn Python library, which standardizes the data by removing the mean and scaling to unit variance, a critical step for ensuring consistency in the input features [23]. These scaled measurements are then processed by a neural network layer consisting of 64 neurons, with ReLU as the activation function and a dropout of 0.5 applied to minimize overfitting and improve model robustness [24]. Table 9.1 illustrates examples of raw sensor readings for each category, showcasing the variability in the input data that necessitates careful preprocessing.

Naturally, both capacitive and inductive sensors sometimes gave incorrect readings. For example, a glass container with a metallic lid could confuse the inductive sensor, as it might react to the metal instead of the glass. Despite these issues, the model was usually able to make the right decision. This reliability came from the large amount of data collected from the sensors, which helped the model learn patterns and reduce the effect of occasional sensor mistakes, making the system more accurate overall.

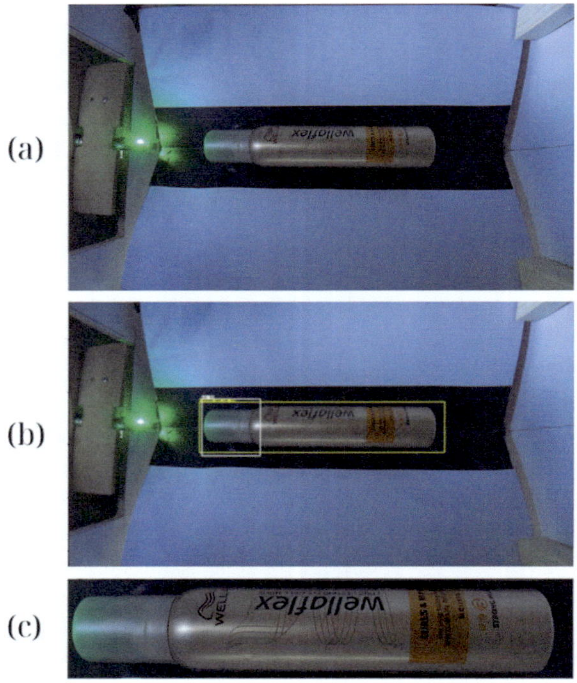

Fig. 9.11 **a** Original Image taken by Picamera. **b** Lid and whole object detected by Tensorflow Object Detection (TFOD) model. **c** Cropped image of the object according to the TFOD bounding box

Table 9.1 Raw sensor readings for a random object of each category

Category	Weight (g)	Capacitive	Inductive
Aluminum	18.03	1	1
Plastic	32.58	0	0
Glass	270.17	0	0
Paper	17.35	0	0
Container with content (plastic)	125.56	1	0

C. *Audio Processing*

For the audio data, after recording, the sound is analyzed to extract a detailed set of features, which includes 7 Mel Frequency Cepstral Coefficients (MFCCs) and their 7 delta MFCCs (first derivatives). These features are well-known for their usefulness in audio classification tasks. MFCCs capture the spectral characteristics of audio signals, while the delta values add information about changes over time. Together, they help the model understand both the sound's structure and its dynamics, making them effective for analyzing a variety of audio types, such as speech and music [25–28].

Each audio recording is sampled at a rate of 48,000 Hz and divided into equal time steps. For each time step, all 14 features (MFCCs and deltas) are calculated, creating a sequence of feature vectors. These sequences are standardized to have a fixed length of 188 time steps to ensure consistency during processing.

The feature sequences are then fed into a neural network, starting with a Long Short-Term Memory (LSTM) layer containing 256 units, designed to handle inputs with a shape of (188, 14). This structure allows the network to capture temporal dependencies within the sequential data, which is essential for understanding audio patterns. LSTMs are particularly effective for audio classification because they can retain context over longer sequences, making them ideal for tasks like speech or music analysis. For example, Zhang et al. demonstrated in their study that LSTMs outperform other methods in classifying urban soundscapes by effectively leveraging temporal features [29]. To further enhance the model's generalization ability and reduce overfitting, a Dropout layer with a rate of 0.5 is applied after the LSTM, ensuring the model remains robust during testing [24] (Figs. 9.12 and 9.13).

D. *Final Model*

All processed data—image, sensor, and audio features—are integrated into a single model. This integration is achieved through a concatenation layer, followed by a dense layer with 256 neurons and ReLU activation, and a dropout of 0.5. The final output layer uses a softmax activation function to classify the input into one of the five predefined categories.

In the following part of the code, executed during training, the structure of the neural network is outlined as we described in previous chapters (Fig. 9.14).

E. *Training Results and Model Evaluation*

This model achieved an accuracy of 95.8% on the test set. Figures 9.15 and 9.16 display the model's loss and accuracy metrics graphically.

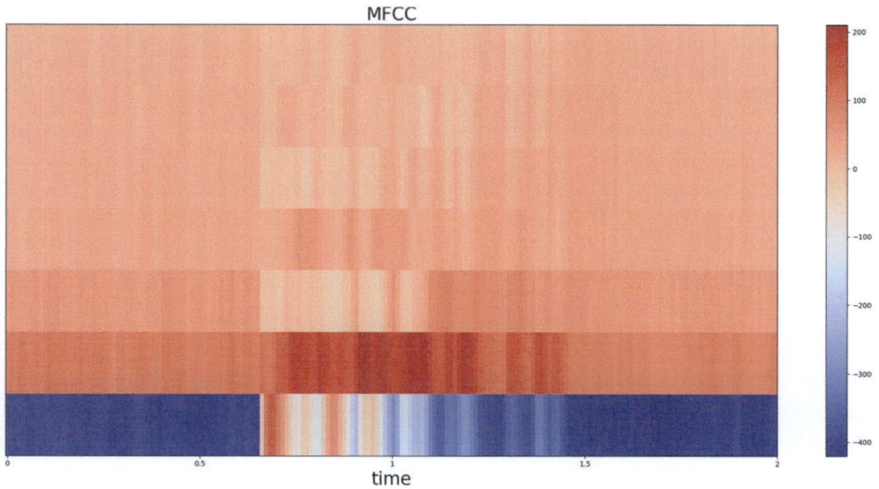

Fig. 9.12 Visualization of MFCCs over time

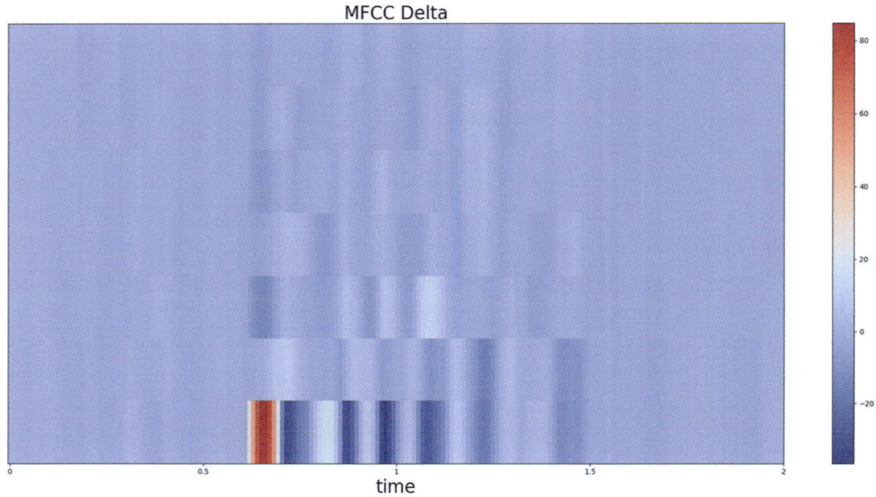

Fig. 9.13 Visualization of Delta MFCCs over time

```
from tensorflow.keras.layers import LSTM, Dropout, Bidirectional, BatchNormalization,GlobalAveragePooling2D
from tensorflow.keras.applications import MobileNetV2,InceptionV3

image_input = Input(shape=(300, 300, 3))
inception = InceptionV3(include_top=False, input_tensor=image_input, weights='imagenet')
inception.trainable = False
image_features = GlobalAveragePooling2D()(inception.output)

sensor_input = Input(shape=(3,))
sensor_features = Dense(64, activation='relu')(sensor_input)
sensor_features = Dropout(0.5)(sensor_features)

audio_input = Input(shape=(188, 14))
audio_features = LSTM(256, return_sequences=False)(audio_input)
audio_features = Dropout(0.5)(audio_features)

merged = Concatenate()([image_features, sensor_features, audio_features])
merged = Dense(256, activation='relu')(merged)
merged = Dropout(0.5)(merged)
output = Dense(num_classes, activation='softmax')(merged)

model = Model(inputs=[image_input, sensor_input, audio_input], outputs=output)
```

Fig. 9.14 Code executed to define the model's structure during the training process

9.5 Future Enhancements

Our current system has demonstrated its effectiveness in accurately classifying waste and providing immediate feedback to users. However, there is always room for growth and improvement. Firstly, we could install more inductive and capacitive sensors at various points on the floor of the bin's chamber to ensure that any entering object makes contact with at least one of them. Furthermore, strategically placed ultrasonic sensors could be added to calculate the maximum thickness of objects. Combined

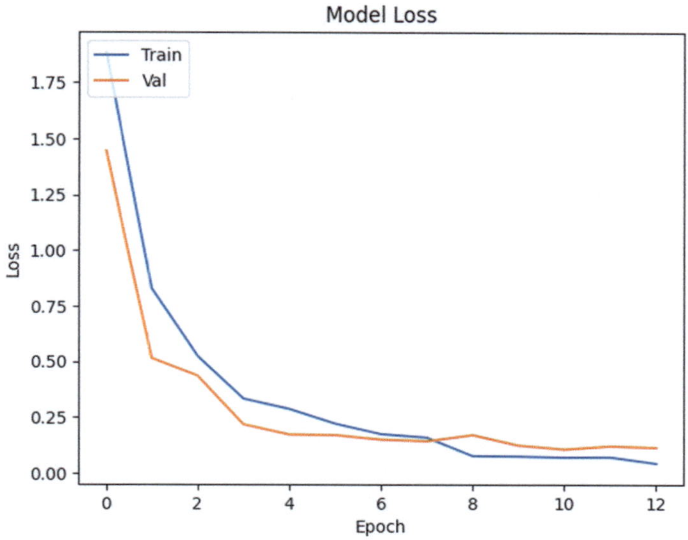

Fig. 9.15 Train and validation loss

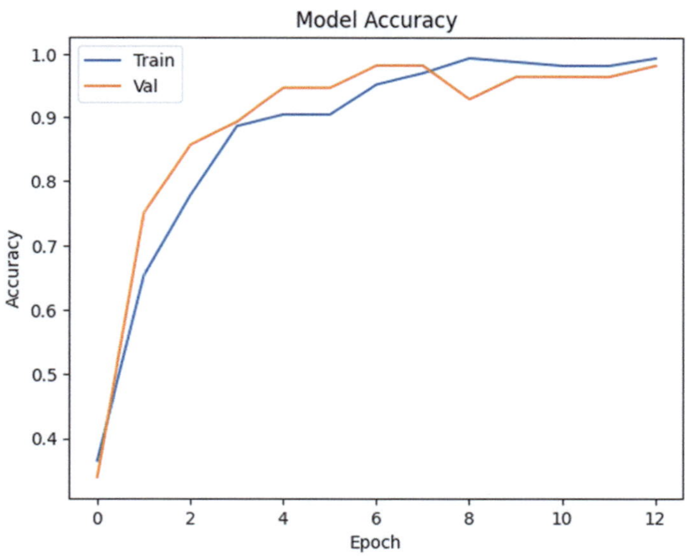

Fig. 9.16 Train and validation accuracy

with the dimensions of the bounding box obtained from the object detection model, this could provide a good estimate of the volume of the object, which could be incorporated into the machine learning model.

Additionally, the use of a larger dataset and the addition of categories such as electronics, textiles, and hazardous materials would enhance the utility and efficiency of the bin. These categories address common waste types that require specific handling and recycling processes, thus improving the sorting accuracy and environmental impact of the recycling system.

Moreover, various automations could be introduced, such as automatic cleaning of the pre-chamber floor, which would serve practical purposes (clearing any residual liquids above an inductive sensor) as well as hygiene. Another useful automation would be the installation of level sensors in the final chambers to alert users and prompt them to empty them.

Additionally, integrating solar power into the system can significantly enhance its energy efficiency and sustainability. Solar panels installed on the bin's surface can power the entire system, reducing reliance on external electricity sources. Energy storage units like batteries can store excess solar energy, ensuring consistent operation during low sunlight. This approach not only lowers operational costs but also aligns with environmentally friendly practices.

Finally, as mentioned in a previous chapter, an automation could be applied to affix a stamp to the waste item before it is separated. This would allow us to utilize the precise categorization information at recycling centers capable of sorting waste based on this stamp. These enhancements have the potential to further optimize the efficiency and accuracy of our smart recycling bin system, contributing to more sustainable waste management practices in the future.

9.6 Conclusion

In this work, we have developed a smart recycling bin that operates efficiently on a Raspberry Pi's modest computational resources. It's equipped with various sensors, aiming to accurately classify waste and enlighten users with immediate feedback. This system stands as an example of how innovative technology can effectively bridge the gap between environmental sustainability and user-friendly waste management. In the future, such devices will become increasingly essential for both outdoor spaces and home environments due to growing environmental concerns and the need to streamline recycling practices. Consequently, ongoing enhancements and widespread adoption of such technologies will play a crucial role in promoting sustainable living and reducing the ecological footprint of waste management.

References

1. A.M. Troschinetz, Twelve factors influencing sustainable recycling of municipal solid waste in developing countries (2005)
2. B. Bell, The battle against recycling contamination is everyone's battle. Waste Management Media Room (2018).
3. F. Waste, Recycling. What is the waste hierarchy (2022)
4. A. Krizhevsky, I. Sutskever, G.E. Hinton, ImageNet classification with deep convolutional neural networks. Commun. ACM **60**(6), 84–90 (2017)
5. P. Sermanet, D. Eigen, X. Zhang, M. Mathieu, R. Fergus, Y. LeCun, Overfeat: integrated recognition, localization and detection using convolutional networks (2013), arXiv preprint arXiv:1312.6229
6. K. He, X. Zhang, S. Ren, J. Sun, Deep residual learning for image recognition, in *Proceedings of the IEEE Conference on Computer Vision and Pattern Recognition* (2016), pp. 770–778
7. X. Li, R. Grammenos, A smart recycling bin using waste image classification at the edge (2022), arXiv preprint arXiv:2210.00448
8. G. White, C. Cabrera, A. Palade, F. Li, S. Clarke, WasteNet: waste classification at the edge for smart bins (2020), arXiv preprint arXiv:2006.05873
9. A.U. Gondal, M.I. Sadiq, T. Ali, M. Irfan, A. Shaf, M. Aamir, … E. Kantoch, Real time multipurpose smart waste classification model for efficient recycling in smart cities using multilayer convolutional neural network and perceptron. Sensors **21**(14), 4916 (2021)
10. R. Rajkamal, V. Anitha, P.G. Nayaki, K. Ramya, E. Kayalvizhi, A novel approach for waste segregation at source level for effective generation of electricity—GREENBIN, in *2014 International Conference on Science Engineering and Management Research (ICSEMR)* (IEEE, 2014), pp. 1–4
11. W. Pereira, S. Parulekar, S. Phaltankar, V. Kamble, Smart bin (waste segregation and optimisation), in *2019 Amity International Conference on Artificial Intelligence (AICAI)* (IEEE, 2019), pp. 274–279
12. J.B. Sigongan, H.P. Sinodlay, S.X.P. Cuizon, J.S. Redondo, M.G. Macapulay, C.O. Bulahan-Undag, K.M.V.C. Gumonan, GULP: solar-powered smart garbage segregation bins with sms notification and machine learning image processing (2023), arXiv preprint arXiv:2304.13040
13. R.Y. Ju, T.Y. Lin, J.H. Jian, J.S. Chiang, Efficient convolutional neural networks on Raspberry Pi for image classification. J. Real-Time Image Proc. **20**(2), 21 (2023)
14. V. Gonzalez-Huitron, J.A. León-Borges, A.E. Rodriguez-Mata, L.E. Amabilis-Sosa, B. Ramírez-Pereda, H. Rodriguez, Disease detection in tomato leaves via CNN with lightweight architectures implemented in Raspberry Pi 4. Comput. Electron. Agric. **181**, 105951 (2021)
15. M. Zamir, N. Ali, A. Naseem, A. Ahmed Frasteen, B. Zafar, M. Assam, … E.A. Attia, Face detection & recognition from images & videos based on CNN & Raspberry Pi. Computation **10**(9), 148 (2022)
16. M.H. Gauswami, K.R. Trivedi, Implementation of machine learning for gender detection using CNN on raspberry Pi platform, in *2018 2nd International Conference on Inventive Systems and Control (ICISC)* (IEEE, 2018), pp. 608–613
17. H. Ewald, H. Krüger, Inductive sensors and their application in metal detection, in *International Conference on Sensing Technology*, New Zealand (2005)
18. W.Y. Du, *Resistive, Capacitive, Inductive, and Magnetic Sensor Technologies* (CRC Press, 2014)
19. D. Wang, FDC1004: Basics of capacitive sensing and applications. Texas Instrum.—Appl. Rep. **12**, 1–12 (2014)
20. https://github.com/tensorflow/models/blob/master/research/object_detection/g3doc/tf2_detection_zoo.md
21. C Szegedy., W. Liu, Y. Jia, P. Sermanet, S. Reed, D. Anguelov, … A. Rabinovich, Going deeper with convolutions, in *Proceedings of the IEEE Conference on Computer Vision and Pattern Recognition* (2015), pp. 1–9

22. T. Hub, Convolutional neural network architectures, in *Principles and Labs for Deep Learning*, vol. 201 (2021)
23. N. Khadka, General machine learning practices using Python (2019)
24. N. Srivastava, G. Hinton, A. Krizhevsky, I. Sutskever, R. Salakhutdinov, Dropout: a simple way to prevent neural networks from overfitting. J. Mach. Learn. Res. **15**(1), 1929–1958 (2014)
25. ResNet based on multi-feature attention mechanism for sound classification in noisy environments
26. X. Liu, M. Sahidullah, T. Kinnunen, Learnable MFCCs for speaker verification, in *2021 IEEE International Symposium on Circuits and Systems (ISCAS)* (IEEE, 2021), pp. 1–5
27. P. Rawat, M. Bajaj, S. Vats, V. Sharma, A comprehensive study based on MFCC and spectrogram for audio classification. J. Inf. Optim. Sci. **44**(6), 1057–1074 (2023)
28. J. Andén, S. Mallat, Multiscale scattering for audio classification, in *ISMIR* (2011), pp. 657–662
29. I. Lezhenin, N. Bogach, E. Pyshkin, Urban sound classification using long short-term memory neural network, in *2019 Federated Conference on Computer Science and Information Systems (FedCSIS)* (IEEE, 2019), pp. 57–60

Chapter 10
Smarts Sensors and Social Sensibility

Francisco V. Cipolla Ficarra

Abstract We present a new quality metric called "4S" (smart sensors and social sensibility), which is based on the Greek notion of "phaneron" to measure the gap between technological reality and the sensation perceived by users in so-called smart cities. The aim is to verify the quality of goods and services aimed at citizens when travelling by traditional public transportation, such as buses, trams, funiculars, subways, taxis, etc., in several European cities. At the same time, the interfaces of interactive systems for users who are inexperienced in the use of collective smart devices, but who are experts in the use of individual smart devices for medical purposes, are analyzed. In addition, a style guide is generated for the communicability of phaneron in the interfaces of personal devices. The aim of the guide is multiple. On the one hand, this tool to reduce digital illiteracy and the generational gap of users, at the same time, to facilitate the use of devices concerning the technological avant-garde. Finally, in this research work, disciplines and areas of scientific knowledge belonging to software engineering, artificial intelligence, design, ergonomics, semiotics, social psychology, sociology, pedagogy and ecology converge. This convergence has allowed the development of an updated heuristic table to determine the qualitative academic fields in the study of smart city metrics.

Keywords Smart cities · Phaneron · Users · Interfaces · Communicability · Urban mobility · Health · Education · Metrics · Quality · Assessment

---

F. V. Cipolla Ficarra (✉)
HCI Lab. – F&F Multimedia Communic@tions Corp., ALAIPO: Asociación Latina de Interacción Persona-Ordenador, Barcelona, Spain
e-mail: info@alaipo.com; ficarra@alaipo.com

HCI Lab. – F&F Multimedia Communic@tions Corp., AINCI: Asociación Internacional de la Comunicación Interactiva, Bergamo, Italy

FMF: Fundation & Academy "Maria Ficarra", Barcelona, Spain

G. A. Tsihrintzis et al. (eds.), *Human-Smart City Interactions and User-Citizen Experiences*, Learning and Analytics in Intelligent Systems 58,
https://doi.org/10.1007/978-3-032-06364-9_10

10.1 Introduction

Nowadays, advances in computing and information technology are present in every human activity, both within and outside their geographical context. These activities are constantly monitored by intelligent or non-intelligent sensors that can communicate with each other, whether fixed or mobile [1, 2]. Sensors emulate the human sensory system. These devices receive external information (chemical or physical stimuli) that is processed internally to generate a response (an electrical impulse).

Hence, we have a diverse range of sensors, such as those related to light, humidity, temperature, speed, contact, sound, ultrasonic, pyroelectric (infrared light radiation), and so on. Therefore, they are basically electromechanical devices that, when incorporated into machines, interact with the natural (air, water, and land) and/or artificial context: Automatic opening of doors and windows; switching on public streetlights; triggering alarms in the event of meteorological disasters; ticket validation in public transportation services, and so forth.

Furthermore, it is in these technological devices that we find the intersection of software and hardware. Both components in sensors are a source of social debate within the global village when the components in the qualitative equation of software and hardware are addressed in emergency situations (earthquakes, floods, fires, etc.) and sensor malfunctions. In short, when we can see that the "real quality" of goods and services is compared with the "theoretical quality" of the sensors offered by hardware manufacturers and software programmers [3].

To this we must add the practicality and ease of use of these devices; the susceptibility and subjectivity of users; and the objectivity or contextual reality in which humans interact with these devices. This equation also includes the notion of social sensitivity within a local or "glocal" and/or "global" context in the face of constant changes arising from information and communication technologies (ICTs).

In this constant metamorphosis, what has remained stable since the twentieth century is the speed of transformation between hardware and software. In other words, hardware progress continues to outpace software. A priori, and in urban settings, these changes are easier for humans to detect or observe with the naked eye.

It is there that the visible or observable constitutes the tip of a kind of iceberg, since currently all technological infrastructure tends to be invisible. Interconnected technological invisibility is a goal pursued in nanotechnology laboratories, through the miniaturization of computing devices in sensors, the enhancement of wireless interconnections between the various equipment used for telecommunications, new materials that facilitate the incorporation of these devices into the human body, and much more [1–5].

However, it is in human perception and the cognitive boundaries between the real and the virtual; objectivity in the face of reality and the susceptibility or subjectivity of users; the priority given to solving local problems over global ones arising from architectural transformations in cities, the displacement of the population from rural areas to cities (or vice versa), the consumption of natural and artificial resources (water, food, electricity, etc.), among others, currently continues to occupy a central axis when conceiving, designing, evaluating, developing, and using new interactive communication systems.

Besides, these systems and sensors tend to incorporate artificial intelligence (AI) [6] but continue to present the typical problems of usability engineering (UE) from the 1980s [7]. This is due to the absence or poor conception of communicability in interactive design, starting in the new millennium [8–10]. The origin of those errors or mistakes that Edward Kit has perfectly identified in software engineering [11] must be located in our day and age, in the academic sector, more specifically, in the human capital, which is completely oblivious or alienated from the reality that surrounds it. This is because many of their interests, projects, research, motivations, etc., have been diverted from the local context to which they belong and physically experience daily. In consequence, these individuals are removed from contextual reality; that is, they live immersed in a kind of mirage of reality. The origin of these deviations must also be investigated in the educational sphere and all its social implications.

10.2 Reality Versus Virtuality: A New or Old Human Dilemma?

The boundaries between the real and the virtual, coupled with the internationalization of interactive design, are a dilemma that has persisted since programmers developed the first interfaces for commercial computer operating systems at the end of the twentieth century [12]. These problems have multiplied exponentially with the various operating systems for smartphones. In this sense, a brief diachronic overview allows us to affirm that since the 1990s, the possibility of users being able to easily modify static and dynamic 2D and/or 3D images with commercial or open source applications has been increasing, not only through computers but also with multimedia mobile phones. This transformative action led some experts in the audiovisual sector to declare "the death of the image" [13].

That is, the disappearance of the validity of an image (analog or digital) as a testament to reality, through photography, film, and video, to name three examples. Consequently, written text was revalued to reinforce the anchoring of images to the reality they described or the function they served in certain environments. It was therefore necessary to include text to describe the function of an icon in an interface [14, 15]. Evidently, the latter reveals flaws in communicability when the user interacts with the system.

Besides, this affirmation of the disappearance of the validity of an image as a testament to reality implied a temporary return in the history of the evolution of interactive design. That is, a return to the origins of the first hypertext systems, where text predominated over images due to the space occupied by data storage (high-resolution photographs, video, music, computer animations, among others), in the offline digital media of the 1980s. Since the democratization of the Internet in the late 1990s, multimedia and hypermedia systems fostered the transformation of each user of interactive systems into a potential content editor [16].

In the new millennium, these users generate millions of photographic and video images to disseminate their messages through their personalized, freely accessible channels and through various social media applications and platforms. These images attempt to represent the real or imaginary world through the emulation and simulation of reality [17–20]. The dilemma between the real and the virtual reappears when many of their creators currently consider that working with a photographic or video camera means capturing reality objectively, when in reality it is not.

A photograph or a frame from a film is a portion of reality, in which components of human subjectivity intervene. Furthermore, this biased view of reality can be manipulated through pixels, bits, qubits, and so on, for various purposes, ranging from commercial to indoctrination or alienation of populations [21, 22].

At the same time, this new dynamic social process aims to foster positive social sensitivity toward human interaction with new technologies, particularly the use of artificial intelligence. This social sensitivity is heightened in certain geographic areas of our planet. For example, the population pyramids of the continents bordering the North Atlantic Ocean. It is contrasted with a kind of diamond-shaped pyramid in the Pacific Ocean regions, from the perspective of the population growth of new generations of users [23].

In both cases, **millions of human beings have begun to freely and democratically access some of the latest technological advances in the so-called: "Quantic-Nanotechnological-Self-Sufficient" era** [24, 25]. For example, in industrialized Asian countries, robots are beginning to supplant humans in everyday, repetitive activities such as serving as waiters in cafes, restaurants, etc., hotel receptionists, citizen information points, and so forth, to which must be added the advantages of agentic AI [26, 27] in carrying out commercial operations, for example. Users belonging to the Zeta or Alpha generations are very favorable to these changes [28]. These changes are perceived more quickly in urban areas than in rural areas, particularly in those countries with large land areas (developed or developing, from an economic perspective).

Hence we can see **the importance of measuring these social changes, starting with the degree of acceptability and enjoyment of interactive systems and sensors by users in the new millennium**. In this sense, the Greek notion of "phaneron" [29] is very useful for measuring the distance between technological reality and the sensation perceived by users in so-called smart cities. Furthermore, from a semiotic perspective, it is a concept that is perfectly suited for the successful design and evaluation of interactive systems with an excellent communicability index. Interested readers can find the demonstrable advantages of applying semiotics to the generation of interactive systems within software engineering over the past 30 years in the following bibliographic references [30–35].

In this work, we will use the notion of phaneron to develop a new metric that combines the four Ss: smart, sensors, social and sensibility, and is oriented toward public passenger transport in Europe (mainly, bus). The main areas of analysis of this research can be schematically represented in Fig. 10.1.

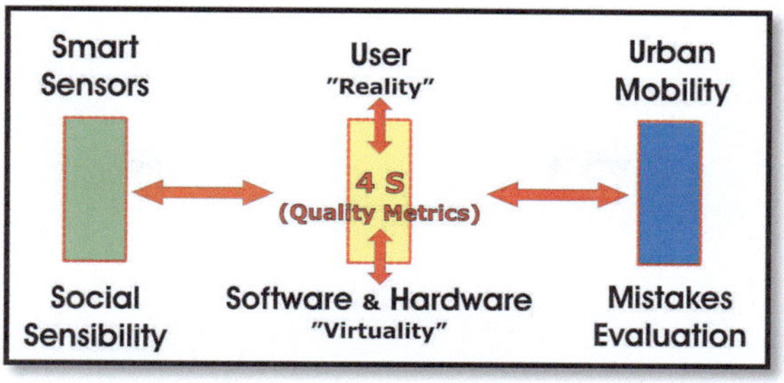

Fig. 10.1 Diagram of 4S (quality metrics for smarts sensors and social sensibility)

10.3 Binary Relationships in Smart Cities: Reality Versus Technological Illusions

In smart cities, there is hypothetically a binary and positive interrelationship between a community's infrastructure (libraries, schools, universities, hospitals, parks, roads, bridges, etc.) and the human capital that inhabits that place. The set of urban planning strategies should favor public services aimed at improving citizens' quality of life. This set of strategies primarily includes the following elements: environmental care, efficient energy use, the constant inclusion of new information technologies, and sustainable mobility. In sustainable mobility for public passenger transport, city councils are adopting various types of solutions to reduce environmental pollution (smog) through buses that run on natural gas, hybrids (electricity and gasoline), hydrogen, electric, and so forth.

Access to this transport service generally has a cost for users, which may vary among different passenger categories depending on age, place of residence, job, and so on. Agreements between city councils, provinces, autonomous regions, etc., and the companies that offer these transport services allow some users to travel free of charge (children, the elderly, the disabled, and so forth) on buses, subways, trains, ferries, and so on. These users usually have special passes that do not need to be validated at ticket machines, for example.

Nowadays, public passenger transport is adopting intelligent computer systems and abandoning mechanical and/or manual systems for access to transport. For example, ticket or pass issuing machines (hourly, daily, weekly, monthly, quarterly, annual, etc.) contain a high-performance personal computer (e.g., an i7) and Ethernet TCP-IP communication, allowing for a reliable and constant connection. These devices allow secure payments with smartphones, using technologies such as contactless, EMV (Europay MasterCard VISA), and keypads. Some of these models in Europe can be seen at the following link: www.pirateando.net/4S. This is the main motivation to measure the efficiency of the new devices being implemented in public transport services and the reaction of passengers to this system modernization.

Social awareness of the intelligent modernization of a system can be distorted by the negative components of commercial advertising, political propaganda, and other hybrid formulas that intersect with the two, with the aim of persuading

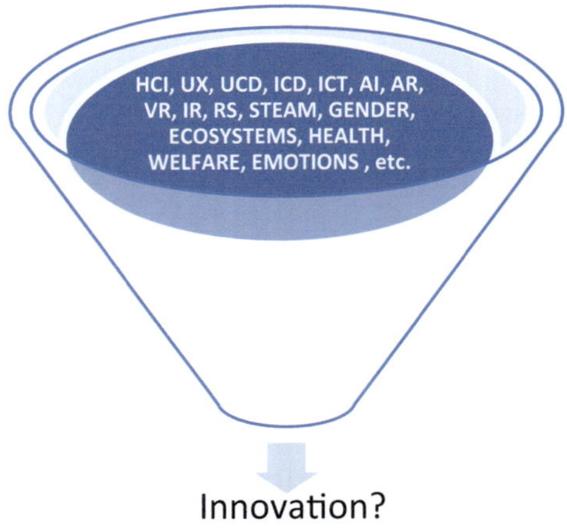

Fig. 10.2 An innovative pleonasm from some fake studies in Ibero-America

users and/or manipulating reality through terms that are currently widely used in educational or scientific vocabulary. Some of these concepts constitute areas of academic study, where users, software, and hardware converge, such as: augmented reality (AR), ecosystems, empathy, ethics, extended reality (XR), gender equality, human–computer interaction (HCI), user experience (UX), information retrieval (IR), interactive smart spaces, recommender systems (RS), STEAM (science, technology, engineering, arts and mathematics), sustainability, technology-enhancing learning and/or teaching, mixed reality (MR), user-centered design (UCD), virtual reality (VR), women's studies, among others (Fig. 10.2). These are concepts where the ambiguity of meanings, vagueness of concepts, and redundancy of notions predominate, if the meaning and signifier of the terms are examined [28].

This triad of ambiguity, vagueness, and redundancy can be summarized in the classic Greek notion of "pleonasm" [29]. Here, it is also necessary to point out that text algorithms generated with artificial intelligence in the triangle of signifier, signified, and referent tend to eliminate the referent [28, 36, 37]. Besides, this elimination seriously harms the intellectual understanding of reality by future users of the next generations of intelligent interactive systems.

One of the purposes of resorting to advertising or propaganda pleonasm in bus services in several European cities is to distract attention from the problems that exist between the actual functioning of ticket machines and the virtual benefits that new technologies are making in smart cities through institutional communication. All these academic trendy terms form a kind of diverse funnel of innovations or trends to follow in the fields of formal, factual, and natural sciences.

However, their results fail to reach the social awareness of city dwellers because they are far removed from their everyday, local realities, or because they fail to solve the operational problems arising from new technologies and the use of sensors. The promotion of these distortions is one of the common denominators of the so-called Omega generation [28, 38]. These deviations from reality have their origin in the "G" factor (G = Garduña—in Spanish, or Gardunia—in English) [28, 39].

10.4 Construction of the 4S Metric for Smart Cities: Human Factors

The importance of the Greek notion of phaneron [29] in evaluating and verifying everything the user infers about reality and mentally imagines that smart cities can offer, through cutting-edge technological devices, which are interconnected with each other, whether or not they resort to artificial intelligence. Aside from this scientifically demonstrable truth (a hodgepodge of pleonastic terms, actions of the Omega generation, and the promotion of the Gardunia factor), today, many of these terms are applied in smart cities to define an intelligent context whose sole theoretical purpose is to improve the quality of life of its inhabitants in their daily activities. Consequently, in urban or rural settings, it is necessary to improve and perfect the telematic infrastructures of goods and services for devices belonging to ICTs, whether public, private, or hybrid. Implicitly, and beneath the visible iceberg of new ICTs, achieving this objective means obtaining data from citizens, that is, potential users.

This data collection for the development, implementation, and operation of new interactive systems with artificial intelligence is coming up against the limits arising from individuals' right to privacy, freedom of expression, the right to truthful data and information, and so forth. Clearly, cities are where the largest amount of data and legal issues are concentrated, given the large population, creating large databases. Legal issues must be resolved prior to the implementation of sensors or automatic payment devices, as they access the banking data of millions of users, through their credit or debit cards, for example [40, 41].

Security breaches in computer networks through cyber-attacks, the application of illegal algorithms with AI to censor investigations conducted by journalists, are some of the illicit acts covered up or legally. Besides, these dishonest actions are internationally protected by groups of lawyers trained through so-called "narco-education" or "education narcotized" [38].

It is a reality that is causing not only serious financial damage but also the emotional distancing of the population from smart cities, ecology, ICTs, AI, industrial automation, robots, university studies related to new technologies, engineering, mathematics, physics, chemistry, and a long etc.

These illegal elements, internationally interconnected and driven by lawyers belonging to the Omega generation, constitute negative human factors that affect the quality of software and the assessment of the veracity of online data and information. All of these harmful elements diminish the positive impact citizens have on the benefits of living in a smart city, undermine the democratic system, and undermine the peaceful coexistence of its inhabitants.

At the same time, this information and data form federated databases, for example. Digital texts, images, audio, etc., are shared within and outside the borders of a province, region, state, or union of states. Furthermore, the illegal actions of these false representatives of impartial justice often remain anonymous, immune, and unpunished, thanks to the misuse of parliamentary or extra-parliamentary laws, as is the case in the EU and the General Data Protection Regulation (GDPR). The current aggravating factor is that it is unlawfully intended to be applied retroactively.

Definitely, **the goal is to destroy social credibility toward justice in order to implement a top-down and autocratic systems in smart communities**. Readers interested in these negative elements (human and social factors) of software engineering, which are expanding with artificial intelligence, can consult the following bibliographic references [36–38, 42, 43]. This is one of the reasons why many dysfunctions of public services (transportation, telecommunications, education, healthcare, social wellness, and so on) within smart cities are not widely known and/or widely disseminated among their inhabitants through social media.

10.5 Paper Versus Bits: A Crossroads for Public Transportation Users

In the 1990s, the digitization of information in paper format (books, magazines, catalogs, etc.), through the first storage media such as CD-ROMs, DVDs, etc., was expected to reduce global paper consumption. However, since the beginning of 2000, paper production has doubled [9]. **Year after year, paper consumption is increasing. Nowadays, the environmental component of smart cities promotes reducing paper consumption at work, home, leisure venues, and so forth**.

On the other hand, paper is still consumed in public transportation. Traditional paper bus tickets, which were validated by mechanical machines, have been replaced by HID (Human Interface Device) ticket validators and readers, for example. These ticket validation and/or reading devices allow passengers on buses, subways, trains, cable cars, ferries, etc., to have a single point to validate any type of ticket, as well as the option to pay for their ticket on the bus, using a single device and interface. Barcodes, QR codes (quick-response), magnetic or electronic strips, and so on, have been progressively included in tickets.

The hypothetical advantages of having a single validation point lie in improving the operational efficiency of the means of transport and streamlining passenger travel on buses. The only exception to the use of paper is payment by credit or debit card, as well as by smartphone and/or smartwatch.

By paying for the ticket, the passenger acquires the right to travel on buses, subways, trains, etc. The way passengers acquire the right to travel on urban public transport varies greatly in cities that can currently be defined as smart cities, or that are moving toward that goal. In some cities on the European or American continents, there are various acquisition methods, and some formats can be found in Appendix 1. A brief summary is as follows:

(1) Purchase the public transport ticket in cash and on paper, paying directly to the driver.
(2) Cash payments are made using special machines that issue tickets and accept coins, but do not issue any remaining cash. Besides, these devices are

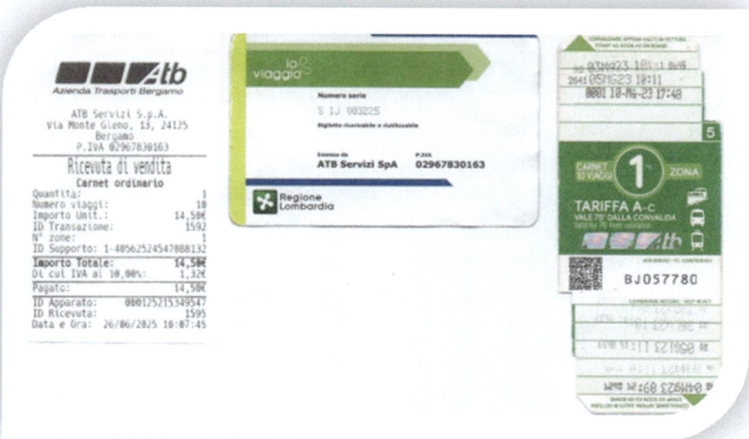

Fig. 10.3 A receipt for the sale of a 10-ride chip card in Bergamo city (left). In the central area there is a paper ticket with chip (Bergamo). Finally, we see a classic 10-ride paper card with printed information such as the date and time. It can be written by the user or driver if the ticket validation machine is not working

programmed to complete the transaction within a few minutes. After these minutes, the machine no longer returns the money the passenger has deposited.

(3) The ten-ride travel card allows passengers to self-enter the day and time of public service use, without having to use the mechanical validation machines (Fig. 10.3—right ticket).
(4) The paper format has a visible magnetic/electronic strip attached for validation.
(5) The magnetic/electronic strip is invisible (chip on paper), allowing the travel card to be refilled once the total number of trips has been used. In other words, it allows for recycling, and therefore, supposedly, there is a saving in paper consumption.

The first option is gradually being phased out as it involves a high level of stress for drivers, who not only have to pay attention to traffic but can also be victims of altercations with passengers or robberies, especially in Ibero-American communities such as Bolivia, Brazil, Colombia, Chile, Paraguay, Peru, Mexico, and so forth. In the second case, many passengers fall victim to misinformation and scams by public transport companies, as they fail to inform them in advance that purchasing a ticket from their machines requires having the exact amount and entering it within a maximum of one or two minutes. The third option is the most convenient when the validators on buses are not working, for example.

In the fourth and fifth options, tickets cannot be folded or brought close to magnetic sources, as they automatically lose their readability, and users are responsible for this loss of information. These users can be fined by the transport company when its inspectors check the travel tickets. Besides, at the time of purchase, purchase tickets are issued (price, location, date, time of purchase, zones of validity, expiration, chip number, and so on) which in some places have the same or more value than the travel ticket itself (undoubtedly lacking purchase information), as is the case with commuter trains in the Italian region of Lombardy, to cite one example. In other words, we are facing a duplication of paper use, in the latter case.

However, with **the use of paper, the user constantly has visual information about the total number of trips available. Per contra, with magnetic/electronic media ideal for smart sensors, the user lacks such information, which can generate a state of anxiety or stress in the passenger**.

While there are devices with options that allow the contents of electronic tickets to be read without consuming the trips, in other modern systems equipped with artificial intelligence this is not possible.

In other words, each time the user passes the ticket over the reader to verify how many trips are available, a trip will be consumed. These real-life cases demonstrate the need to measure the positive and negative factors in the use of these devices in cities that are or are becoming "smart cities." These are studies that confront subjectivity and objectivity in the use of a system.

10.6 Measurement of Social Subjectivity

Historically, the "subjectively pleasurable" use of a system was a component of usability [44]. Consequently, it was part of what Nielsen called the "practical acceptability" of the system, along with usefulness, cost, compatibility, reliability, and so forth; contrasting it with "social acceptability" [7, 44]. In the new millennium, usability has been superseded by the notion of communicability with the rise of interactive communications between humans, smart devices, robots, and so on, interconnected through the Internet, intranets, and extranets, for example. **Communicability brings together, among other notions, "practical acceptability" and "social acceptability"** [36, 45]. That is, they have overcome the stage of separation between the two.

The problem that persists and that is related to the present work is the subjective aspect of the user and that is shaping social sensitivity to certain malfunctions of technological devices in smart cities. Hence, many transport users believe that the objectivity/veracity/invariability of data and information remains in paper form, while subjectivity/virtuality/volatility resides in data and information in digital form.

Social subjectivity is an issue that can be professionally resolved through the tools, methods, and techniques of social communicators, sociologists, social psychologists, and philosophers. In other words, it is not a problem that can be solved in university departments or faculties of mathematics, physics, chemistry, fine arts, technology-enhanced learning/teaching, computer architecture technology, computer engineering and systems, computer science and automation, software engineering, electronics and telecommunications, among others. However, it is evident that these faculties and/or departments address this problem by generating epistemological distortions within the sciences, especially when they resort to HCI, UX, UCD, IR, and RS, to name five examples. The resolution from a social perspective and that of new information and communication technologies requires authentic expert and ad hoc trained figures, such as the "communicability specialist." This is a professional who stands at the intersection of formal, natural, and factual sciences. Readers interested in the genesis and evolution of the communicability specialist for interactive systems can consult the following bibliographic references [12, 25, 46].

Normally, when faced with a problem of social subjectivity, the communicability professional can resort to mental representational operations, such as enumeration, description, comparison, distinction, classification, definition, and so forth. That is, they can answer the first rhetorical question: "What is there?" Consequently, they can identify the problem by searching for common, contradictory, atypical, etc. elements; limiting the temporal and geographical spaces. Finally, they can establish links between causes and effects, seek analogous theories or situations to better understand the facts, and formulate other rhetorical questions, such as: "What to do?", "How to do it?", "Why do it?", and "What to do?."

In this research work, answering this set of questions involves generating quality attributes and metrics, establishing objectives to be achieved; using methods, techniques, and instruments; generating a set of tasks; selecting the users who will participate in the experiment; and collecting, synthesizing, and monitoring the results obtained. In our case, this experimentation begins with the technique of direct observation applied to real users. The actions examined begin the moment the user interacts with fixed and mobile devices to access and travel on public transport in several European smart cities.

Regarding the objectivity of observation, Mario Bunge [47] argues that one of the fundamental characteristics of the scientific method is its objectivity, as it seeks to achieve factual truth, regardless of the values and beliefs of the researcher. Some argue that it is a criterion of a single person, although observation is a process in which the habits, expectations, practical skills, and scientific knowledge available to the observer are of great importance and are constantly interacting. Accordingly, direct observation is a deliberate search for data and information. The observer uses their senses to observe facts and realities of the context, recording what happens. These actions include users in their real context, where they normally carry out their activities. In short, it is an intentional perception of reality since it is done with a specific objective, such as the collection of data and information [48].

The observation methods used by the communicability expert can be summarized as follows:

1. It is a structured or systematic observation,
2. Non-participant, and
3. Carried out in real life.

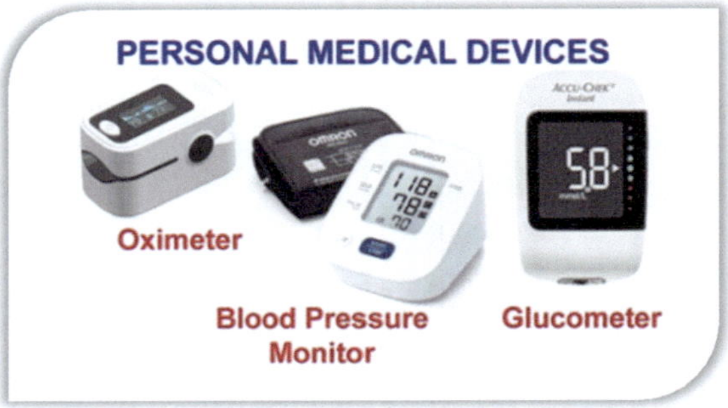

Fig. 10.4 Medical devices used independently by members of the group

Structured observation allows for establishing the objectives of the observation; delimiting and defining the field of observation, also establishing the most relevant aspects; selecting the instruments and techniques to be used; and accurately recording the data and initial information obtained. Non-participation refers to the observer remaining detached or external to the situation being observed. Finally, it is carried out in real life, meaning that the events are collected naturally and as they occur. In other words, there is no prior preparation or manipulation.

Once the data and initial information are obtained, the communicability expert begins to establish the tasks to be performed by the users who will participate in the various phases of the experiments. In this research project, the experimentation was carried out with a team of 14 adult (7 women and 7 men) bus users (10 buses—different lines), ranging in age from 50 to 60. None of them had prior computer training or experience using personal computers. However, all of them independently used medical devices to measure blood pressure (a blood pressure monitor), blood sugar (a glucometer), and blood saturation level (an oximeter). Some examples can be seen in Fig. 10.4.

This group of users participated in the viewing of a set of science fiction films related to AI, home automation, ICTs, interfaces, robotics, smart cities, and so on. The study universe consisted of a total of 30 films, which are listed in Appendix 2. The audiovisual material was selected through a random process (drawing) from a total of 120.

The purpose of the viewing was to identify interfaces and the interactive design components that predominated within them: text, numbers, photos, maps, three-dimensional and/or two-dimensional animations, videos, icons, predominant colors (primary, secondary and tertiary), and the screen format (rectangular, square, circular, spherical, and cube). In addition, participants had to determine whether or not there was interaction between the user and the interface. If so, the modality used was determined: tactile/vocal/visual/mental. This activity motivates group members and has positive effects, as it allows the group to unite and facilitates the acquisition of knowledge about interfaces and their main components.

10.7 The Importance of Interfaces in Smart Cities

Attention to interfaces arose with the rise in the use of personal computers in the home and the development of the first operating systems. At first, users were interested in programming using computers like the Commodore Amiga and some versions of the BASIC (Beginners' All-purpose Symbolic Instruction Code) language, and even went so far as to program some computer games.

Most users interacted with the computer and its peripherals (printers, floppy disk drives, magnetic tapes, and so forth) through a cathode ray tube, the keyboard, and by typing commands. In the 1980s and 1990s, there was a marked difference between users who used computers for work and/or entertainment purposes. The former generally used IBM PC (personal computers) or IBM-compatible computers. The latter were more oriented toward design tasks, for example, acquiring computers with better graphics cards or greater processing memory capacity, such as Apple's Macintosh (Mac).

In this context, the common denominator among the different brands was the reliability of the operating system and the design of the first interfaces [15, 49]. These interfaces would evolve graphically and functionally as new versions of processors, graphics cards, keyboards, mice, joysticks, and so on, were marketed. Given the periodic advancement of software and hardware, the first style guides were generated for the internationalization of operating system interfaces in MS-DOS/Windows (Microsoft) and System, Mac OS, Mac OS X, etc. for Macintosh. These guides facilitated the work of thousands of computer system designers and programmers, who had to address the diverse cultures of users located throughout the world.

Besides, software production costs decreased, as they shared the same operating systems, commercial applications (spreadsheets, word processors, presentation programs, email tools, and so on), and programming languages, adapted to different cultural systems. All of this helped to considerably reduce errors resulting from a lack of knowledge of the customs and practices of other peoples. Through these guides, the first strategies for software adaptability were established, in multiple languages, to facilitate the understanding and sharing of applications related to work, study, research, and so forth, in offices, town halls, businesses, industries, universities, schools, etc. These communication tactics were also used in the first commercial interactive multimedia products and offline support, conforming to the five principles of usability engineering, enunciated in the 1990s [7, 44].

On the other hand, all these positive aspects of design aimed at diverse cultural systems began to decline. The decline of these style standardization guides in the southern European Mediterranean began when these interactive multimedia systems were required to operate exclusively on the Internet.

User disorientation when navigating interactive content in numerous commercial systems (online and offline), regardless of the purpose (education, tourist information, entertainment, and so on) and the quality of the metaphors used in the interfaces, originated with the emergence of web page designers and content creators—popularly known as "creatives" but basically, with originality equal to zero—from the fine arts, music, anthropology, psychology, philosophy, etc., in addition to programmers trained in faculties of industrial engineering, telecommunications, physics, mathematics, and so forth. Team members noted a lack of synergy between programmers and creatives. There was a wide educational gap between the two groups, coupled with a resistance to change and a lack of experience in teamwork.

This was one of the reasons, at the beginning of the new millennium, why many European multimedia companies dedicated to developing applications for teaching languages, mathematics, computer science, and so on, closed their operations. The only exceptions to this restriction in the audiovisual and multimedia sector were those firms focused on the video game industry. Global interface design in video games constitutes a kind of "kernel" for their commercial success, regardless of the user's age, gender, or geographical location. Per contra, not all potential users of the various applications and computing devices available in smart cities interact with video games on a daily basis. Therefore, this is not the ideal solution for achieving positive social awareness of smart cities.

Both the ergonomic design of smart city devices and the interactive design of the software in the presentation of data and/or information on the different screens tend toward minimalism. **Minimalism, communicativeness, and the universality of interactive design are the best solutions for users in smart cities, who lack prior experience using technological devices, with or without artificial intelligence**.

Some components of universality in interactive multimedia design have been successfully incorporated into the interfaces of transportation devices. These are visible elements in rectangular (horizontal or vertical) or square interfaces. A summary of these elements, in Western countries, is as follows: Traffic signs, such as arrows, or shapes (triangles, circles, squares, and so forth); flags to indicate different languages; primary/secondary/tertiary colors; typography (Arial or Times); classic symbols that indicate acceptance (√) or denial (X); emoticons; signage to indicate actions to users, among others. Regarding this last component, such as the use of innovative icons on smartphones or watchphones, these must respond to some type of universal coding so as not to generate ambiguity or vagueness in communication with the end user. This is one of the reasons why these images are accompanied by text, which may require prior language selection to correctly interpret the text's meaning and function. Text serves as an anchor in the relationship between meaning and signifier, eradicating ambiguity and/or vagueness. This language change not only involves additional time when interacting with the ticket vending machine or ticket validation machine, but also the possibility of inadvertent errors on the part of the user when interacting with these public transportation devices.

On the other hand, the ability to access information in multiple languages denotes the presence of the quality attribute known as "richness." This attribute has always been positive in teaching and/or learning processes based on interactive multimedia systems, hypermedia, augmented reality, expanded reality, among others. Richness increases the efficiency and satisfaction of an expert user when faced with the IT services present in smart cities [50–53]. However, if the user is inexperienced and the system has a wide range of navigation styles, multiple paths to access the same data, and an abundance of text, graphics, maps, animations, videos, sounds, music,

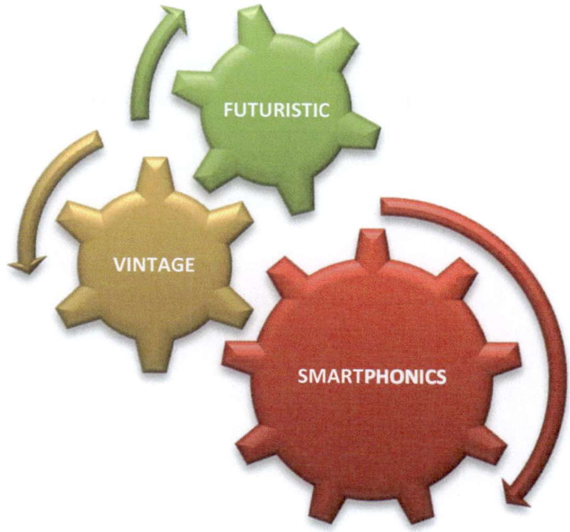

Fig. 10.5 A new classification of the interfaces for public transport in the smart cities

etc., all of this can impair the mnemonic aspect. That is, it is very likely that the user will not remember the steps followed to access a piece of information or will become disoriented when interacting with a system for the first time. Subsequently, dynamic media (animations, videos, sound effects, guided photo tours, etc.) that are part of richness must be deactivated to avoid distractions. These distractions, when viewed by inexperienced users, can lead to inadvertent errors and, consequently, discourage the use of the interactive system.

The interfaces analyzed in several European cities related to public transport have been divided into three large groups, which we can define as "sui generis." These groups are: "Vintage," "Smartphonics," and "Futuristic" (Fig. 10.5).

The first group "vintage" contains the first information point interfaces from the 1990s, characterized by horizontal rectangles representing computer keys. These were 2D but emulated three-dimensionality through edge shading, for example. These interfaces were located in libraries, hospitals, tourist offices, etc. Computers and systems operated on totem poles for informational purposes. In just a few minutes, the user could find out what they were physically looking for and how to easily navigate to a location, for example. The predominant interactive design in these interfaces was determined by a minimalist style.

The second "smartphonics" interfaces are those currently found on traditional smartphones or more modern ones such as 4G/LTE/4G+/LTE Advanced, or 5G, 5G+/5G UW/5G UC. All of them have a horizontal rectangular format, similar to a credit card. The interfaces contain a multitude of new, unconventional icons that are used to activate novel applications. These icons do not always allow a user experienced in new technologies to correctly infer or understand their functionality, especially when interacting with them for the first time.

In conclusion, there are those that we call "futuristic." These interfaces feature transparency of materials (glass) over the textures or background colors of the previous two. These background transparencies support the components of static and dynamic media related to data and information. Usually, they use three-dimensional computer graphics for navigation or presentation of information on the screen. In this type of interface, data and information are displayed dynamically and in 3D: holography, videos, animations, equations, scrolling text (depending on the language, letters may be written left to right—i.e., Japanese; right to left—i.e., Castilian, and so forth). The format can vary from horizontal or vertical rectangles, cubes, spheres, etc., occupying large spaces on walls and even dividing the spaces of a large room. It is an interface where richness is almost always present as an attribute of quality. In consequence, there is a combination of dynamic and static images that make up the content or navigation elements within these interfaces. These interfaces are more present in science fiction films than in the reality of smart cities. At the same time, this type of interface, from a hardware and software perspective, is gradually being implemented in real-life settings such as fairs or exhibitions, museums, theme parks, and so on. These events and spaces are designed to present the latest advances in ICT to all users. In our study universe, interfaces from the first category predominate. Some excellent examples are in Fig. 10.6 (Stansted) and Fig. 10.7 (London). Other images can be seen at the following link: www.pirateando.net/4S.

It features a minimalist design based on 3D-like keys available in rows/columns, using three colors: red, gray, and white. Some navigation icons have a faded background, and the letters are in the foreground.

It has the same minimalist design as in Fig. 10.6, but uses blue for the 3D-like keys. Additionally, it allows you to select 15 languages, 14 of which have their corresponding flags at the bottom of the interface: UK, France, Germany, Greece, Italy, Poland, Spain, Portugal, Turkey, Japan, China, India, Pakistan, Bangladesh, and Arabic (Fig. 10.8).

Fig. 10.6 This is a classic interface from the 1990s, located at the Stansted (UK) airport

Fig. 10.7 This is also a traditional 1990s interface for paying for subway tickets with credit or debit cards (seven in total)

Fig. 10.8 Fifteen languages for buy a ticket in London

Analyzing the interfaces and the results obtained from the experiments with the 14 users has allowed us to generate an initial style guide (Appendix 3), which can serve as a basis for interface models for public use and for users with no computer experience (Fig. 10.9).

Universality, minimalism, simplicity, and social cross-culturally are the four cardinal points that comprise the proposed style guide for interfaces for all audiences, without physical and/or mental disabilities. The participants in the

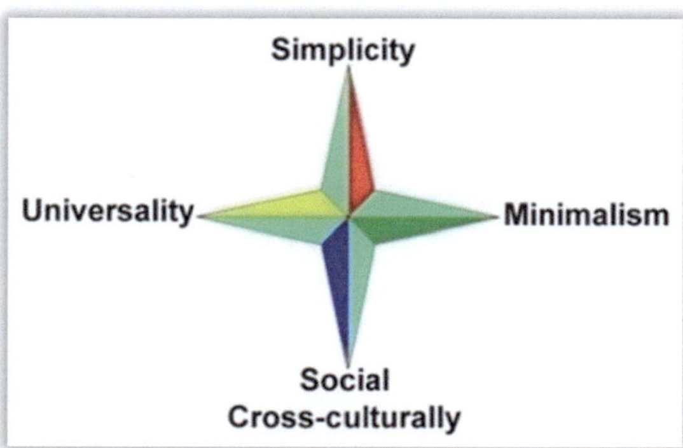

Fig. 10.9 Convergence of the four main components for design and evaluation of the interfaces

experiments daily use basic health devices such as oximeters, blood pressure monitors, and glucometers (excluding models related to continuous glucose monitors), among others. The order in which these devices are presented already indicates the degree of difficulty involved in their use. The first (oximeter) is universal and requires no operations other than placing it on the fingers. The second, depending on the model, is placed on the pulse or arm to automatically measure blood pressure. The last, however, requires greater user concentration, as they must insert a drop of blood onto the test strips on the device's reader. In all of them, the ergonomics and design of the plasma screens (reading the results) show common denominators and/or isotopies, whether iconographic and/or numerical.

10.8 A Concave Deltoid for a New Metric

The social aspects of human sensitivity to smart sensors in public passenger transport lead this work to take the form of a concave deltoid when establishing each of the steps followed to generate the 4S metric. Figure 10.10 shows two vertices: "A" and "B." They represent the social and technological aspects, respectively. The knowledge and experience of the heuristic evaluator in communicability allow us to address this problem from these two perspectives. In addition, the phaneron will measure the distance between the two, once the experiments with users have been conducted. Therefore, in Fig. 10.10, the set of activities can be seen as those carried out individually by the communicability expert (areas 1, 2, etc.) and those carried out in groups with users (areas 4, 5, and so on). Below, and in parentheses, each of the activities carried out in the deltoid concave is briefly listed.

First, the expert determines the technological reality oriented toward public transportation services equipped with smart sensors, such as paying for a bus ticket with a smartphone, or automatically opening doors when boarding, or exiting a funicular. In a smart city, public transportation sensors are distributed throughout the city network and provide real-time information to the user and simultaneously to the companies involved in that service, through video cameras, Wi-Fi, plasma screens at the various bus stops, and so on.

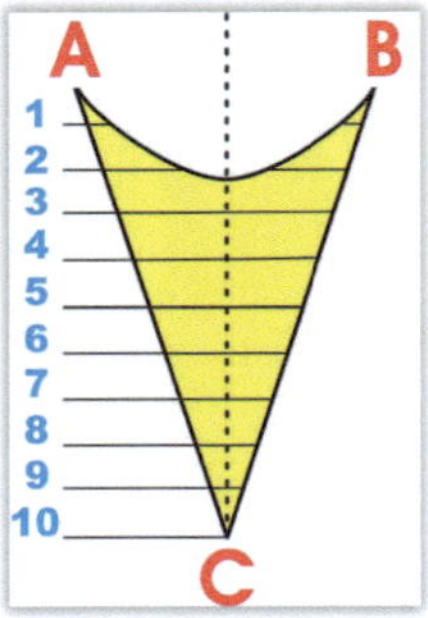

Fig. 10.10 A concave deltoid for heuristic assessment

The evaluator will conduct an initial inventory of these devices, their functioning, and passenger reactions to them, especially when malfunctions occur due to user error or failures due to the devices' malfunction or interactive design. Based on these observations, they can begin to develop a set of hypotheses (1). The evaluator's direct observations have been made in the following cities: Barcelona (Spain), Bergamo (Italy), Madrid (Spain), Milan (Italy), London (UK), Luxembourg (Grand Duchy of Luxembourg) and Vienna (Austria), from October 2024 to May 2025. The state-of-the-art compiled already allows them to establish some premises and hypotheses to be verified with heuristic evaluation, using users (2). User experiments have been conducted in the following cities: Barcelona and Bergamo.

Consequently, the examiner begins to compile a list of the detected errors, establishes quality attributes, and metrics (3). This operation belongs to software engineering and is located precisely at the intersection of function-oriented metrics and people-oriented metrics, within the square defined by Pressman and Maxim [54]. Readers interested in the development of quality metrics and attributes for the heuristic evaluation of interactive systems can consult the bibliography. These references demonstrate how they have been rapidly developed for different uses, with excellent results, and within the framework of "low cost and high quality."

After, he/she has the elaboration of the set of tasks (4) that potential users will perform when interacting with devices for public transport services. Selection and compaction of the user group (5). Users must watch films in the laboratory and answer closed-ended questionnaires related to the main components of the interfaces.

Experimentation with users in a real-life context, i.e., interacting directly with the machines to purchase tickets and validate them on buses (6); compilation of the results obtained and generation of tables and/or graphs (7); verification of the hypotheses made by the evaluator (8); organization of the conclusions and future lines of research for smart cities (9), and establishment of a first style guide for design, in order to facilitate the communicability of the phaneron in smart city interfaces for inexperienced and adult users (10).

This is a first guide that is oriented towards mobility within public transport services and personal use. Some instruments, methodologies, techniques and stratagems for heuristic assessment have already been described or referenced above.

10.9 Inspecting the Technological Components and the Reactions of Human Capital of Smart Cities

First, each European city in our potential study universe was assigned a number, as shown in the following list:

1. Andorra
2. Berlin
3. Copenhagen
4. Ljubljana

5. London
6. Luxembourg
7. Madrid
8. Malta
9. Milan
10. Oporto
11. Oslo
12. Rome
13. San Marino
14. Vienna
15. Zagreb

Subsequently, a random draw was used to select smart cities (only five). This is the set of cities that participated in a random selection, through a lottery: London, Luxembourg, Madrid, Milan, and Vienna. The objective of working with these first cities was to conduct an initial test to detect functional problems, observe user behavior, formulate hypotheses, and establish strategies to continue the study, using user experiments.

Next, two cities Barcelona (Spain) and Bergamo (Italy) were incorporated into the study since the users for the public transportation experiments reside in one of those cities and have no problems traveling to the other city being analyzed. The users do not have physical and/or mental disabilities. All of them speak and understand the languages of those two cities perfectly, in addition to English. The users of both groups have always been the same throughout the entire experimental period. Lastly, we can see the complete list of potential cities where the communicability expert would conduct direct observation of users when purchasing or validating transportation tickets.

These requirements have allowed the group from smart city (Barcelona) to evaluate the technological devices and services aimed at public transportation in smart city (Bergamo), and vice versa. This strategy has yielded reliable results, with reduced costs, and in a short time.

In the complete collection of technological devices analyzed within the cities related to public transportation, the following technological, functional, ergonomic, communication, and interactive design characteristics are highlighted:

(a) **Internalization of Design and Ergonomics**: The operation, format, and physical structure of the ticket vending machines are practically standard and similar across the various countries analyzed. Reading contactless cards allows for simple, reliable, secure, and fast recharging. The use of these cards reduces the stress user face at these machines when they have to insert coins and/or bills while facing a long line of passengers boarding transport. Many of these cards are ergonomically designed to provide efficient access for people with reduced mobility, and even feature Braille on the keypads. They commonly feature TFT (Thin Film Transistor) and multi-touch screens (10 or more touches), which enhance user experience and enhance the color quality and brightness of the images. Some models allow ticket validation using barcode technology via an

Fig. 10.11 This is a negative example of the integration of technology in smart cities, where priority is given to the machines, not their users

infrared reader for paper tickets, QR codes, and barcodes. Other components include a thermal printer for receipts and an IP intercom with SIP (Session Initial Protocol). Assistance or help can be requested by pressing a button. In short, these are devices that contain a wide range of sensors, ranging from temperature measurement to motion detection in the event of vandalism, including the activation of an alarm speaker. Some of these sensors refer to whether they are placed inside or outside the stations. Figure 10.11 shows how a bus stop has been completely designed to protect the ticket vending machine but not the passengers. For example, there is no bench to sit on in case of adverse weather conditions: intense heat, hail, snow, etc. The architectural solution is wrongly focused on the technology, not the user. In other words, it is the complete opposite of the basic paradigms of industrial design and/or the adaptation of technologies to people.

The electronic, automatic, and other components of the device may have an attached icon representing its functionality. However, there are exceptions, as these may be omitted or accompanied by bilingual text. That is, the local language appears first, followed by English or another language. Different fonts, smaller letters, and different colors are commonly used, to cite a few examples of visual communication. In the interfaces or screens of these devices for public transport, there is a widespread trend in Europe to use only four or five languages: English, French, Spanish, and German. Clearly, there are exceptions, such as the one seen in Fig. 10.8, where the quality attribute "richness" is present due to the wide range of languages represented (Figs. 10.12, 10.13, 10.14 and 10.15).

(b) **Interaction Time**: In the following Fig. 10.16 for the metro in Madrid or Milan, we can see that there are lines in front of the ticket issuing machines. Some machines may be separated based on the end user. That is, whether they are for local users or tourists/foreigners. The average time spent is 3–7 min for each user, for both types of machines. In contrast, the average time spent on the 16 figures was 2–5 min, and there was no physical division between local and tourist/foreign users. Times can exceed 10 min at the tourist-only machines located at the airport metro entrance (Barcelona, Spain), for example. The difficulty adolescents, young adults, and adults (tourists) have in interacting with this

Fig. 10.12 The users can see only the flags of four languages/dialects beginning with local/regional (Catalan), national (Spanish), English, and French, in the Barcelona Metro

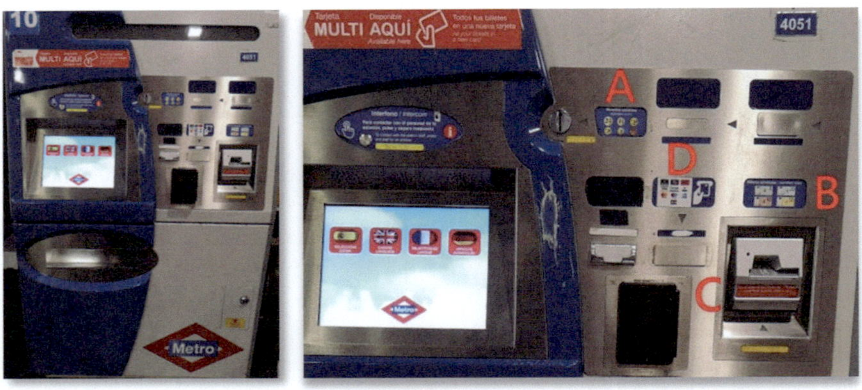

Fig. 10.13 These ticket vending machines in Madrid ("Adolfo Suárez" Madrid-Barajas Airport—Metro/Subway) support the four main languages that are currently the common denominator in many European cities: English, French, Spanish and German

technological device slows down the flow of passengers to the metro. Therefore, for a communicability evaluator, it is beneficial to observe the interaction and reaction of tourists with public transport ticket machines located at airports, train stations, ports, etc. Many of these users are interacting with these devices and their corresponding interactive systems, sensors, keyboards, etc., for the first time. Therefore, through a short period of direct, non-participatory, and objective observation, it is possible to detect and verify common errors among different types of users.

Fig. 10.14 This is a ticket validation machine on the bus (City of Bergamo) that supports the same foreign languages as the ticket vending machine in Madrid

Fig. 10.15 This is special vending machines on the bus (Vienna, Austria) because supports three visible foreign languages: English, Italian, French and have a special key with more languages: Czech, German, Hungarian, Russian, and Spanish

(c) **Payment Methods**: One of the characteristics of smart cities is to reduce paper consumption, and therefore, cash payments. Therefore, some machines only accept electronic payments, via smartphones, watches, credit or debit cards, etc., as seen in Figs. 10.14 and 10.17. Others, however, simultaneously accept electronic and cash payments (coins and bills). However, in this last option, the information regarding the coins or bills accepted by the system is not

Fig. 10.16 Difficulty interacting with the system leads to lines forming in front of the ticket vending machines (Barcelona—left, and Milan—right), where people have to ask for help directly from the transport company staff (if present) or from other users who have purchased a metro or commuter train ticket, for example

always specified. Furthermore, an additional problem with paying in cash is the remaining amount, that is, the refund when the amount paid exceeds the cost of the service.

Although these machines are frequently monitored, there may be periods when they run out of coins or bills to issue the remaining amount. These devices can also be configured to accept certain bills when the purchase exceeds certain limits. For example, Fig. 10.13 in Madrid (Airport Subway) states that the ticket vending machine will only accept 50-euro bills (B) when the purchase of tickets

Fig. 10.17 This ticket vending machine in Milan accepts the following coins (0.05, 0.10, 0.20, 0.50, 1 and 2 euro) and bills (5, 10, 20, 50 and 100 euro)

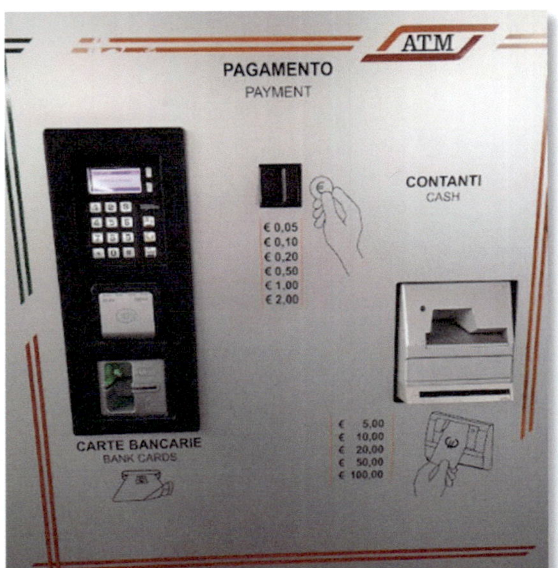

exceeds 22 euros (C). The English text is practically unintelligible to the user due to its size and the label red color. The same figure represents the coins accepted by the system, that is, almost all euro coins, except for 1 and 2 cents (A). This information is more readable in Milan city (see Fig. 10.14).

Finally, smart cities tend to increase the use of electronic payments as it facilitates passenger flow during peak hours. Not all senior users have smartphones or carry credit/debit cards. Others resist the use of electronic payment methods due to the lack of reliability regarding the use of their personal data by the public or private institutions that manage them. Despite this contextual reality, currently, some public transportation systems in smart cities only accept payment by bank card (Barcelona—Fig. 10.17).

(d) **Managing Malfunctions**: Although fixed ticket vending machines have options to communicate with an assistant in case of problems, this service is not guaranteed 24 h a day, seven days a week, in some smart cities. Nor can the assistant immediately resolve on-site problems arising from malfunctioning sensors related to reading contactless cards, as new international software and hardware companies have begun to provide this service, such as Apple Pay/VAS, Samsung Pay, Google Pay/Smart Trap, and so forth. New technology multinationals are adding to traditional financial services cards such as Mastercard or Visa, to name a couple of international examples. Furthermore, these vending devices have a standard limit for issuing travel tickets in analog formats (paper, cardboard, etc.), which can range from 500 to 1,000. The same applies to those that accept bills and/or coins of all programmed denominations. For example, depending on the device manufacturer, the bill handling capacity can vary between 1,000 and 1,500 (Fig. 10.20).

With devices for purchasing or validating tickets on public transport, a malfunction quickly reveals the level of responsibility of the business operator of the service:

(1) Ticket Purchase: In Fig. 10.18 (Bergamo city) a small, mobile ticket vending machine has the following written in red and within a yellow box at the bottom of the front: "If the ticket vending machine is out of order, tickets must be purchased before boarding." That is, the responsibility for the malfunction does not fall on the service provider, but rather on the user, who will be fined by inspectors for not having a ticket before boarding the bus. This is the penultimate sentence of the text on the front. There is another sentence before it that reads: "Ticket vending machine does not return coins and does not give change." However, it does not specify the time available to the user to purchase the ticket, since after that time, the machine does not return the money inserted. Both texts, written in red, do not stand out visually since the device externally has the same shade of red. In addition, the text is written in bold for the local language, while in English it is written normally. Lastly, the location of these machines at the entrance of the bus causes that many passengers do not have time to read all the illegible writing, as they must allow other passengers to board.

(2) Ticket Validation: Figs. 10.18 and 10.19 contain two posters in English and the local language (Italian), which state what to do if they are not working,

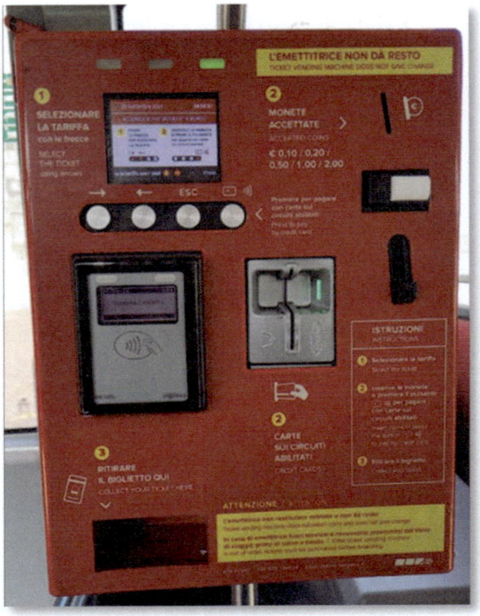

Fig. 10.18 Device that accepts different payment methods in the bus, but the information sent to the user may be unintelligible (Bergamo city)

resorting to the strategy of the rhetorical question: "This validator is out of service? Use the others on board. If none is working, buy the ticket at the automatic vending machine on board or through the (acronym of the transport company) Mobile app." One answer, again resorting to rhetoric, is: "What should passengers do when public transport is so crowded that they cannot try the other validators?" Once again, these messages are written at the bottom of the posters. In short, with these two phrases and the adhesive posters, the company managing the monopolistic public service in some smart cities makes the users responsible for malfunctions, which result in them being unduly penalized with fines (see the cost of these fines at the following link: www.pirateando.net/4S). Besides, there is a kind of hidden scam or abuse targeting passengers who insert their coins into the vending machines on public transport without knowing in advance how much time they have available to complete the transaction (Figs. 10.21, 10.22 and 10.23).

These are some negative human factors within the management of software and hardware in smart cities. This is a negativity that denotes the lack of corporate responsibility and the lack of oversight by those who are the concessionaires of the public transport service. This reality fosters negative social factors in the corporate communication of city councils and transport service companies, generating a high level of social rejection toward other smart sensors within smart cities. Readers interested in other technological components (positive and negative) for users, with their corresponding examples, can consult the following link: www.pirateando.net/4S. The set of observations made by the communicability expert, summarized below, have formed quality attributes, which are included in the new metric of this research work: "4S." The attributes that comprise the new quality metric are:

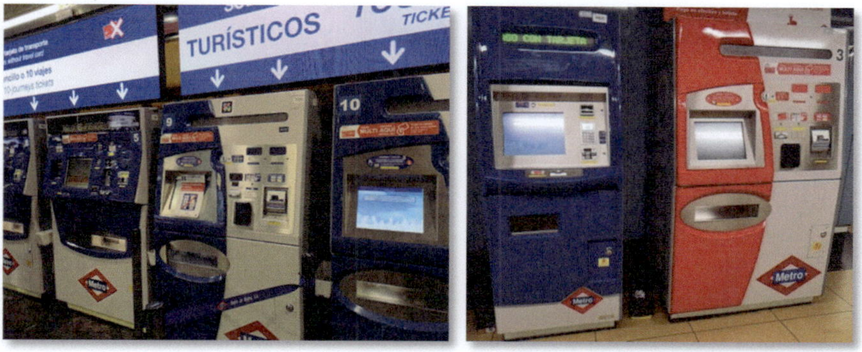

Fig. 10.19 Ticket vending machines #9 and #10 (right) only for tourist tickets, in Madrid. Furthermore, we can see another two machines: One for credit or debit card payment (blue color) and other machine for cash and card payment (red color)

Fig. 10.20 This is a machine in the urban bus (public service) that only accepts card payments (Barcelona city)

- Communication Transparency [CT],
- Absence of Pleonasm [AP],
- Minimalist Information [MI],
- Mnemonic Structure [MS],
- Rich(ness) Communication Media [RCM],
- Technological Acceptability [TA],
- Technological Operability [TO] (mobile and/or fixed: ticket issuing and validating devices),
- Functional Inference [FI],
- Contextual Connectivity [CC] (local GIS—geographic information system, Google Maps, etc.),

Fig. 10.21 Validator out of service: English text

Fig. 10.22 Validator out of service: Italian text

- Motivational Promotion [MP],
- Responsible Service Management [RSM], and
- Undue Penalties [UP].

These attributes denote a bi-unique, orthogonal relationship with social sensitivity and technologies (sensors) in smart cities.

Fig. 10.23 Penalty poster in Barcelona bus

10.10 Users: Set of Tasks

The set of tasks to be performed by users includes the compilation of forms. The forms primarily have multiple-choice responses. Binary responses, i.e., yes or no, have also been included. The content of these responses does not vary depending on the smart cities where the data is collected (Barcelona or Bergamo). The acronyms that appear in the figures and their corresponding data are indicated in brackets. The average of the results obtained is represented as percentages (%). Besides, in the heuristic evaluation of the systems and devices related to public transportation analyzed, the evaluator and/or user uses a stopwatch to record the time consumed in some of them. Below, the tasks to be performed by users on smart city buses are briefly described; the results can be seen in Appendix 4.

The set of tasks [T] is divided into two subsets. The first involves interaction with a fixed ticket vending machine outside the public transportation vehicle. The second is mobile and takes place inside the means of transport. Some tasks are repeated in both subsets, for example, those related to the interactive design of interfaces. Therefore, although these are interrelated technological devices within smart cities, they will be accompanied by the two acronyms of the software and hardware evaluated (external and fixed: EF, or internal and mobile: IM), combined with artificial intelligence, human and social factors, semiotics, communicability, interactive design, among other components:

T01: Check whether the ticket vending machine is on and without any error message on the interface [EF – IM].

T02: Determine whether the terminal has a touch screen or plasma screen with buttons. Also, determine whether any type of audiovisual content is activated when interacting with that screen. For example, a virtual guide for purchasing tickets, figures with facsimiles of the different tickets, etc. [EF – IM].

T03: Recognize different sensor sounds in the case of successful actions (ticket purchase) or errors when paying with cards, smartphones, watchphones, and so forth, inside public transportation [IM].

T04: Confirm the shape of the interface. That is, whether it is a rectangle (horizontal or vertical), a square, or another format. Add to this analysis the topological distribution of the options or keys on the screen. In other words, they are horizontal, vertical, or a combination of both [EF – IM].

T05: Establish the predominant colors in the interface components (background, letters, titles, icons, maps, photos, and so on): Primary (blue, yellow, and red), secondary (green, violet, and orange), tertiary (yellow-orange, red-orange, red-violet, blue-violet, blue-green, and yellow-green), and neutral (black, white, and grayscale) [EF – IM].

T06: Examine the interface's main font, whether it is classic (Arial, Times, Verdana, etc.) or special (artistic or decorative) [EF – IM].

T07: Navigate within the system using virtual keys (2D and/or 3D), or if they are real, that is, physical and external to the screen. Also, verify whether the physical keys contain the Braille alphabet [EF – IM].

T08: Include the interface of the ticket vending machine examined within one of the following subsets: vintage/smartphonics/futuristic [EF – IM].

T09: Confirm the existence or absence of a minimalist style, as well as the absence of ambiguity and/or vagueness between the meaning and significance of the texts, images, and audio.

T10: Investigate the possibility of paying in cash. If so, verify the maximum denomination of the bills accepted and the coins not accepted. Introduce bills and coins that the system does not accept to verify whether they are automatically returned [EF – IM].

T11: Examine electronic card payments ("classic"—Visa, Mastercard, American Express, etc., and "modern"—associated with multinational software and hardware companies), the maximum number of cards accepted, and whether the system issues a receipt for the payment made [EF – IM].

T12: Contact the virtual assistant (intercom) in English, requesting help paying with a smartphone or smartwatch [EF].

T13: Pay attention to updated audiovisual information about the route and the stops the vehicle makes [IM].

T14: Request additional information from the driver in English regarding the route of the vehicle and a tourist monument or museum [IM].

T15: Investigate the synchronization between the time and real-time information about the route the vehicle is taking and what is announced on the screens at the stops [IM].

T16: Measure the time difference between ticket issuance when purchased in the local language and in English [EF – IM].

T17: Verify the maximum number of languages for interaction between the user and fixed/mobile technological devices [EF – IM].

T18: Investigate whether there is additional information within the means of transport regarding the operation of ticket validation or purchase devices inside the unit, solutions to possible malfunctions, sanctions imposed on passengers, etc. [MI].

These tasks have been carried out progressively during the six-month evaluation period, with the same users, always using the same urban transport lines, and the same machines installed outside or inside metro stations, buses, airports, and so on. However, the results obtained with the 4S metric applied to public passenger service in the two cities studied denote the presence of negative factors usually referred to as "human factors" in software engineering. These factors impede the harmonious expansion of the use of smart sensors, for example, in smart cities. This means that the notion of "phaneron" is represented not as a gap, but rather as an abyss, considering the distance between technological reality and the sensation perceived by public passenger transport users.

Public transport and users who a priori tend to value the ecological aspects of cities, avoiding environmental pollution, reducing paper consumption, promoting the use of smart sensors, etc. For example, traffic sensors can be used to predict routes that can be traveled without so much congestion, facilitating the remote operation of traffic lights, and reducing the concentration of carbon dioxide and particulate matter in the air. Monitoring these last two elements requires sensors designed to measure environmental pollution variables. The use of these sensors is linked to citizens' health and quality of life. This is the source of a positive social motivation and assessment toward them.

The social equation that citizens of some smart cities make is that less traffic in the city theoretically implies better quality of the air they breathe. Contrarily, when one of the cities examined in this research is located 10 min away by city bus from the busiest (low-cost) passenger airport in northern Italy (Orio al Serio—Bergamo, Italy), the data collected by smart sensors can do little or nothing to change the situation and improve air quality for citizens in smart cities.

10.11 Results Obtained: Apocalyptic and Integrated

The concepts of semiotician Umberto Eco [55] allow us to divide city residents and those facing new technologies into two large groups: Integrated and Apocalyptic. The first group includes users belonging to different generations of the population who favorably accept all technological advances, since they can hypothetically benefit them and improve their quality of life. The second group is those who resist changes arising from information technology, industrial automation, e-commerce, and so forth. Some express themselves openly, while others do so in a camouflaged manner.

A classic example of the latter are the members of the Omega generation, who slow the advance of technological innovations by spreading fake news, professionally discrediting pioneers in the fields of social communications and information technology, systematically plagiarizing colleagues, and promoting narco-education or narco-illegality with AI, to cite a few examples, which can be expanded upon in the following references [39, 42, 43] and there is a new map and table (Appendix 5).

In this research, the users of the experiments have detected that "apocalyptic" concepts predominate in the city of Bergamo (Lombardy, Italy), while "integral" concepts are more present in the city of Barcelona (Catalonia, Spain). Nowadays, urban transportation in many cities involves maximizing technological advances stemming from electronic payments, cybersecurity, artificial intelligence, metaverse, and so on, including the distribution and management of smart sensors throughout cities.

In opposition, technological and human dysfunctions, which generate incidents in public, private, and hybrid services, automatically increase social unrest. This negative reaction is also observable in education, healthcare, city halls, etc., and transportation services are no exception. Besides, the COVID-19 pandemic has led to a population shift to non-urban areas. Therefore, in some parts of Europe, the number of citizens who daily use and combine various means of public transport, such as trains, buses, subways, funiculars, ferries, and so forth, to get to work has increased.

Hence, it is necessary for this combination of means to operate efficiently and on pre-established schedules. This efficiency must be guaranteed from the different ticket acquisition methods to ticket validation, as expressed by 92% of users (average) in the two cities evaluated: 89% (Barcelona) and 95% (Bergamo), and independently of the total number of inhabitants in a city.

It has also been found that the proper functioning of smart sensors in transport devices can favor their continued expansion in the most diverse structural areas of the city. Consequently, the so-called technological operability must be easy to remember or mnemonic (in the study universe, it reached a high percentage), intuitive for all types of users (tourists were those who reported the most problems: 74%), and oriented toward constant updating.

That is, a positive response to services significantly and motivationally encourages users to expand all the advances that service managers can offer them through cutting-edge devices, with or without artificial intelligence. On the other hand, motivation is initially high among users and will remain high over time when malfunctions are nonexistent. However, if service administrators and guarantors do not assume their responsibilities in the face of malfunctions (95%), and only implement fines for passengers (99%), the number of apocalyptic incidents clearly increases (in parentheses are the final results in percentages, achieved through experimentation with users). This is one of the causes of the formation of social movements against telecommunications advances in mobile telephony for smart cities, such as 5G or 6G.

In the interactive systems of the devices analyzed, it has been observed how minimalist design favors the absence of pleonasm, for example. Contrarily, the communication transparency or the seriousness of the corporate communication of the companies that provide these services to citizens in smart cities is equal to zero (i.e., Bergamo).

This result is achieved by the lack of indication of the time available to the user to insert coins and purchase a ticket on the bus (Fig. 10.15), or the problems derived from the design of corporate communication, whether in analog format (read: paper, stickers) texts on the front and/or side of devices, etc.), or on digital media (videos). In both media, a lack of textual legibility has been detected due to errors in mixing different fonts, inadequate combinations of background colors or textures, variable arrangement of visual resources on different screens during navigation, the artistic use of navigation icons or keys to access information searches in databases, and the small size of signs inside the means of transport.

The richness of communication media has been detected mainly outside the study universe and the borders of the EU. It is represented in Fig. 10.7 from London (UK), by the use of a large number of languages in a ticket vending machine. Within the study universe, access to different languages on this type of device is practically similar, ranging from 4 to 5 languages at most. From the standpoint of activating and deactivating dynamic and/or static media, which favor the presence of the wealth attribute, it has been verified that this is not widely diffused, with the exception of maxi-tablets (live-touch).

These devices are found at some bus stops and have an interface that exceeds approximately 40/50 inches. They are arranged in a vertical rectangular format. Through them, real-time information on urban mobility (AVM: Automatic Vehicle Monitoring) is accessed, which indicates via Wi-Fi the schedules, the total number of free seats on public transport, etc. Besides, the maxi-tablets allow users to locate the city's tourist attractions, calculate the time to reach them, calculate bus routes, etc. Among other features and depending on the model, they have USB ports to recharge smartphones, for example.

These equipments are combined with other systems related to contextual and real-time connectivity (AVM). Paper data is generally outdated with digital data when considering the variable of transport arrival times at the bus stop, subway, or train station. Surface transport service delays frequently exceed the punctuality of posted bus stop schedules. This is an inconvenience that prevents good connections with other modes of transport. Furthermore, it is a source of workplace conflict due to employee punctuality. This leads to employees having to drive cars, bicycles, scooters, and so on, increasing urban traffic problems. Undoubtedly, re-motivating people to choose public transport again is not an easy task in a smart city, due to the accumulation of negative user experiences.

When users exchanged cities to experience a different evaluation scenario than their everyday one, these negative experiences were confirmed. Some of them include: Reducing bus frequency during the summer and winter vacation periods for schools, colleges, universities, and so on; Not increasing the vehicle fleet if there are large crowds of passengers outside of school start or end times; Leaving neighborhoods without transportation service on weekends; Not providing service at night; Hiding bus line numbers through advertising on the sides or back of buses, among others that can be found at the following link: www.pirateando.net/4S (Fig. 10.24).

This contextual reality of the Bergamo city, for example, converges towards a rejection by broad sectors of the population of the advances of smart cities. For their part, users of the experiment who participated in the evaluation of the Barcelona city confirmed the impossibility of paying in cash for public transportation. If they do not have a ticket before using the service, they question using a debit or credit card, for example, due to the annual maintenance costs of these cards at financial institutions; the continued elimination of financial branches and/or ATMs for recharging bank cards; the difference in costs between tickets purchased at vending machines and on the bus (the latter option being more expensive); the potential demagnetization of the

Fig. 10.24 The bus service in Luxembourg is an excellent European example about the visibility bus route numbers because the users can see the numbers. The external advertising is equal to zero

card; problems arising from personal safety in confined spaces crowded with passengers; and the high cost of fines for passengers (triple compared to other European cities).

All users highlighted the lack of a uniform visual style across the various modes of transport. That is, there are no isotopies in the different categories of interactive design of the various devices and their interfaces, which contain smart sensors. This inadequacy was also noted in the institutional communication of those responsible for the transport service. The new subset of negative elements was easier to detect, with the exchange of analysis scenarios and users, in our study universe. One way to partially and significantly overcome these problems has been for riders to interact in English with transportation drivers when requesting information about the route (i.e., 100% in Barcelona) or assistance via the intercom. While some drivers did not speak perfect English, almost all of them were able to provide the information requested by the passenger. However, the assistance provided via the intercom sometimes had a better level of English, yet riders had difficulty operating the devices to purchase tickets (36%). This value is lower than the first. All of this suggests that older adults in both cities prefer human, face-to-face communication to anonymous telecommunications. Therefore, human capital that interacts with passengers in a friendly manner increases social sensitivity, even in situations of technological device malfunction, which have been evaluated in this study.

10.12 Lessons Learned and Future Research

The results obtained highlight the importance of measurement in the expansion of smart sensors and the human perception of these changes in so-called smart cities and the evolution [56–60]. It is not an easy task because it combines the material aspects of the context with human elements of a subjective, emotional, or sentimental nature. However, already in ancient Greece, Plato and Aristotle argued that the tangible world

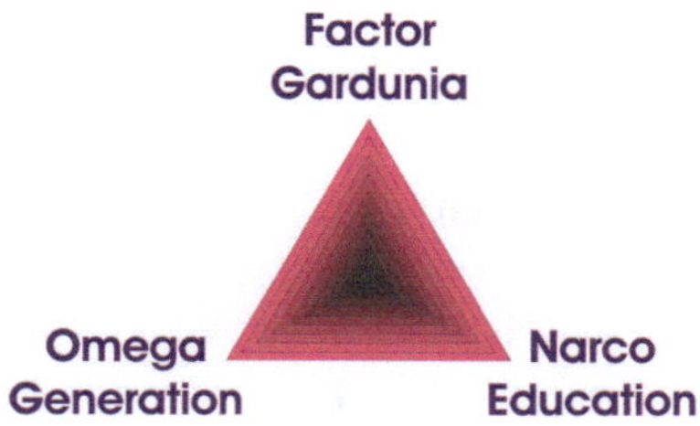

Fig. 10.25 A triad of elements that negatively promote the divide between social sensitivity and smart sensors

was constantly changing and that reality could be best understood through immaterial elements. In other words, Aristotelian causes and Platonic forms.

The problem today is that reality can be manipulated to such an extent that human beings no longer distinguish between the real and the virtual, with all the advances coming from computer graphics (rendering), for example, extended reality [61]. Hence the importance of studies related to communicability, interactive design, ergonomics, software engineering, hardware standardization, artificial intelligence, the miniaturization of new information technology devices, social psychology, sociology, pedagogy, and ecological context, among others, are carried out by genuine professionals. **The primary objective of these studies should be to elevate the quality of life, not the alienation of human beings**.

Simultaneously, **the notion of Phaneron, in this work, has denoted an excessive distance between the subjective aspects of users and the use of technologies that include smart sensors.** Furthermore, it has connoted that this abyss has its human epicenter in a bidirectional triad, formed by the following three components: Gardunia factor, Omega generation, and narco-education (Fig. 10.25).

This triangulation in the academic sector has demonstrated its presence and expansion in countless examples during the pandemic (Covid-19), ranging from the awarding of engineering, bachelor's, master's, and/or doctoral degrees in record time, with the minimum requirements in the final projects or theses, and with virtually no academic controls in the final evaluation; the assignment of lifelong, finger-appointed jobs in the educational, healthcare, government, judicial, and other fields; the approval of international R&D projects among friends; the distribution of financial resources from national, regional, provincial, and municipal tax collection to solve overseas problems, and so on (Appendix 5). To this uncomfortable reality or truth, we must add the new, veiled attacks on the social sciences, coming from professionals in the natural, formal, and factual sciences (humanities computer scientists, mathematicians, physicists, chemists, artists, lawyers, and so forth) who lack proven experience in systems analysis and/or computer programming, for example.

These are destructive actions typical of the members of this triad. In the third decade of the new millennium, they are currently abandoning the sectors of humanities computing, recommender systems, information retrieval, interaction design and children (IDC), HCI, UX, UCD, STREAM, etc., and who have hypothetically solved all the problems of the disabled, the elderly, children, women, blind person, etc., in their interdisciplinary research. The new goals they have set for subjugation and manipulation are: human sciences and welfare innovation; human well-being from a physical perspective (health); the isolation of people and the reduction of the cognitive capacities of new generations through podcasting in education, the narcissism of static and dynamic images on social media, the misuse of artificial intelligence, ethics to eradicate the truth on the Internet, associations, organizations, foundations, and so on, to commit certified crimes, and the legal profession to spread illegality and muzzle journalists, and a very long etc.

Consequently, these members of the triad are opposed to the modernization of science and universal education, democratic access to truthful information, the right to be informed while respecting the freedom of thought and expression of all human beings, among many other principles, rights, and obligations that citizens of planet Earth have progressively acquired over the centuries, and which theoretically should be present in smart cities [62]. In other words, their fundamental objective is not the social well-being or the common good of the community that finances their activities, either directly or indirectly. In short, they act unethically to keep members of the human capital isolated and separated from each other, given the spread of the latest technological developments that seek to democratize, such as the Internet at the end of the last century, and today, perhaps the effective use of artificial intelligence.

It is precisely in these actions that we see the gap, which continues to grow exponentially, as already pointed out by Phaneron's notion. The results obtained in the experiments carried out verify this reality. For example, there is no longer an interrelation between the Universities located in a community and the real problems of their inhabitants' daily lives, such as public transportation. These institutions are located in a specific area, but much of their research, funded by local residents, serves to solve problems for multinationals (companies, industries, laboratories, and so forth), SMEs, startups, angel investors, data scientists, researchers, professors, etc., who are thousands of kilometers away from the headquarters of their academic units.

This is one of the reasons why, when a ticket vending machine (Fig. 10.11) located in front of the main hospital in a provincial capital in Lombardy, a foreign language (German, Spanish, French, and English) is selected, for example, English, the information appears in the local language (Italian), as can be seen in Fig. 10.23. Then, in the rest of the screen navigation to complete the operation, there are texts only in English or Italian. The same occurs with the other foreign languages. These are classic mistakes of usability. These errors were detected and resolved in the university context with heuristic evaluations of interface design in the 1990s. Therefore, in the new millennium, they should be avoided since they are interactive devices and systems for public use.

That is, thousands of local users and tourists detect these errors from the beginning of the interaction, but they appear invisible to those at local academic institutions who should point them out to service managers for immediate correction. Axiomatically, we are dealing with lifelong employees at academic institutions, who first access a job and then study, and not the other way around. So much so that in the city where these machines with mistakes from the last century are installed, there are unexpectedly adult individuals (over 40 years old, for instance) with no educational, research, or training experience, and so on, with a simple bachelor's degree (i.e., computer science), in addition to volunteering for rare diseases, religious unionism and a tractor driving license, which are sources of a sui generis metamorphosis on the European continent. In record time, uneducated individuals first become experts in usability engineering, multimedia publishing, hypermedia communication, artificial intelligence, pedagogy, podcasting, etc., and then become professors of "Specialized Teaching and Pedagogy for Children with Disabilities," directors of PhD programs and PhD theses (even if they do not have an university doctorate or a master's degree), deans of humanities and social sciences departments, censors of international scientific publications, and a long etc. This transformation takes place within a rainbow of topics of interest, covering all the sciences. Obviously, **this is a clear example of the power of this triad that distorts, from the university sphere, the proper functioning of smart cities, now and in the future, since they are "deforming" them for future generations, interested in new technologies**.

This example, which originated in the new millennium, combined with the results of this work, has led to the creation of an updated heuristic table to determine the qualitative academic fields in the study of smart city metrics. This additional activity is due to the fact that many errors detected by users could already be avoided, as they are historical failures (more than 30 years old) and documented in an extensive and international bibliography. This should be one of the objectives of research within universities: to avoid constantly repeating errors already resolved due to inexperience or inadequate training of the academic community, before accessing these jobs. In other words, these educational institutions should have and demonstrate a greater interest in the well-being of the local population than in the militarization of international education through inhumane behaviors of permanent staff, ranging from mobbing to stalking, including bossing. The main components of the heuristic evaluation table are in Appendix 6. This table is fundamentally based on the binary presence metric (yes = 1 and no = 0). It allows us to easily and quickly detect the quality of university human capital, particularly those who hold roles as teachers, researchers, thesis supervisors, R&D project managers, international interlocutors, etc. The remaining components can be consulted at the following link: www.pirateando.net/4S.

Additionally, interested readers can consult the temporal evolution of these components and the psychological, sociological, pedagogical, labor, and so forth variables in the following references [63–70], spanning more than four and a half decades of research on human and social factors in the fields of software, education, and the modernization of science (Fig. 10.26).

10.13 Conclusion

Through the new metric and its inherent quality attributes, it has been effectively established that there is a segment of the population that does not accept devices with smart sensors due to failures, errors, or mistakes primarily found in software engineering and its main components. Added to this is the human/

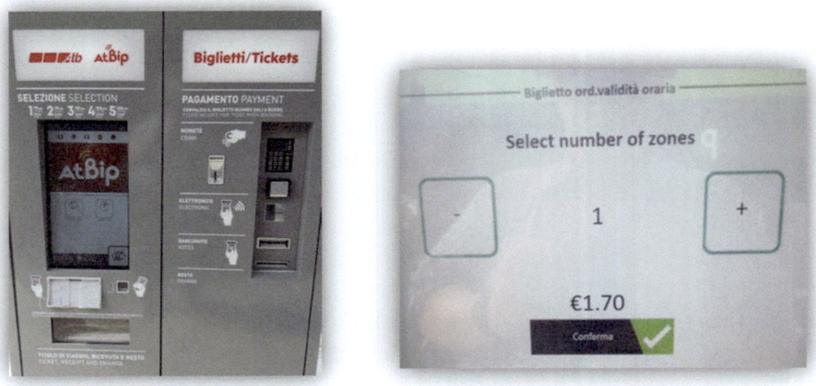

Fig. 10.26 A ticket vending machine (left) that occupies all the available space at the stop (without seats for passengers—Fig. 10.11), where the interactive system interfaces present classic mistakes of the heuristic evaluation of usability in the last century. After the English selection, the information on the screen has text in local language (right)

social factor. That is, human behavior that, from engineering and other areas of scientific knowledge, distorts reality within smart cities, between what a service should be (theoretical perspective) and what it actually is (practical perspective). For example, poor management of malfunctions in interactive systems guarantees for the protection of users' personal data online, structured and illegally endorsed disinformation to evade citizens' right to objective information, censorship to impede freedom of access to truthful information, among other factors.

At the same time, today, the majority of humanity in developed and emerging economies is mechanically and/or involuntarily witnessing a new technological and social revolution. It is the advance of industrial automation and robotics, humanoid or otherwise. Little by little, this new reality is rapidly being incorporated into the productive spheres of goods and services, being most visible in large cities. Therefore, it is necessary to constantly measure the convergence between these technological advances and social sensitivity toward them.

Since ancient times, measurement has been the basis of the human cognitive pyramid, upon which reflections are made about one part of reality with another to enumerate and classify the points of convergence and divergence between the theoretical and the practical, that is, between what should be and what actually is. Without the action of measurement, it would be impossible to observe the context, learn, and conduct experiments, for example. In our case, we worked with users with no computer experience who regularly walk the streets of smart cities, whether above ground or below ground. These users lack university education and lack the latest technological advances in computing or mobile multimedia telephony. Furthermore, for health reasons, they constantly monitor three pieces of data from their medical devices (oximeter, blood pressure monitor, and glucometer). These data are collected autonomously.

However, all of these users possess an exceptional cultural background, which they acquired through 20th-century social media such as the press, radio, and television. At that time, one of the primary functions of these media was not only information and entertainment, but also training. This wealth of information and knowledge accumulated over time allows these generations to navigate interaction with intelligent machines and their interactive systems, whether national or international. Now, the source of rejection detected with the phaneron is the dysfunctions of moving around public transportation services.

These days, with few exceptions, the dysfunctions fall mainly on the users of the service and not on those who manage them. This is one of the reasons why the adult population feels vulnerable to the changes introduced in smart cities. That is, a constant state of helplessness when they don't know how to respond to these malfunctions, which range from interface design to the institutional communication of transportation companies. Consequently, there is a widespread feeling of helplessness. This leads to an outright rejection of everything related to new technologies and telecommunications. The exceptions about this widespread rejection are those interrelated with health and personal well-being. Quantitatively, these users represent a portion of the population within smart cities that will grow exponentially in the coming decades. Hence the importance of the first style guide developed in this work to improve the communicability of the phaneron in device interfaces.

There is a revaluation of analog media such as paper, where a group of users who participated in the experiments considered themselves autonomous in determining the time available to make a trip, for example. A priori, this is something trivial, but it is not within the current universe of study. They believe they have lost their freedom of movement by having to rely on technological tools to access information that was theirs by right, and which they have involuntarily given away with the use of chip-on-paper. Furthermore, **contrary to the claims of smart city authorities based on the premise of "safeguarding the environment," users have found that paper is still used in European public transport, with or without chip-on-paper, and its consumption has even increased**. For example, with the issuance of payment receipts when purchasing or recharging a cash travel card for buses, metro, trains, etc.

This also occurs when making a payment with contactless cards, whether debit or credit. Authorities do not encourage paper recycling. There is no economic benefit for the user if they recharge their old chip-on-paper card. Added to all this is the fact that in some places, corporate inconsistency or roguery has taken root in the programming of ticket readers, since consulting the residual trip information on paper chip cards means consuming a trip. At the same time, in other cities (i.e., London, Barcelona, etc.), other transportation management models continually monitor the movement of the elderly, disabled, and chronically ill, among others. These individuals have special passes but are required to validate them each time they board a bus, subway, train, and so on. This not only slows down the flow of passengers, but also makes users feel violated in their right to freedom of movement, without being constantly monitored regarding where they have been, where they are, and where they may be going.

Fig. 10.27 High quality service of urban mobility (free for all) in an European smart city: Luxembourg

A simple way to solve the main problems detected in the interrelationship between users and sensors in smart cities related to urban mobility is to implement the Luxembourg model. That is, free public transport for all, every day of the week. Another way is to eradicate monopolies in public services. In some EU cities, this model is focused solely on the sector with the highest number of passengers, such as students. By increasing the number of companies providing the same service, competition generally directly increases service quality (Fig. 10.27).

Acknowledgements The author would like to thank Maria V. Ficarra, Teresa Rivera-Díaz, Carmen Tolosa-Flores, Montse Ferrer, Rita Cattaneo and Roger Ferrer for their help in the experiments; Giselda Verdone and Miguel C. Ficarra for the collaboration in the labs. A special thanks to Carlos, Mary, Luisa, Amélie, and University of Parma.

Appendix 1

Some ticket formats in European smart cities:

See Figs. 10.28, 10.29 and 10.30.

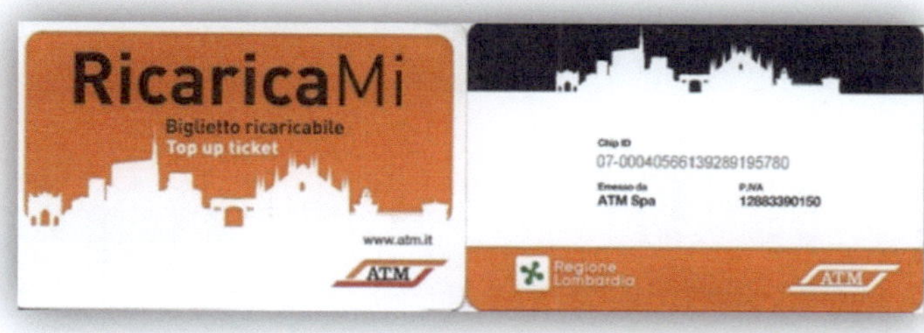

Fig. 10.28 A recyclable paper chip card in Milan, Italy

Appendix 2

Films:

- A-X-L (2018)
- Chronicle (2012)
- Dredd (2012)
- Edge of Tomorrow (2014)
- Geostorm (2017)
- Ghost in the Shell (2017)
- Infinite (2021)
- Jupiter Ascending (2015)
- Limitless (2011)
- Lucy (2014)
- Matrix Resurrections (2021)
- Moonfall (2022)
- Oblivion (2013)
- Quantumania (2023)
- S1mOne (2002)
- Source Code (2011)
- The Machine (2013)
- Upgrade (2018)
- Virtual Revolution (2016)
- Voyagers (2021)

Fig. 10.29 A standard (left) Italian train ticket, which includes a QR code, the amount paid, the validity date, and the route. A paper chip card for train travel (right), with its corresponding paper receipt. The receipt is more valuable than the travel card. The receipt image displays, among other information, the date, fare, origin and destination of the trip, and the validity period

Appendix 3

Communicability of Phaneron in the Interfaces: Towards a First Style Guide

This first guide covers the interactive design category known as layout/presentation of information, content, and navigation. It requires the following:

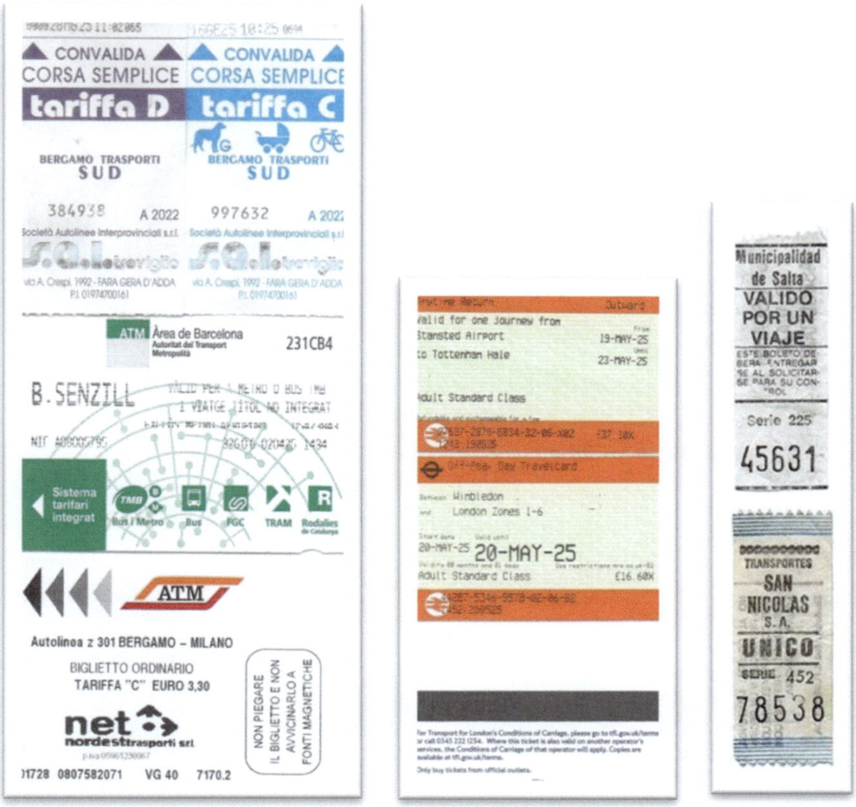

Fig. 10.30 A classic paper ticket for commuter bus trips (different fares depending on the zone), with the date and time printed on it, for example (left area). Besides, a single-ride ticket with a non-recyclable magnetic strip for buses, metro, suburban trains, and trams in the city of Barcelona. Last, the single-ride ticket for bus between Bergamo–Milano. Besides, we can see travel tickets for trip by train, metro, bus, etc. in London (central area), with a magnetic strip. The design is almost identical, except that the second one features the logo of the city's public transport network. Both indicate the cost, validity, duration, etc. A traditional Latin American model of a ticket paid in cash to the bus driver (right area)

- Use bright plasma screens in a rectangular (horizontal, preferably) or square format. The former is preferable as it emulates television monitor screens. Our study universe is made up of a generation of people who have spent many hours in front of the television screen. The new generations are accustomed to consuming audiovisual content on smartphones; therefore, rectangular screens should be formatted vertically.
- Use larger numbers than the rest of the elements that make up the device's interface. In the case of travel card readers, the OK icon or similar is generally larger than the total number of trips available to the user (95% of the cases studied).

- Promote circular shapes in the elements that make up the interface. Users reject vertical or linear shapes, such as right angles at the ends of photographs, videos, and so forth, or in some fonts.
- Use neutral or primary colors, primarily.
- Insert emoticons to visually summarize the values of the results.
- Use simple shapes (traffic signs) for navigation, such as triangles or arrows to advance or rewind screens.
- Eliminate company logos as navigation icons.
- Include sounds before displaying the final results. Avoid unpleasant sounds when operations are in progress. Both successful and unsuccessful operation sounds should have the same volume.

Readers interested in other style guides developed following some of the aforementioned cardinal points and communicability, with positive results in their application over time and in the context of education, tourism, ecology, home automation, manufacturing, and so on, can consult the following Refs. [31, 33, 35].

Appendix 4

Results of Heuristic Evaluations:

See Figs. 10.31, 10.32, 10.33 and 10.34.

The high rates of task completion in both cases indicate that they received adequate explanations and fully understood the tasks they were required to perform in the city assigned to the experiment (Fig. 10.35).

References: Communication Transparency [CT], Absence of Pleonasm [AP], Minimalist Information [MI], Mnemonic Structure [MS], Richness Communication Media [RCM], Technological Acceptability [TA], Technological Operability [TO] (mobile and/or fixed: ticket issuing and validating devices), Functional Inference [FI], Contextual Connectivity [CC], Motivational Promotion [MP], Responsible Service Management [RSM], and Undue Penalties [UP].

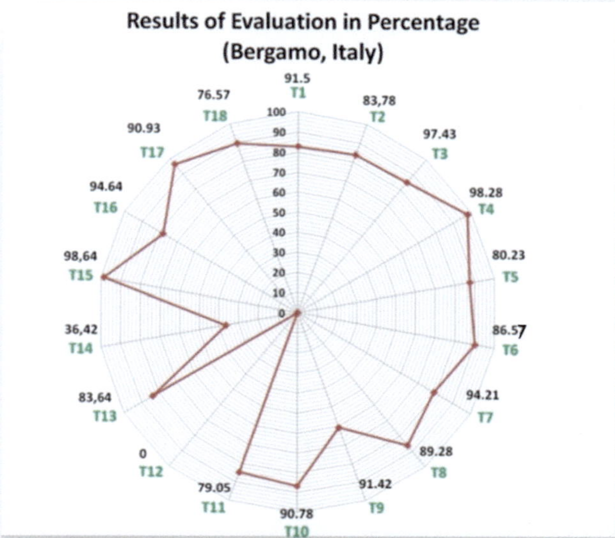

Fig. 10.31 Graph of the average results obtained from the tasks performed by users from Barcelona who interacted with smart devices using public services in Bergamo city. A total of 10 randomly selected urban lines

User	T1	T2	T3	T4	T5	T6	T7	T8	T9	T10	T11	T12	T13	T14	T15	T16	T17	T18
BCN W1	71	90	87	98	90	81	55	74	40	100	87	0	77	0	95	88	100	95
BCN W2	85	97	95	100	87	100	68	90	55	90	89	0	90	45	100	72	100	100
BCN W3	72	69	88	90	73	92	71	73	39	98	91	0	85	20	100	91	77	90
BCN W4	89	95	99	100	96	98	90	94	80	94	86	0	64	30	100	85	100	97
BCN W5	77	83	75	100	98	79	86	72	91	88	90	0	80	0	90	94	100	100
BCN W6	81	90	94	96	89	85	91	83	63	70	78	0	79	74	100	61	100	93
BCN W7	90	78	67	100	84	100	88	95	41	82	90	0	83	0	100	49	89	100
BCN M1	88	93	71	100	90	95	79	88	56	93	99	0	58	55	100	96	100	90
BCN M2	65	54	63	100	71	79	91	76	83	79	87	0	89	0	96	56	100	44
BCN M3	91	95	100	100	93	96	97	99	89	86	95	0	96	0	100	84	92	63
BCN M4	90	83	92	100	77	87	68	99	58	62	78	0	82	40	100	67	100	100
BCN M5	89	79	80	91	94	89	76	100	67	86	86	0	97	63	100	74	100	93
BCN M6	73	90	83	100	97	94	81	83	38	90	73	0	91	88	100	89	90	100
BCN M7	99	77	95	100	93	89	67	79	52	97	59	0	100	95	100	86	99	90
Average	83	84	85	98,21	88	90,29	79	86	61	87	85	0	83,64	36	99	78	96,21	89,64

Fig. 10.32 Table of the two main groups of users (women and men), indicating the percentage of activities they were able to complete correctly

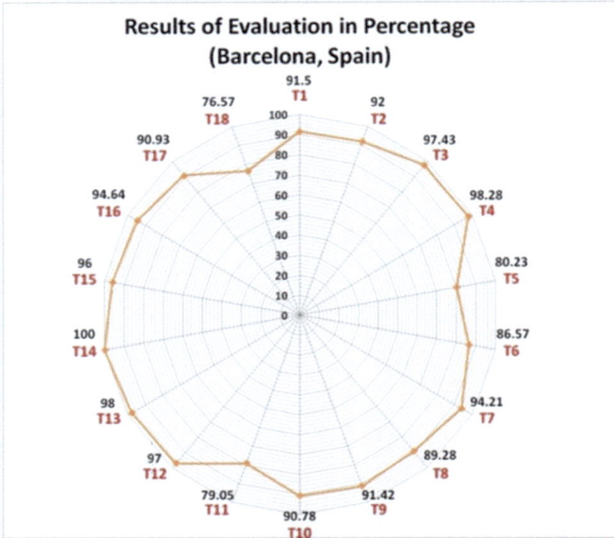

Fig. 10.33 Graph of the average results obtained from the tasks performed by users residing in Bergamo. They interacted with a total of 10 smart devices on 10 bus lines in Barcelona city. The bus lines were randomly selected by lottery

User	T1	T2	T3	T4	T5	T6	T7	T8	T9	T10	T11	T12	T13	T14	T15	T16	T17	T18
BG W1	80	90	95	95	67	79	81	95	89	99	73	95	99	100	100	92	79	82
BG W2	90	93	100	99	99	94	100	89	97	100	79	100	100	100	100	99	85	70
BG W3	100	99	100	100	92	98	100	95	99	100	84	100	100	100	100	100	88	66
BG W4	95	98	100	100	79	95	98	89	100	97	81	100	100	100	100	98	90	84
BG W5	98	93	100	100	90	100	94	96	100	100	92	100	100	100	100	94	100	90
BG W6	89	81	100	97	63	57	89	93	82	49	53	83	90	100	85	79	99	85
BG W7	91	100	100	100	78	99	97	100	94	100	85	98	100	100	96	99	100	71
BG M1	78	86	89	96	59	65	100	57	77	83	88	100	95	100	81	85	100	67
BG M2	95	90	100	100	92	91	100	98	95	100	94	93	100	100	100	97	100	78
BG M3	93	88	94	99	67	89	93	92	90	88	70	97	100	100	100	93	78	69
BG M4	100	100	100	100	83	95	100	100	94	100	88	100	100	100	100	100	100	96
BG M5	100	92	99	100	90	98	91	96	95	100	90	100	100	100	100	100	95	77
BG M6	100	95	97	100	84	100	89	94	99	100	89	100	100	100	99	100	90	95
BG M7	72	83	90	90	75	52	87	56	69	55	47	95	88	100	83	89	69	42
Average	91,5	92	97,43	98,29	80,23	86,57	94,21	89,29	91,43	90,79	79,5	97,21	98	100	96	94,64	90,93	76,57

Fig. 10.34 The table shows the total number of activities that users residing in Bergamo were able to successfully complete in Barcelona

Appendix 4

International Octopuses—Map and Table:

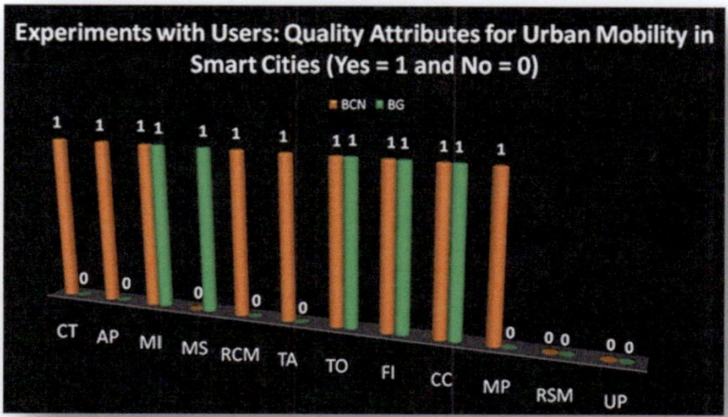

Fig. 10.35 Verification of the quality attributes for urban mobility in smart cities

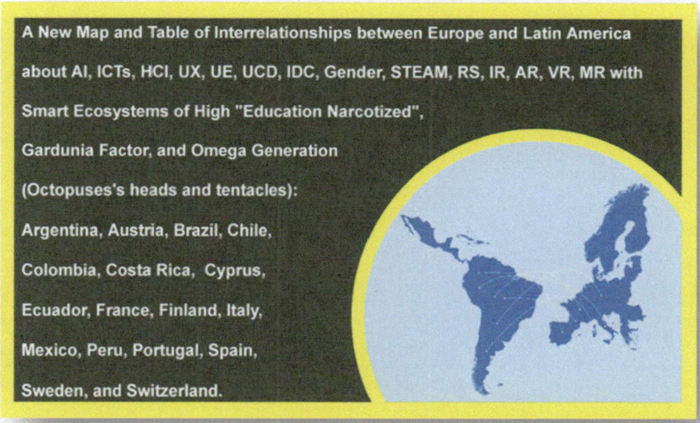

Fig. 10.36 First list of the countries analyzed between Latin America and Europe

A new analysis of the nodes and links that converge in some of the main cities and universities (Octopuses's elements (O): H = Head and T = Tentacle). In the last column, the total number of triangles represents the intensity of the Gardunia factor, Omega generation and Narco-education (minimum is equal to 1 and maximum is equal to 5) (Fig. 10.36).

Country and Cities	O	University Acronyms	Total
Spain (S)			
S01. Barcelona	H	UPC, UPF –ICT, URL –La Salle, UB	▲ ▲ ▲ ▲ ▲
S02. Ciudad Real	T	UCLM	▲ ▲ ▲ ▲
S03. Donostia	H	UPV/EHU	▲ ▲ ▲ ▲ ▲
S04. Granada	H	UGR	▲ ▲ ▲ ▲ ▲
S05. La Laguna	H	ULL	▲ ▲ ▲ ▲ ▲
S06. La Rioja	H	UNIR	▲ ▲ ▲ ▲ ▲
S06. Leganes	H	UC3M	▲ ▲ ▲ ▲ ▲
S07. Lleida	H	UdL	▲ ▲ ▲ ▲ ▲
S08. Madrid	T	URJC	▲ ▲ ▲ ▲ ▲
S09. Palma de Mallorca	H	UIB	▲ ▲ ▲ ▲ ▲
S10. Salamanca	H	USAL	▲ ▲ ▲ ▲ ▲
S11. Seville	T	US	▲ ▲ ▲ ▲
S12. Valencia	H	UPV and UV	▲ ▲ ▲ ▲ ▲
S13. Zaragoza	H	UZ	▲ ▲ ▲ ▲ ▲
Portugal (P)			
P01. Algarve	H	UAlg	▲ ▲ ▲ ▲
P01. Aveiro	H	UA	▲ ▲ ▲ ▲ ▲
P02. Lisboa	T	UL	▲ ▲ ▲ ▲ ▲
P03. Porto	T	UPT	▲ ▲ ▲ ▲
Italy (I)			
I01. Bari	H	POLIBA	▲ ▲ ▲ ▲ ▲
I02. Bergamo	H	UNIBG	▲ ▲ ▲ ▲ ▲
I03. Brescia	T	UNIBS	▲ ▲ ▲ ▲
I04. Bolzano	H	UNIBZ	▲ ▲ ▲ ▲ ▲
I05. Milan	H	IULM, POLIMI, UNIMI, UNIMIB, UNIBOCCONI, and UCSC	▲ ▲ ▲ ▲ ▲
I06. Modena	T	UNIMORE	▲ ▲ ▲
I07. Naples	T	UNINA	▲ ▲ ▲
I08. Pisa	H	UNIPI, SSSUP, and ISTI	▲ ▲ ▲ ▲ ▲
I09. Rome	H	UNIROMA2	▲ ▲ ▲ ▲ ▲
I10. Trento	H	UNITN and FBK	▲ ▲ ▲ ▲ ▲
I11. Turin	H	UNITO	▲ ▲ ▲ ▲ ▲
I12. Udine	H	UNIUD	▲ ▲ ▲ ▲ ▲
France (F)			
F01. Grenoble	T	UJF	▲ ▲ ▲
F02. Toulouse	H	UNIV-TOULOUSE –ICS/IRIT	▲ ▲ ▲ ▲ ▲
Finland (N)			
N01. Espoo	H	AALTO	▲ ▲ ▲ ▲
N02. Tampere	H	TUNI	▲ ▲ ▲ ▲▲

Sweden (W)			
W01. Stockholm	H	SU	▲▲▲▲
Switzerland (Z)			
Z01. Lugano	H	USI	▲▲▲▲▲
Austria (U)			
U01. Graz	H	TU-GRAZ	▲▲▲▲▲
Cyprus (Y)			
Y01. Nicosia	H	UCY	▲▲▲▲▲
Colombia (O)			
O01. Armenia	T	UNIQUINDIO	▲▲
O02. Cali	T	UNAD	▲▲▲
O03. Medellin	T	UDEMEDELLIN	▲▲▲▲
O04. Popayan	H	UNICAUCA	▲▲▲▲▲
Argentina (A)			
A01. Buenos Aires	H	UNLaM	▲▲▲▲
A02. Corrientes/Chaco	T	UNNE	▲▲▲
A03. Rosario	T	UNR	▲▲▲
A04. La Plata	H	UNLP –Lifia	▲▲▲▲▲
A05. San Juan	T	UNSJ	▲▲▲▲
A06. Santiago del Estero	T	UNSE	▲▲▲▲▲
Chile (H)			
H01. Santiago de Chile	H	USACH and PUC	▲▲▲▲▲
H02. Valparaiso	H	UPLA, PUVC and UV	▲▲▲▲▲
Brazil (B)			
B01. São Paulo	H	MACKENZIE	▲▲▲▲▲
B02. Rio de Janeiro	H	PUC-RIO	▲▲▲▲▲
Peru (P)			
P01. Lima	H	PUCP	▲▲▲▲▲
Mexico (M)			
M01. Aguascalientes	T	UAAG	▲▲▲▲▲
M02. Mexico DF	T	ITAM	▲▲▲▲
M03. Monterrey	H	ITESM / TEC	▲▲▲▲▲
M04. Puebla	T	BUAP	▲▲▲
M05. Zacatecas	T	UAZ	▲▲▲▲
Ecuador (E)			
E01. Quito	T	PUCE	▲▲▲▲
E02. Loja	H	UTPL	▲▲▲▲
Costa Rica (R)			
R01. Cartago	T	TEC	▲▲
R02. Heredia	T	UNA	▲
R03. San José	T	UCATOLICA	▲▲▲

The violet colors are the most dangerous, as they are destructive of scientific knowledge and all those authentic democrats who, in a modest and honest way, care for the general well-being of human beings through education and the modernization of science.

Appendix 6

First heuristic table to determine the qualitative academic fields in the study of smart city metrics:

1. Academic certification of university degrees, with their corresponding study programs, transcripts of course grades, and final projects leading to the degree.
2. Certified work experience in the educational field prior to hiring.
3. Qualification to teach at a third level (university).
4. Academic qualification to conduct research, direct projects funded by public entities (municipal, provincial, regional, state, and European), and supervise master's, doctoral, and postdoctoral theses. In other words, a bachelor's degree holder is not a doctor and cannot direct doctoral theses, in Europe.
5. Opposition to the phenomenon of "meteoric" degrees and/or access to lifetime positions. For example, obtaining two or more bachelor's/engineering degrees, master's degrees, and doctorates in a single decade, while developing high-level business, commercial, and other activities: management or general management.
6. Secularism in the professional practice of teaching.
7. Absence of conflicts of interest between teaching and potential other interests related to commercial, industrial, union, religious, volunteer work, etc. This component must be maintained throughout a teaching career.
8. Periodic psychiatric examinations for acts of psychological vandalism such as: workplace harassment, professional identity theft, promotion of hatred and racism toward other or former colleagues, the generation of enslaved students, cybercriminals, etc., the promotion of narcissism on social media, the constant broadcasting of lies for the purpose of rapid career advancement, defamation and professional discrediting to eliminate potential competitors for the same lifelong position, incentivizing the censorship of scientific publications, among other ethically deplorable actions.
9. Theoretical and practical demonstrations to improve the quality of life of local citizens, through results obtained individually and within their work group.
10. Annual and external verification (auditory with external organizations or associations) of the work carried out autonomously.

Through an analysis of the state of the art of human capital in universities, with these ten main points, we can better understand the current gap between users, their use of new technologies in smart cities, and the real/social involvement of higher education institutions in solving potential problems, such as those detected in this research.

References

1. V. Tomer et al., *Nanosensors for Smart Cities* (Elsevier, Amsterdam, 2020)
2. L. Neckermann, *Smart Cities, Smart Mobility: Transforming the Way We Live and Work* (Troubador Publishing, London, 2018)
3. A. Kent, H. Kim, S. Sabri, *Smart Cities for Technological and Social Innovation: Case Studies, Current Trends, and Future Steps* (Academic Press, London, 2020)
4. R. Khatoun, S. Zeadally, Smart cities: concepts, architectures, research opportunities. Commun. ACM **59**(8), 46–57 (2016)
5. H. Zhu et al., Integrating IoT-sensing and crowdsensing with privacy: privacy-preserving hybrid sensing for smart cities. ACM Trans. Internet Things **3**(4), 1–30 (2022)
6. C. Kirwan, Z. Fu, *Smart Cities and Artificial Intelligence: Convergent Systems for Planning, Design, and Operations* (Elsevier, Amsterdam, 2020)
7. J. Nielsen, The usability engineering life cycle. IEEE Comput. **25**(3), 12–22 (1993)
8. F.V. Cipolla-Ficarra, *Handbook of Research on Software Quality Innovation in Interactive Systems* (IGI Global, Hershey, 2021)
9. F. Cipolla-Ficarra et al., *Technology-Enhanced Human Interaction in Modern Society* (IGI Global, Hershey, 2017)
10. F. Cipolla-Ficarra, *Handbook of Research on Interactive Information Quality in Expanding Social Network Communications* (IGI Global, Hershey, 2014)
11. E. Kit, *Software Testing in the Real World: Improving the Process* (Addison Wesley, New York, 1995)
12. F. Cipolla-Ficarra, M. Cipolla-Ficarra, Computer graphics and mass media: communicability analysis, in *Proc. First International Conference on Advances in New Technologies, Interactive Interfaces, and Communicability (ADNTIIC, 2011)*, ed. by F. Cipolla-Ficarra (Springer, Heidelberg, 2011), pp. 182–192
13. R. Debray, *Vie et mort d'image: Une histoire du regard en Occident* (Gallimard, Paris, 1992)
14. F. Cipolla-Ficarra, A user evaluation of hypermedia iconography, in *Proc. Compugraphics* (GRASP, Paris, 1996), pp. 182–191
15. Apple: Macintosh Human Interface Guidelines (Addison-Wesley, Boston, 1992)
16. F. Cipolla-Ficarra, M. Ficarra, A. Kratky, Computer graphics for students of the factual sciences, in *Proc. First International Symposium Communicability, Computer Graphics and Innovative Design for Interactive Systems (CCGIDIS 2011)*, ed. by F. Cipolla-Ficarra (Springer, Heidelberg, 2012), pp. 79–93
17. F. Cipolla-Ficarra, J. Alma, M. Cipolla-Ficarra, Digital television and senior users: design evolution or involution? in *Technology-Enhanced Human Interaction in Modern Society*, ed. by F. Cipolla-Ficarra (IGI Global, Hershey, 2017), pp. 143–158
18. J. Bolter, M. Engberg, B. MacIntyre, Media studies, mobile augmented reality, and interaction design. Interactions **20**(1), 36–45 (2013)
19. W. Sherman, W. Craig, *Understanding Virtual Reality: Interface, Application and Design* (Morgan Kaufmann Publishers, San Francisco, 2003)
20. J. Searle, *The Construction of Social Reality* (MIT Press, Cambridge, 1995)
21. S. Jones, *Young People and Politics in the UK: Apathy or Alienation?* (Palgrave MacMillan, London, 2007)
22. D. Patai, N. Koertge, *Professing Feminism: Education and Indoctrination in Women's Studies* (Lexington Books, Lexington, 2003)
23. L. Weber, *Demographic Change and Economic Growth: Simulations on Growth Models* (Physica-Verlag, Heidelberg, 2010)
24. F. Cipolla-Ficarra, Mobile phones, multimedia and communicability: design, technology evolution, networks and user issues, in *Mobile Phones Technology, Networks and User Issues*, ed. by M. Barnes, N. Meyers (Nova Publishers, New York, 2011)
25. F. Cipolla-Ficarra, E. Nicol, M. Ficarra, Communicability era: new professionals for interactive systems, in *Quality and Communicability for Interactive Hypermedia Systems: Concepts and Practices for Design*, ed. by F. Cipolla-Ficarra (IGI Global, Hershey, 2010), pp. 204–227

26. A. Stavrou, J. Voas, General and agentic AI, and the challenges of Xplainable reliability. IEEE Comput. **58**(6), 69–72 (2025)
27. A. Tiwari, Beyond automation: the emergence of agentic urban AI. Automation **6**(3), 29 (2025)
28. F. Cipolla-Ficarra, MEAU: a method for the evaluation of the artificial un-intelligence, in *Handbook on Artificial Intelligence-Empowered Applied Software Engineering*, vol. 2, ed. by M. Virvou, G.A. Tsihrintzis, N.G. Bourbakis, L.C. Jain (Springer, Heidelberg, 2022), pp.251–301
29. W. Nöth, *Handbook of Semiotics* (Indiana University Press, Bloomington, 1995)
30. F. Cipolla-Ficarra, A method that improves the design of hypermedia: semiotics, in *Proc. of the International Workshop on Hypermedia Design*, ed. by S. Fraïssé et al. (Springer, Heidelberg, 1995), pp. 226–227
31. F. Cipolla-Ficarra, Evaluation and communication techniques in multimedia product design for on the net university education, in *Multimedia '96*. ed. by B. Urban (Springer, Vienna, 1996), pp.151–165
32. F. Cipolla-Ficarra, M. Cipolla-Ficarra, Interactive systems, design and heuristic evaluation: the importance of the diachronic vision, in *Proc. New Directions in Intelligent Interactive Multimedia* (Springer, Heidelberg, 2008), pp. 625–634
33. F. Cipolla-Ficarra, *Communicability Design and Evaluation in Cultural and Ecological Multimedia Systems. ACM Multimedia 2008* (ACM Press, New York, 2008), pp. 1–8
34. F. Cipolla-Ficarra, MECEM: metrics for the communications evaluation in multimedia, in *Proc. of 6th International Conference on Information Systems Analysis and Synthesis (ISAS 2000)* (International Institute of Informatics and Systemics, IIIS, Orlando), pp. 23–31
35. F. Cipolla-Ficarra, A set of rules and strategies for UNSAM virtual campus, in *Proc. International Conference on Human-Computer Interaction (HCI '09)* (Springer), pp. 101–110
36. F. Cipolla-Ficarra, Software and innovation: detecting invisible high-quality factors, in *Handbook of Research on Software Quality Innovation in Interactive Systems*, ed. by F. Cipolla-Ficarra (IGI Global, Hershey, 2021), pp. 1–40
37. F. Cipolla-Ficarra, Challenges and limits of the scientific education and new media, in *Assessment Methods and Success Factors for Digital Education and New Media*, ed. by F. Cipolla-Ficarra (IGI Global, Hershey, 2022), pp. 157–180
38. F. Cipolla-Ficarra, Machine learning and human unlearning, in *Extended Selected Papers of the 14th International Conference on Information, Intelligence, Systems, and Applications. IISA 2023*, ed. by N. Bourbakis, G.A. Tsihrintzis, M. Virvou, L.C. Jain, Lecture Notes in Networks and Systems, vol. 1093 (Springer, Heidelberg, 2024), pp. 72–118
39. F. Cipolla-Ficarra, The "G" factor in the web, new technologies, and education, in *Handbook of Research on Software Quality Innovation in Interactive Systems*, ed. by Cipolla-Ficarra (IGI Global, Hershey, 2021), pp. 437–463
40. S. Ahmed, A. Hassanien, M. Elhoseny, *Security in Smart Cities: Models, Applications, and Challenges* (Springer, 2018)
41. D. Rawat, K. Ghafoor, *Smart Cities Cybersecurity and Privacy* (Elsevier, Amsterdam, 2018)
42. F. Cipolla-Ficarra, V. Ficarra, Motivation for next generation of users versus Parochialismin software engineering, in *Proc. International Conference on Advances in New Technologies, Interactive Interfaces, and Communicability. ADNTIIC 2011*, ed. by Cipolla-Ficarra et al. (Springer, Heidelberg, 2011), pp. 124–133
43. F. Cipolla-Ficarra, Revaluing the strategies and narrative techniques of Pinocchio in the social media network, educational and job context, in *Assessment Methods and Success Factors for Digital Education and New Media*, ed. by Cipolla-Ficarra (Hershey, IGI Global, 2022), pp. 384–406
44. J. Nielsen, *Usability Engineering* (Academic Press, London, 1993)
45. F. Cipolla-Ficarra, M. Cipolla-Ficarra, Universability and communicability in computer animation, in *Advanced Research and Trends in New Technologies, Software, Human-Computer Interaction, and Communicability*, ed. by Cipolla-Ficarra (IGI Global, Hershey, 2014), pp. 131–142

46. F. Cipolla-Ficarra et al., *Handbook in Advances in Dynamic and Static Media for Interactive Systems: Communicability, Computer Science and Design* (Blue Herons, Bergamo, 2011)
47. M. Bunge, *The Science: Your Method and Your Philosophy* (Siglo, Buenos Aires, 1981), p. XXI
48. F. Cipolla-Ficarra, The resolution of the problem of objectivity in a method of evaluation for interactive applications. ACM SIGWEB Newslett. **5**(2), 8 (1996)
49. T. Fernandes, *Global Interface Design: A Guide to Designing International User Interfaces* (Academic Press, London, 1995)
50. P. Nicopolitidis, M. Obaidat, *Smart Cities and Homes: Key Enabling Technologies* (Morgan Kaufmann, Burlington, 2016)
51. C. Stimmel, *Building Smart Cities: Analytics, Ict, and Design Thinking* (CRC Press, Boca Leon, 2015)
52. M. Maheswaran, E. Badidi, *Handbook of Smart Cities: Software Services and Cyber Infrastructure* (Springer, 2019)
53. F. Cipolla-Ficarra, Evaluation heuristic of the richness, in *Proc. International Conference on Information Systems Analysis and Synthesis, ISAS* (Orlando, 1999), pp. 23–30
54. R. Pressman, B. Maxim, *Software Engineering: A Practitioner's Approach* (McGraw-Hill Education, New York, 2015)
55. U. Eco, *Theory of Semiotics* (Indiana University Press, Bloomington, 1979)
56. H. Jahankhani, S. Dadwal, G. Bowen, *Technology and Talent Strategies for Sustainable Smart Cities: Digital Futures* (Emerald Publishing Limited, Leeds, 2023)
57. M. Lytras, A. Housawi, B. Alsaywid, *Smart Cities and Digital Transformation: Empowering Communities, Limitless Innovation, Sustainable Development and the Next Generation* (Emerald Publishing Limited, Leeds, 2023)
58. A. Tsokas, P. Pardalos, S. Rassia, *Artificial Intelligence, Machine Learning, and Optimization Tools for Smart Cities: Designing for Sustainability* (Springer, 2023)
59. Z. Allam, Z. Allam, *The Metaverse and Smart Cities: Urban Environments in the Age of Digital Connectivity* (Elsevier, Amsterdam, 2024)
60. V. Tecim, S. Kahyaoglu, *Artificial Intelligence Perspective for Smart Cities* (CRC Press, Boca Leon, 2022)
61. H. Wang, H. Dong, A. El-Saddik, Immersive multimedia communication: state-of-the-art on extended reality streaming. ACM Trans. Multimed. Comput. Commun. Appl. **21**(7), 1–33 (2025)
62. T. Vinod-Kumar, *E-Democracy for Smart Cities* (Springer, Heidelberg, 2018)
63. F. Cipolla-Ficarra, *Persuasion On-Line and Communicability: The Destruction of Credibility in the Virtual Community and Cognitive Models* (Nova Science Publishers, New York, 2010)
64. A. Serrano-Gómez, Corrupción en la Universidad. La ley y sus efectos negativos (Madrid, Dykinson, 2015)
65. M. Hirigoyen, *Les Narcisse: Ils ont pris le pouvoir* (La Découverte, Paris, 2019)
66. F. Cipolla-Ficarra et al., *Handbook of Computational Informatics, Social Factors and New Information Technologies: Hypermedia Perspectives and Avant-Garde Experiences in the Era of Communicability Expansion* (Blue Herons Bergamo, 2011)
67. F. Cipolla-Ficarra, E. Nicol, M. Cipolla-Ficarra, *Vademecum for Innovation through Knowledge Transfer: Continuous Training in Universities, Enterprises and Industries. Innovation through Knowledge Transfer* (Springer, Heidelberg, 2010), pp.139–149
68. F. Cipolla-Ficarra et al., *Handbook of Advantages and Recent Developments in Human-Computer Interaction and Quantum Computing Applied to Nature, Society and Education* (Blue Herons, Bergamo, 2021)
69. F. Cipolla-Ficarra et al., *Handbook of Interaction Techniques and Technologies Applicable to Learning and Teaching: Changing Relations between New Media, Users, Contents and Evaluation of Interactive Systems* (Bergamo, Blue Herons, 2021)
70. F. Cipolla-Ficarra, V. Ficarra, M. Cipolla-Ficarra, New technologies of the information and communication, in *Proc. International Conference on Advances in New Technologies, Interactive Interfaces, and Communicability. ADNTIIC*, ed. by Cipolla-Ficarra et al. (Springer, Heidelberg, 2011), pp. 71–84

Part V
Digital Public Services

Chapter 11
A Framework for Autonomous Crowd Management Through Reinforcement Learning and Digital Twins

Chairi Kiourt, Zoi Lygizou, Anestis Koutsoudis, and Dimitris Kalles

Abstract Effective crowd management in dynamic urban environments is critical for public safety, efficient mobility, and enhanced citizen experiences. Traditional methods, such as static planning and manual interventions, often fail to adapt to the real-time and unpredictable nature of human behavior during large-scale events or emergencies. This chapter proposes an innovative framework for autonomous crowd management that integrates Reinforcement Learning, Digital Twins, Multi-Agent Systems, and Intelligent Internet of Things devices. Our approach leverages RL for adaptive decision-making, MAS for decentralized collaboration, and DTs to create virtual replicas of urban environments for real-time monitoring and simulation. The integration of multimodal IoT data streams enhances situational awareness and enables dynamic system responses to evolving crowd conditions. By prioritizing human-in-the-loop mechanisms, the proposed framework ensures transparency, ethical oversight, and trust in automated decisions. This work demonstrates a dynamic, scalable, and proactive crowd management system that not only addresses immediate operational challenges but also provides a roadmap for future research in smart urban environments.

Keywords Human behaviors · Reinforcement learning · Multi-agent systems · Digital twin · Crowd management

C. Kiourt (✉) · A. Koutsoudis
"Athena" Research Centre, University Campus at Kimmeria, Xanthi, Greece
e-mail: chairiq@athenarc.gr

A. Koutsoudis
e-mail: akoutsou@athenarc.gr

Z. Lygizou · D. Kalles
School of Science and Technology, Hellenic Open University, Patras, Greece
e-mail: zoi.lygizou@ac.eap.gr

D. Kalles
e-mail: kalles@eap.gr

G. A. Tsihrintzis et al. (eds.), *Human-Smart City Interactions and User-Citizen Experiences*, Learning and Analytics in Intelligent Systems 58,
https://doi.org/10.1007/978-3-032-06364-9_11

11.1 Introduction

With the growing complexity of urban environments, managing crowds has become a critical challenge in ensuring public safety, facilitating efficient mobility, and enhancing overall citizen experiences [1]. Whether in densely populated areas, large-scale events, or during emergency situations, the demand for effective crowd control has never been more critical [2].

Traditional methods [3, 4], such as static planning or human interventions, often struggle to cope with the dynamic, real-time nature of crowd behaviour [5] and have several limitations in managing crowds effectively [6]. For example, human operators rely on manual monitoring and decision-making, which are susceptible to delays and errors, especially in fast-evolving situations. Usually, they are based on limited, biased and oversimplified what-if scenarios, since predicting all possible situations is impossible and it may be difficult to consider hitherto unseen situations. Traditional methods struggle to manage large, complex crowds effectively, as human capacity to monitor and control vast areas simultaneously is limited. Human operated surveillance cameras are not ubiquitous, data is not collected in real-time and communication is usually done verbally.

The rise of smart city technologies—such as simulation models [7], sensing technologies [8], digital twins [9] and reinforcement learning [10] methods—offers new opportunities to address such challenges. Yet, finding a balance between public safety, privacy, and resource efficiency in crowd management continues to be a challenge, which is exacerbated by the need to respond in real-time and adapt to rapidly changing conditions without requiring manual intervention.

In the Crowd Management (CM) framework, Machine learning (ML) has emerged as a transformative approach in addressing complex challenges, with Reinforcement Learning (RL) [11] playing a pivotal role in adaptive decision-making and real-time optimization. By leveraging large datasets, ML techniques can model, predict, and optimize crowd (agent) behavior [2, 12] in dynamic urban settings. Supervised Learning (SL) methods are used to analyze historical crowd movement patterns and identify potential risk zones, while Unsupervised Learning (UL) facilitates clustering and anomaly detection, such as overcrowding or panic situations [13]. RL, a subset of ML, goes a step further by training agents to make adaptive, real-time decisions in non-stationary environments. These agents mainly simulate human behaviors [2] by learning optimal strategies through interaction with their environment, focusing on objectives like minimizing congestion, maximizing evacuation efficiency, or ensuring public safety. The integration of ML and RL with real-time data streams from IoT devices and sensors has further enhanced the ability to provide intelligent, data-driven solutions for crowd management [14]. Together, all these methods enable the creation of systems that are not only predictive but also proactive in addressing urban challenges.

Multi-agent systems (MAS) offer a decentralized and adaptive approach to managing and simulating complex crowd dynamics by enabling collaboration and coordination among multiple autonomous agents [15]. These systems consist of individual agents, each capable of perceiving its environment, making decisions, and interacting with other agents to achieve shared objectives. In the context of crowd management, MAS can simulate and control large-scale behaviors, such as guiding people towards evacuation routes [2] and optimizing pedestrian flow [16]. By employing techniques like Multi-Agent Reinforcement Learning (MARL), these systems learn to address challenges collaboratively [17], trying to enhance safety. The flexibility and scalability of MAS make them an indispensable tool for managing crowds in urban environments, where real-time adaptability (non-stationarity) is crucial for success.

By integrating ML, MAS, IoT, and other advanced approaches, Digital Twin (DT) technologies [18] have transformed crowd management by offering real-time guidance through the utilization of live data, virtual replicas, and predictive simulation models. These technologies empower city planners and authorities to monitor, predict, and manage crowd dynamics in real-time, enhancing decision-making, improving public safety, and optimizing urban mobility [18]. By integrating data from IoT sensors, cameras, and other sources, digital twins provide a dynamic visualization of crowd behavior, helping to predict and mitigate issues such as overcrowding, congestion, or safety hazards during large-scale events or emergencies. These virtual environments allow for testing various scenarios, such as evacuation strategies or public space utilization, without disrupting real-world operations. The ability of digital twins to adapt to real-time data inputs ensures that crowd management strategies remain responsive and proactive, enhancing both public safety and operational efficiency.

The motivation of this book chapter is to address the significant limitations of traditional crowd management approaches, which struggle to cope with the growing complexity of modern urban environments. These traditional methods, characterized by limited situational awareness and the inability to perform real-time monitoring and communication, are often inadequate for managing large-scale, dynamic, and non-linear crowd behaviors. In emergency scenarios or densely populated urban spaces, the lack of real-time data and adaptive decision-making leads to delays, inefficiencies, and increased risks to public safety. Moreover, traditional systems often rely on manual intervention, pre-defined scenarios, and static models that fail to account for the unpredictability and fluidity of human behavior. This book chapter is driven by the need to overcome these challenges through the integration of cutting-edge technologies, including reinforcement learning, digital twins, multi-agent systems, and intelligent IoT devices. By leveraging these innovations, the proposed framework aims to enhance real-time adaptability, improve operational efficiency, and ensure safer, smarter, and more responsive crowd management strategies.

The main contribution of this book chapter lies in the development of an innovative and comprehensive framework for autonomous crowd management, which integrates advanced technologies such as reinforcement learning (RL), digital twins (DT), multi-agent systems (MAS), and intelligent IoT devices. Unlike traditional methods, the proposed framework offers a dynamic, real-time, and scalable solution capable of addressing the challenges posed by complex, non-linear, and unpredictable crowd behaviors. This work emphasizes the synergy between RL and DT to simulate and manage intricate human dynamics, leveraging MAS for collaborative and competitive decision-making. Additionally, it introduces the integration of multimodal IoT data streams for enhanced situational awareness and real-time adaptability. By prioritizing human-in-the-loop decision-making mechanisms, the framework ensures ethical oversight, transparency, and trust. Ultimately, this book chapter not only bridges gaps in existing crowd management methodologies but also provides a roadmap for future research and implementation in dynamic urban environments.

The remainder of the book chapter is organized as follows: Sect. 11.2 highlights the key challenges in urban crowd management, emphasizing the limitations of traditional methods and the need for innovative approaches. Section 11.3 provides an in-depth overview of reinforcement learning for crowd simulation, detailing its applications in modeling human behaviors, multi-agent systems, and trust dynamics. Section 11.4 explores the integration of digital twins in urban environments, focusing on their architecture, 3D digitization, semantic segmentation, and the role of multimodal intelligent IoT nodes in real-time monitoring and management. Section 11.5 proposes a general framework for autonomous crowd management, combining RL, MAS, IoT, and digital twin technologies, along with a discussion of its scalability and adaptability. Finally, Sect. 11.6 concludes the chapter, summarizing the findings and discussing future research directions to advance crowd management in dynamic urban settings.

11.2 Challenges in Urban Crowd Management

Crowd management involves a set of measures, methods, technic and technologies designed to facilitate people's movement and enjoyment [1], and it can be defined as the systematic planning and supervision of the organized movement and assembly of people with the goal of avoiding the rapid movement of groups and critical crowd densities that could lead to chaotic states and panic outbreaks [3]. Crowd management is a difficult task with many challenges:

- **Unpredictable human behavior**: [19] Crowd behavior is highly dynamic and difficult to predict, particularly during large-scale events or emergencies. Unlike regulated systems like traffic (vechicles and driving regulations), pedestrian movements are fluid and can change without notice, making it challenging to maintain order and safety.
- **Anonymity and behavioral risks**: Crowds foster anonymity, a sense of belonging, and perceived freedom, which can reduce individual responsibility, leading to a loss rationality and respect for rules. This often results in behaviors such as pushing, vandalism, or other forms of risky and antisocial actions, especially under heightened emotions or extreme enthusiasm.

- **Trade-offs between safety, privacy, and resources**: Striking a balance between ensuring public safety, respecting individual privacy, and efficiently utilizing available resources remains a critical challenge. Implementing surveillance and monitoring systems often raises concerns about privacy and ethical considerations.
- **Complexity of large crowds**: Managing vast, heterogeneous crowds across modern large urban areas or during significant events poses logistical and operational difficulties. Human operators struggle to monitor and coordinate effectively at scale.
- **Need for real-time monitoring systems**: The increasing complexity of modern urban environments necessitates real-time, intelligent, and adaptive systems to respond dynamically to changing crowd conditions (non-stationarity) [20]. Traditional systems, reliant on human intervention, are often too slow and error-prone for such tasks.
- **Limitations of traditional methods**: Traditional crowd management techniques rely on pre-defined scenarios, limited situational awareness, and manual communication methods, which are insufficient for addressing real-time and large-scale challenges.
- **Personalised support**: Crowds are composed of individuals with varying needs, preferences, and behaviors. Catering to these differences in real time, especially in large-scale events or emergencies, is highly complex.
- **Data integration and real-time processing/monitoring**: Managing and integrating diverse data streams from IoT devices, sensors, and other sources into a unified system such as the digital twin remains a significant challenge. Real-time processing of this data to model dynamic crowd behavior and provide actionable insights requires robust computational infrastructure and advanced algorithms.
- **Multimodal data for effective crowd management**: Effectively managing crowds requires integrating multimodal data, such as location parameters, weather conditions, and environmental factors like hunger, thirst, and road blockages. Combining these diverse data types into a cohesive framework presents challenges in data collection, synchronization, and real-time analysis, especially in dynamic urban environments.
- **Scalability and adaptability**: Designing systems that can scale to accommodate large, diverse crowds across different urban settings is complex. Additionally, these systems need to adapt dynamically to changing conditions, such as emergencies or unexpected crowd behaviors, while maintaining efficiency.
- **Accuracy versus efficiency trade-off**: Achieving a balance between high accuracy in crowd simulations and the efficiency of algorithms is a persistent challenge. Computationally intensive methods like reinforcement learning and digital twin simulations often struggle to meet the demands of real-time applications, especially for large-scale urban environments.

In this work, we propose a comprehensive framework that addresses these challenges through innovative approaches, leveraging advancements in digital twins, reinforcement learning, and multi-modal data integration. By tackling issues such as scalability, real-time monitoring, and behavioural unpredictability, the framework aims to enhance the effectiveness and adaptability of crowd management systems in dynamic urban environments.

11.3 Overview of Reinforcement Learning for Crowd Simulation

Reinforcement Learning (RL) [11] is a machine learning mechanism where an agent learns to make decisions by interacting with its environment (S_t). The agent receives feedback (R_t) in the form of rewards or penalties based on its actions (A_t) and uses this feedback to improve its decision-making policy over time. This abstracted architecture is depicted in Fig. 11.1. Unlike supervised learning, RL does not require labelled data; instead, it focuses on learning optimal strategies through trial and error.

RL applications can generally be categorized into tabular approaches and approximation-based approaches (commonly related to as Deep Reinforcement Learning, or DRL). Tabular methods rely on maintaining explicit tables of state-action values, which makes them computationally efficient for simple and discrete environments. However, these methods become impractical as the complexity and scale of the environment increase. In contrast, DRL leverages neural networks to approximate state-action values, enabling RL to handle high-dimensional and continuous spaces. This flexibility allows DRL to be effectively applied to complex real-world problems, such as crowd management, where the environment is dynamic, multi-agent, and non-linear.

In dynamic environments like crowd management, RL is particularly valuable due to its adaptability and real-time decision-making capabilities [21]. Crowds exhibit complex, non-linear behaviour influenced by factors such as movement patterns, environmental constraints, and individual behaviours and preferences [2]. RL agents can navigate such complexities by continuously updating their strategies in response to changes in the environment. For example, RL can be used to optimize evacuation routes in emergencies by minimizing congestion and ensuring the safety of individuals. Additionally, RL agents can predict and mitigate potential risks, such

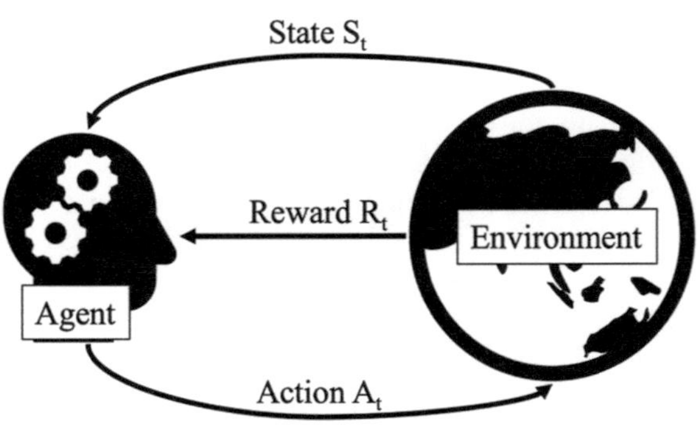

Fig. 11.1 Reinforcement learning agent's main architecture

as overcrowding or bottlenecks, making them an essential tool for effective crowd management in unpredictable and dynamic scenarios.

11.3.1 Human-Level Behaviours Modelling

Reinforcement Learning models are increasingly utilized to simulate real-time human behaviours, such as predicting actions and dynamics. In crowd management, they are particularly effective in modelling and optimizing crowd flows and behaviours. In the realm of human locomotion, Deep RL (DRL) has been utilized to model neuromechanical control, allowing for the generation of human-like movements without direct reliance on experimental motion data [22]. This approach enhances the predictive capabilities of simulations in novel situations, which is beneficial for applications such as assistive device development and rehabilitation treatments. In robotics, RL has been applied to enable autonomous navigation through dense crowds [23]. By integrating trajectory prediction algorithms with RL-based collision avoidance methods, robots can anticipate pedestrian movements and adjust their paths accordingly, resulting in smoother and safer navigation.

For crowd evacuation scenarios, RL-based models have been developed to simulate agent behaviours in three-dimensional environments [24]. By incorporating human-like perception and decision-making processes, these models effectively reproduce navigation and obstacle avoidance behaviours observed in real-world crowds. Furthermore, RL has been employed to model environment-sensitive crowd behaviours under the guidance of situational information sharing [25]. This method improves the efficiency of crowd evacuation simulations by enabling agents to adapt their behaviours based on shared experiences and environmental cues. Additionally, RL-based frameworks have been developed to model crowd-oriented behaviours by embedding emotion models, allowing agents to adapt their actions based on psychological factors [26]. These applications demonstrate RL's capability to model complex human behaviours and adapt to dynamic environments. The process of shaping RL agents' behaviour involves adaptive feedbacks to guide more effectively the learning, thereby enhancing the agents' ability to emulate complex human behaviours. For instance, the study of Sidiropoulos et al. [27] discusses methods for incorporating dynamic rewarding algorithm into the RL process, enabling agents to adapt more effectively to dynamic environments. By leveraging such techniques, RL agents can better predict and manage crowd behaviours, leading to more efficient and responsive crowd management strategies.

Despite RL's advancements, challenges persist in modelling complex human dynamics, achieving scalability for large crowds, incorporating psychological factors, and ensuring real-time adaptability in rapidly changing environments. Overall, the application of RL in simulating real-time human behaviours and managing crowd flows offers a dynamic and flexible approach to understanding and controlling complex crowd dynamics. By learning from interactions within the environment, RL models can predict and influence crowd behaviours, contributing to more effective crowd management strategies.

11.3.2 Collaboration and Competition in Multi-agents Systems

On one hand, single-agent RL allows agents to make decisions based on observations, but interactions with other agents can turn the environment into a non-stationary problem. This dynamic adaptation makes it challenging to achieve a global optimum. Multiagent Reinforcement Learning (MARL) overcomes the limitations of single-agent RL taking into consideration the existence of other entities modeled as agents, such as human, autonomous vesicles etc.. MARL is advantageously placed to cope with the complex issues that arise in dynamic environments because it allows agents to interact and coordinate their actions to achieve a common goal, which ultimately results in enhanced system performance and efficiency [28].

In a Multi-Agent System, agents can either *collaborate*, *compete*, or a combination of both (*mixed*), depending on the environment and objectives [29]. In a *collaborative behaviors* [30], all agents work together towards a common goal, requiring coordination and communication to complete tasks efficiently. On the other hand, in a *competitive behaviors*, agents pursue conflicting goals, often maximizing their own rewards at the expense of others, with little or no communication. A *mixed behaviors* [31] introduces a hybrid approach, where some agents cooperate while others compete, leading to dynamic relationships that can shift between collaboration and competition based on the context. The complexity of interactions within MAS makes them highly effective for addressing real-world challenges that involve both cooperation and competition. A prime example is crowd management [2], where individuals often need to collaborate to ensure safety and efficiency. However, in certain high-stress or survival situations, competition may arise as people act in their own self-interest. MAS through MARL can simulate and optimize these dynamic behaviors, trying balancing collaborative and competitive actions to better understand and manage crowd dynamics in complex environments.

Figure 11.2 presents an indicative structure of a Multi-Agent Reinforcement Learning (MARL) system, where multiple agents interact simultaneously with the environment. Each agent operates independently, receiving its own observations, rewards, and selecting its own actions, while also considering the potential impact of other agents' actions. This parallel interaction enables agents to learn from both their individual experiences and the broader system dynamics, fostering a *cooperative and/or competitive* behaviours. As a result, the agents can adapt to complex scenarios more effectively, improving the overall system's performance and adaptability. The structure also highlights how each agent contributes to shaping the environment, potentially leading to a more dynamic and responsive learning process compared

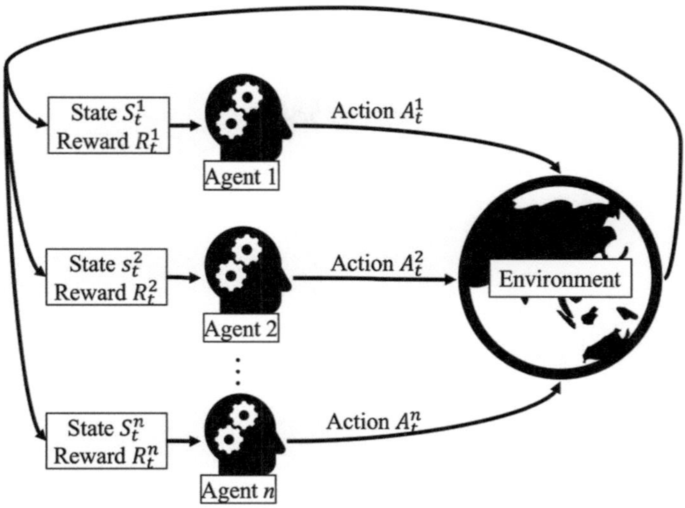

Fig. 11.2 Multiagent reinforcement learning main architecture

to single-agent systems. This coordination of actions is essential for solving problems in environments that are too intricate for isolated decision-making, making MARL an essential tool in areas like autonomous vehicles, smart grids, and disaster management.

Figure 11.3 depicts an example of mixed behaviors in a MAS within a crisis scenario. In this case, multiple autonomous agents (e.g., humans) exhibit both collaborative and independent behaviors, with some cooperating while others may compete. The orange agents, connected by green arrows, represent a collaborative relationship, such as that between a parent and child. However, these agents may compete against other individuals (connected by red arrow) for survival in an emergency situation, such as a fire, where access to an exit is limited to only a few people. This dynamic interplay of cooperation and competition highlights the complexity of human behavior in high-stress environments.

Recent advancements in MARL have enhanced pedestrian simulation and crowd management. In 2014, Martinez-Gil et al. introduced MARL-Ped, a model-free RL framework where agents used environmental data for navigation, solving problems like congestion and pathfinding, though with limited adaptability for dynamic behaviors [32]. Later, they improved the framework with Q-learning and iterative learning strategies, making it more scalable [33]. In 2018, Lee et al. applied deep RL for agent navigation using depth maps, though agents occasionally failed to avoid obstacles [34]. Hildreth and Guy [35] introduced a cooperative MAS algorithm using decentralized communication for coordinated navigation, integrating collision avoidance and optimization techniques. The authors in Wang et al. [36] propose a path planning method based on improved MARL, where agents in an evacuation scenario form groups and select a leader, typically the agent closest to the exit and most familiar

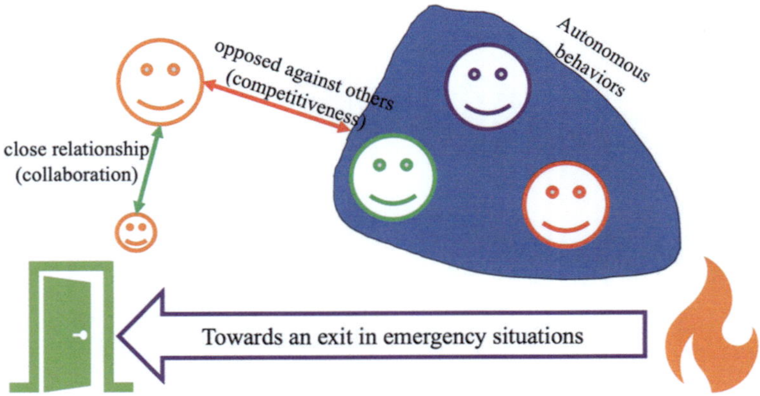

Fig. 11.3 Mixed behaviors in MAS (crisis use-case)

with the environment. The leader uses the MARL algorithm to choose the best evacuation path, guiding the group to safety, while a bulletin board stores and shares the empirical knowledge gained during the learning process. In [2] the authors employed Proximal Policy Optimization (PPO) and Soft Actor-Critic (SAC) algorithms to train agents to behave like humans in fire crisis simulations, yielding promising results. They utilized realistic perception mechanisms, such as "3D Ray Perception Sensors", which closely mimic human perception processes. Another important research direction of the last years is the Mobile Crowd Sensing (MCS) which has emerged as a powerful tool for performing complex sensing and processing tasks, leveraging the capabilities of smart mobile devices. To address the uncertainties and dynamic environments faced by participants, the Intelligent Crowd algorithm [37] uses MARL to optimize real-time decision-making, allowing participants to select the best sensing options and earn rewards based on their contribution to the overall crowdsourced data. These and many other research directions showcase the growing potential of MARL in modeling complex crowd behaviors.

MARL faces key challenges in modeling dynamic behaviors where agents must balance cooperation and competition. A major challenge is creating strong relationship behaviors within MAS, such as those seen in crowd management, where agents collaborate for safety but may compete in high-stress situations. A critical challenge in MARL is the development of algorithms, such as MA-POCA [38], that can model these shifting behaviors effectively, balancing cooperation and competition. Additionally, ensuring scalability and real-time adaptability in large-scale applications, such as disaster management and autonomous vehicles, requires efficient coordination and learning among agents in complex settings.

11.3.3 Trust in Multi-agent Systems

MAS research shows that agents must voluntarily enter relationships to collaborate effectively; otherwise, efforts may fail. Trust is crucial for promoting collaboration and is defined from various perspectives [39]. Trust between agents in MAS is similar to human trust. In the domain of urban crowd management, trust in MAS is paramount for ensuring effective and reliable operations. This section delves into two critical aspects of trust: (a) the trust in agents that simulate diverse human behaviors and the (b) trust in agents related to the Internet of Things (IoT). By exploring these dimensions, we aim to highlight the importance of accurate behavior simulation, robust data integrity, and secure communication in building and maintaining trust within MAS. Understanding these factors is essential for developing systems that can adapt to dynamic environments, make informed decisions, and ultimately enhance public safety and efficiency in urban settings.

11.3.3.1 Trust in Agents Simulating Different Human Behaviors

Trust in MAS is crucial, especially when agents are designed to simulate human behaviors [40]. These agents must accurately reflect the diverse and dynamic nature of human actions to ensure effective crowd management. Trust is built on several key factors, including, behavioral accuracy, behavioural relationships, negotiation, coommunication and adaptability. Trust in agents is established through their ability to accurately simulate a wide range of human behaviors, from everyday actions to responses in emergencies. This accuracy is essential for predicting and managing crowd dynamics effectively [41, 42].

Agents must adapt to changing environments and scenarios, learning from interactions with other agents and continuously improving their behavior models to remain relevant and effective, just as in real-life situations. Current research focuses on autonomously establishing such behaviors through MARL algorithms, like MA-POCA [38]. MA-POCA enhances coordination and cooperation among agents by allowing them to learn from both their own and others' experiences, developing trust and collaborative strategies crucial for managing complex environments. In crowd management, agents can predict and respond to others' movements, ensuring more cohesive and efficient management. Another perspective on trust is introduced by Tweedale and Cutler [39], discussing the Trust, Negotiation, Communication (TNC) model. This model enhances MAS by fostering trust between agents, helping them dynamically organize within teams or distributed systems. The TNC model equips agents with mechanisms to build formal trust networks through cooperation or collaborative associations, emphasizing the importance of trust for effective communication and teamwork in MAS.

Incorporating trust mechanisms into MAS is essential. Trust can be modeled similarly to human relationships, such as the trust between a parent (*agent*) and child (*agent*) (Fig. 11.3) in an emergency situation. This relationship is characterized by high levels of reliability, predictability, and mutual support. Moreover, integrating social dimensions into agent interactions can significantly enhance trust. Agents can be programmed to prioritize interactions with familiar or 'trusted' agents, similar to how humans rely on known relationships in stressful situations. This approach not only improves the reliability of the system but also enhances its resilience by fostering a network of supportive and cooperative agents. By simulating such relationships, agents can better manage critical situations, ensuring that vulnerable individuals receive the necessary attention and protection.

To further build trust, agents must be transparent in their decision-making processes [43]. Clear documentation and understanding of how agents are programmed to behave in various situations help build trust among users and stakeholders. Regular validation and testing against real-world scenarios are necessary to ensure that agents' behaviors remain accurate and reliable. This process helps identify and correct any discrepancies between simulated and actual behaviors. Conducting verification and validation of multi-agent systems is crucial for maintaining transparency and reliability [44].

11.3.3.2 Trust in Agents Related to IoT

In this section, we treat IoT nodes (Sect. 11.4.3) as agents because they operate autonomously, possessing their own abilities and mechanisms for decision-making, collaboration, and communication with other nodes. These IoT agents can independently gather data, process information, and execute actions based on their programmed objectives and environmental interactions. By functioning as autonomous entities, IoT nodes can dynamically adapt to changing conditions, optimize their performance, and coordinate with other agents (other IoT nodes) to achieve collective goals. This autonomous behavior is crucial for managing complex systems, such as smart cities, where real-time data processing and adaptive responses are essential for efficient operation and enhanced user experiences. Additionally, treating IoT nodes as agents allows for the implementation of advanced algorithms, such as MARL, to further improve their decision-making capabilities and collaborative efforts.

A critical aspect of this approach is agent trust. Trust between IoT agents is essential for ensuring reliable and secure interactions. Each agent must be able to trust that the data it receives from other agents is accurate and that the actions of other agents are aligned with the collective goals. This trust is built through mechanisms such as direct and indirect trust assessments, reputation systems, and continuous validation of agent behaviors. By incorporating trust mechanisms, IoT agents can more effectively collaborate, share information, and make decisions that enhance the overall performance and reliability of the system. This trust-based approach not only improves the efficiency of IoT networks but also ensures that they can adapt to and manage complex, dynamic environments more effectively.

Traditional methods like cryptographic and access control are insufficient for handling internal attacks in open, dynamic, and heterogeneous MAS. Trust and

Reputation Models (TRM) have been proposed to address this by assessing entities' trustworthiness through data collection [45]. TRMs integrate Direct Trust (DT) from an agent's interactions and Indirect Trust (IT) from third-party recommendations. However, DT can be unreliable due to feedback sparseness and malicious attacks, and combining DT and IT is complex. In Samuel et al. [46] each agent has a Balanced Trust Incentive (BTI) unit that is used to interact with other agents. BTI is increased if the agent's credibility assessment is positive, which indicates that the agent is honest. Conversely, BTI will be reduced if credibility assessment is negative, which shows that the agent is not honest. TRMs face limitations such as the inability to share trust information across applications, leading to repeated reputation building. A proposed framework facilitates the exchange of trust and reputation information [45]. Misbehaving devices can attack IoT devices for personal profit, but filtering recommendations through social relationships can mitigate this [47]. The Social Internet of Things (SIoT) allows devices to form social relationships, enhancing trust and collaboration [47, 48]. TRMs can be centralized, decentralized, or blockchain-based. Centralized models are prone to single points of failure and manipulation, while decentralized models face data sharing and privacy issues. Blockchain-based models offer security benefits but require high resource consumption and face scalability challenges [46, 49, 50]. Edge/fog computing enhances blockchain use in IoT by providing more storage and computation resources, supporting scalability, and offering real-time trust services [51]. This approach addresses data trading security and trustworthiness in contexts like the Internet of Vehicles (IoV) [51]. In Guleng et al. [52] the authors present a multi-agent trust management scheme for decentralized vehicular networks. A Q-learning approach is employed to estimate the indirect trust of nodes that are not directly connected to a trustor node, in which estimation is conducted by averaging the evaluation reports from multiple nodes.

In conclusion, the primary challenges in establishing trust among IoT agents revolve around ensuring reliable and secure interactions in dynamic and heterogeneous environments and entities. Traditional methods like cryptographic and access control are insufficient for internal threats, necessitating advanced TRMs. These models must balance direct and indirect trust assessments, manage feedback sparsity, and mitigate malicious attacks. Additionally, the integration of social relationships and blockchain technology, while promising, introduces complexities related to resource consumption and scalability. Addressing these challenges is crucial for enhancing the efficiency, adaptability, and security of IoT networks.

11.4 Digital Twins for Urban Environments

A digital or cyber twin is a virtual representation of a real-world entity, system, process, or other abstraction, which synchronizes with its physical counterpart [53]. Digital twins are created by integrating real-time data from the environment to represent, validate, and simulate the physical entities' present and future behaviour [54]. As the virtual asset behaves similarly to the physical one, it can be used to simulate actions and consequences, as well as forecasting of the physical entity evolution [55]. Around 2002, Grieves presented a project for a Product Lifecycle Management Centre, in which the basic principles of the Digital Twin (DT) were present [56]. In 2015 Grieves [57] defined the DT as a "*virtual representation of what has been produced*" and indicates that comparing the DT "*to its engineering design (allows) to better understand what was produced versus what was designed, tightening the loop between design and execution*". The recent advances in technology and network infrastructures widened the horizons of the DT concept and the significant cost reductions to the IoT and sensing technologies, leading, increasingly, towards the consideration of a technological framework applicable to the civil protection and human management approaches [58]. Nowadays, most of the DT approaches focus on the management and monitoring of artificial entities, such as manufacturing/production-related approaches, etc. There are also some attempts trying to introduce the human in the DT procedure and are treated as equal/similar entities, without focusing in-depth on their specialized needs and capabilities [59]. Through AI, DTs are evolving into powerful, dynamic, and automated tools. AI appears as a strong candidate to handle all generated data, and to improve the automation and smartification process.

The DT architecture is composed of three main elements:

(a) *The physical world*: includes the physical objects or elements (in this case, the crowd) we want to manage, as well as sensors, actuators, and processing capabilities that collect and process real-world data,
(b) *The virtual world*: comprises the digital twin itself, along with associated databases, machine learning models, and data processing techniques for simulating, analysing, and predicting behaviours and
(c) *The connectivity layer*: serves as the bridge between the physical and virtual worlds, facilitating communication through a variety of protocols and interfaces to ensure seamless data/knowledge exchange and synchronization.

This architecture enables continuous monitoring, visualization, and analysis of the physical world in real time, enhancing decision-making and operational efficiency [54, 60]. In the context of urban spaces, digital twins model a city's dynamic environment by combining data from multiple sources like citizens, IoT data from smart devices and sensors, and urban mobility [54]. Digital twins enable city planners, engineers, and decision-makers to better visualize and comprehend urban dynamic by collecting and analysing real-time data on issues such as weather conditions, traffic patterns, energy consumption, and public safety. This integration improves resource management, infrastructure planning, and quality of life for the citizens.

Figure 11.4 depicts the main structure of the technological approach of a digital twin, having in the core the AI-digital twin platform. In the left-hand side of the graph, we can see a simulation system (*Virtual World*) encompassing simulation systems, machine learning (ML), multi-agent systems (MAS), and dynamic crowd simulation technologies. These components model and predict behaviour within simulated environments. On the right-hand side, the (*Physical World*) is depicted, powered by a Multimodal IoT Network that collects and processes real-world data in real time. This network integrates information from sensors, pedestrian behaviour, and

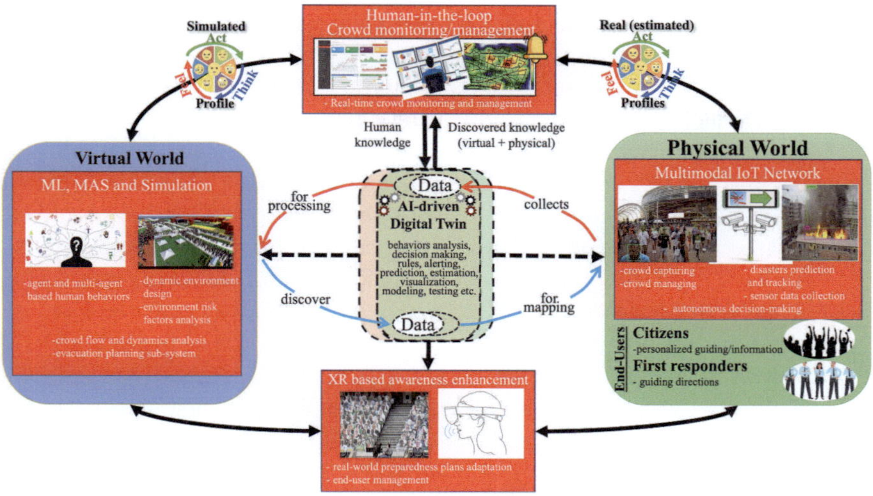

Fig. 11.4 AI-driven digital twin architecture

environmental conditions, enabling accurate and up-to-date insights into the physical environment. At the top of the figure, a Human-in-the-Loop Command and Control System is showcased, which leverages processed data from the Virtual and Physical Worlds along through *The Connectivity Layer* with the risk-assessment, management, and alerting mechanisms. This system facilitates real-time monitoring and decision-making, ensuring effective crowd management. Finally, the incorporation of XR-based technologies enhances situational awareness and improves responsiveness for first responders. This holistic approach ensures a seamless connection between simulated and real environments for robust crowd management and safety.

11.4.1 3D Digitisation of Urban Environments

Urban environments (UEs) are inherently dynamic, with elements such as shifting traffic patterns, on-going construction activities, seasonal vegetation changes, and human interactions constantly altering the landscape [61]. These changes present unique challenges during the digitisation process, as data collected at different times or from varying perspectives may no longer represent the current state of the environment. Today, UEs 3D digitisation is considered a common task, due to technological advancements, but still challenging [62]. Within the UE framework such approaches involve capturing spatial data of cities, including buildings, streets, parks, and infrastructures. It is the process of creating 3D digital representations of physical spaces by utilising multimodal terrestrial and aerial data collection approaches such as Light Detection and Ranging (LIDAR) and/or photogrammetry. The first provides precise mapping through laser scanning, while the latter creates 3D models from image sequences. In addition, LIDAR excels at capturing detailed elevation data and penetrating vegetation to map underlying terrain, while photogrammetry provides high-resolution colour information [63]. In both cases, unmanned aerial, and ground vehicles (*UAVs and UGVs*) enable faster, more efficient, and consistent data collections [64]. The production of accurate 3D models of UE relies on a pipeline that involves apart from the initial partial raw data collection (*vast sets of multimodal datasets including point clouds and/or images*), their pre-processing including filtering and fusion, georeferencing and post-processing to create applicable digital derivatives for tasks such as urban planning, heritage preservation, disaster management and response planning, optimised infrastructure development, smart-city initiatives, construction progress tracking, etc. The 3D visualisation of these derivatives, combined with diverse georeferenced data streams, do form valuable 3D digital twins which serve as robust testbeds, enabling efficient and sustainable approaches to monitoring, planning, decision-making, scenario simulation, and idea testing [65].

More specifically, the data collection phase is primarily affected by the scale and the complexity of the area to be captured. Weather conditions, data collection sensors range, line-of-sight and environmental obstructions may also affect this phase. Whatever data collection approach is applied, the main scope is to capture the UE as accurately as possible and as complete as possible. Thus, aerial and terrestrial data collection scenarios need to be designed and executed based on the specifications of UAVs and UGVs. The utilisation of such semi-automations has a significant impact on the efficiency of implementing the data collection but also for the consistency of the captured data in terms of resolution, sampling, and accuracy [66]. Furthermore, equipping these vehicles with multimodal sensors, including multispectral and thermal cameras along with the ability to accurately repeat the data collection scenarios, enhances the application domains of the 3D derivatives. Repeatability is an important feature that is highly related to the need for longitudinal studies or even data accuracy improvement over multiple sessions. Furthermore, the data collection phase must address other potential challenges such as data storage requirements, processing bottlenecks and vast data handling. In addition, within the broader domain of 3D digitisation, a diverse range of technologies exists, each offering distinct advantages and limitations [67]. This underscores the fact that no single approach can be considered a panacea for all challenges in the field. The two primary data collection approaches (LIDAR and photogrammetry) are still considered complementary. In many cases data fusion techniques can be utilised to improve the quality and accuracy of the produced digital replica of a UE. In fact, LIDAR and photogrammetry data can be fused. However, effective data fusion has some strong prerequisites. These are a detailed planned data collection pipeline that addresses the unique challenges of integrating multimodal datasets. A key consideration is the alignment of the captured data at different times and/or from different sources. The ability to precisely synchronise sensors and robust georeferencing methods in-situ or/and in a post-data collection processing phase is also such a strong prerequisite. Additionally, sensor calibration is equally important as the process contributes to resolution, scale and orientation consistency while reducing possible discrepancies during the post-processing phase.

UAVs and UGVs equipped with multiple sensors play a pivotal role but still as they also face numerous challenges performing the actual data collection phase and thus it is important for them to collect apart from the actual data, various *paradata* (*data related to the actual performance of the data collection*) which will ensure data traceability during the fusion process [68]. It should be noted that a fusion process comes with its own pre- and pro-processing steps such as noise filtering, outlier removal, data normalisation as well as fusion validation. The latter relies on benchmarking against standards such those provided by the Open Geospatial Consortium and the European INSPIRE directive or through the comparison against independent measurements. These independent measurements rely on residuals (e.g. *photogrammetric targets play the role of ground control points*) that are distributed within the digitised environment. The positions of such residuals may be measured with GNSS topographic stations. Additionally, UAVs and UGVs can be used in combination with a GNSS-based positioning technique, the Real-time Kinematic (RTK) that provides centimetre-level accuracy in real-time navigation and georeferencing. Hence, comparing the residuals geolocation between the two types of measurements contributes to validating the quality of the produced data. This approach is also applicable when LIDAR or photogrammetry are not fused applied separately. It should be noted that RTK introduces the use of a base station which is used to correct positional errors caused by atmospheric disturbances, satellite orbit inaccuracies and other error sources. RTK significantly contributes to data collection scenarios repeatability and thus an aerial or terrain path can be followed with minimal deviations. Furthermore, if synchronous aerial and terrestrial data collection is a requirement since an urban environment is a living case-study and a continuously changing environment then state-of-the-art integration of UAVs and UGVs in collaborative scenarios can deliver the required [69]. Again, AI-supported path generation algorithms can optimise the trajectories of UAVs and UGVs, ensuring efficient coverage, minimizing redundancy, and dynamically adapting to environmental changes or obstacles during data collection. Given all the above, it is obvious that the ability to achieve a 100% complete digital replica of a UE is inherently challenging. Based on the requirements, a UE digitisation project often led to compromises in data completeness or precision. Temporal mismatches caused by UE's dynamic nature are inevitable and thus different states, of a large-scale environment, are usually visible on the collected data [70]. The limitations in sensor modalities may introduce data inconsistencies while scalability challenges including data handling and computational constraints are always present when working on vast and complex terrains.

Apart from the data collection, data fusion and georeferencing of the raw data, the post-processing phase plays also a critical role in transforming or sampling on these data to produce digital derivatives that meet the requirements of other application specific algorithms or tools. It is a phase where heavy computational and analytical steps are applied. Some of the major directions on this stage are automated/semi-automated/manual feature extraction, classification and annotation. Automated techniques often rely on machine learning algorithms to identify and segment features such as roads, buildings, or vegetation from point clouds or mesh models [71]. Tools like QGIS or Pix4D can support semi-automated processes, allowing users to refine results interactively. Manual annotation, while labour-intensive, is crucial for ensuring accuracy in sensitive or complex cases, such as classifying historical landmarks. Other popular post-processing tasks involve point cloud to 3D mesh conversion through algorithms such as Poisson reconstruction, mesh decimation, edge preservation and retopology, texture mapping resolution reduction, geometry details baking on normal and displacement maps, scalable 3D rendering in simulations by utilising solutions such as different *level-of-detail* (LOD) 3D models. These processes are fundamental in producing high-quality digital twins that balance fidelity with computational efficiency, making them suitable for applications like VR/AR.

11.4.2 Semantic Segmentation of 3D Replicas

Semantic segmentation of digital replicas, particularly for 2D and 3D urban environments, plays a crucial role in modern urban planning, environmental monitoring, and disaster management. Semantic segmentation leverages advanced machine learning and computer vision techniques to classify various elements, such as buildings, building's material, vegetation, and roads, into meaningful categories. In this section we introduce state-of-the-art methods for feature-based segmentation, discussing the challenges in processing high-dimensional data from multiple sources, and highlighting the innovations within our proposed framework.

Semantic segmentation of urban areas has significantly progressed through the use of advanced multispectral data, LiDAR, and machine learning approaches. Yan et al. [72] integrated airborne laser scanning data with high-resolution multispectral satellite imagery for urban land-use classification, distinguishing buildings, trees, and streets. Popescu et al. [73] utilized UAV-mounted LiDAR to classify water, vegetation, and soil based on LiDAR intensity values. Gibril et al. [74] employed supervised models, such as random forests, support vector machines, and decision trees, to perform land use/cover classification of aerial imagery. Similarly, Pan et al. [75] used machine learning methods and multispectral imagery to detect asphalt pavement potholes and cracks. Additional contributions, such as those by Jurado et al. [76], Weinmann and Weinmann [77], explored the fusion of spatial and multispectral data for unsupervised classification of natural materials and the semantic interpretation of urban areas, respectively. As highlighted by Kerdegari et al. [78] and Kemker et al. [79], the scarcity of pixel-level annotated datasets remains a challenge for urban imagery segmentation. Building upon these studies, Koutsoudis et al. [80] introduced a cost-efficient and innovative pipeline that leverages multilayer texture information, a low-cost multispectral camera, and advanced machine learning methods to address existing limitations. Their results (Fig. 11.5) were impressive, providing detailed insights into the segmented elements of the urban environment, which can aid in disaster response and crowd management.

Figure 11.5 illustrates the transformation of a large urban area through semantic segmentation. The left image showcases the digitized representation of the urban

Fig. 11.5 (left) 3D digitized large urban area, (right) semantic segmentation of the area

environment of the olds town of Xanthi, Greece, highlighting its intricate structures, such as densely packed buildings and surrounding vegetation. In contrast, the right image demonstrates the semantic segmentation outcome, where individual features of the area, such as buildings, building elements (e.g., rooftops), vegetation, and roads, are distinctly classified and color-coded for visualization purposes. This segmentation highlights the granularity and precision, enabling enhanced analysis and decision-making for urban planning and monitoring tasks.

Despite advancements, several challenges persist in semantic segmentation of urban environments. Processing high-dimensional data from diverse sources, such as multispectral imagery and LiDAR, requires complex fusion techniques. Limited availability of pixel-level annotated datasets and the computational demands of large-scale urban analysis further hinder progress. Additionally, accurately segmenting dense urban areas with overlapping features and diverse materials remains difficult. Addressing these challenges is essential for improving applications in urban planning, disaster management, and crowd monitoring.

11.4.3 Multimodal Intelligent IoT Nodes

In crowd management, digital twins rely heavily on data from IoT devices, sensors and other data sources to monitor, predict and manage crowd behavior. Integrating all these diverse data sources, real-time, into digital twins has a revolutionary impact, allowing for improved situational awareness, proactive decision-making, and real-time crowd control [53]. The integration of IoT technologies into digital twin approaches offers powerful solutions for real-time monitoring, analysis, prediction, and control of physical systems [81]. In the context of crowd management, these capabilities are especially crucial for dynamic and complex environments, including transportation systems, healthcare facilities, large events, smart city development, and other densely populated areas.

Personal IoT devices like smart-phone (cameras), wearable devices equipped with GPS or biometric sensors and smart smartwatches or tickets for tracking attendees in several events, collect and transmit data from various points in the physical environment [82]. Their role in crowd management is to provide real-time location and movement data, enable tracking of crowd density and flow patterns, and support monitoring of individual health metrics (e.g., heart rate) in stressful events. The heterogeneous IoT sensor data can be utilized to virtualize and visualize cities, facilitating hazard prevention and remote management of workplace safety issues [60].

Robotic systems like UAV technologies, integrated with various sensors, has transformed how society collects data in a variety of applications [83]. UAVs can track traffic patterns in real time, gathering data on congestion, accidents, and road closures, resulting in optimized routes, shorter travel times and improved overall transportation efficiently. In crowd management strategies, UAVs play a crucial role by providing an aerial perspective for monitoring [84]. This point offers significant advantages, as it remains unaffected by the dynamics of the crowd or the crisis being observed, ensuring consistent and comprehensive surveillance even in challenging scenarios.

Social media plays a crucial role in aggregating data for digital twins, particularly in profiling crowds [85] by delivering real-time updates on crowd sentiment, movement, and activity patterns. The widespread accessibility of social media on mobile devices has resulted in an enormous influx of georeferenced socioeconomic data, enabling more intricate urban modelling, simulations and prediction of behaviours. [27] This data supports improved decision-making and management of urban environments, fostering sustainability while addressing the preferences and needs of citizens for more responsive and inclusive city planning [86].

Figure 11.6 depicts a Multimodal Internet of Things (IoT) network, established in the physical words, integrated into an AI-driven Digital Twin (Fig. 11.4) framework designed for advanced crowd monitoring and management. The network is composed of distributed IoT nodes strategically placed throughout urban environments to collect and analyze real-time data. These nodes are equipped with a variety of sensors capable of capturing environmental conditions, human activity, and crowd density, providing a comprehensive view of the physical world. In addition to their sensing capabilities, the IoT nodes are equipped with sirens and speakers, enabling direct communication with the crowd for alerts or guidance during critical situations. The nodes are also equipped with solar panels, making them energy-efficient and autonomous, ensuring continuous operation without reliance on external power sources. Communication between IoT nodes and the AI-driven Digital Twin platform is facilitated through high-speed 5G networks, ensuring seamless data transmission and low latency for real-time decision-making.

The collected data is transmitted to the AI-driven Digital Twin platform, which processes and integrates the information to deliver actionable insights. By combining multiple data modalities, such as visual data from cameras, acoustic signals, and environmental metrics, the system ensures thorough and accurate monitoring of dynamic urban settings.

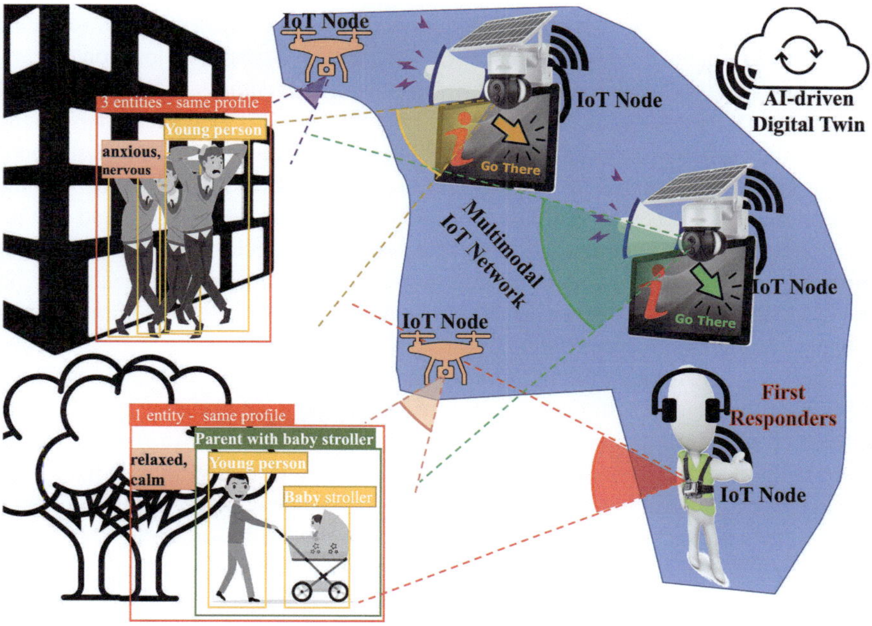

Fig. 11.6 Multimodal IIoT network

11.4.4 Simulation

Over the recent years, with the increases in computational speed of modern processors, it has become feasible to employ multi-agent approaches in order to simulate large crowds [2]. Simulation systems can now choose from many available algorithms to model such crowds [87, 88]. It has also become possible to simulate crowds interacting with vehicles [89], which play an important role in emergency situations (e.g. ambulance reaching injured people at the heart of a large crowd).

Over the years, various innovative systems have been developed to address large-scale crowd simulation challenges. Early advancements, such as Van Den Hurk and Watson's [90] multi-layered flocking system, extended Reynold's model [91] to handle complex navigation but faced limitations with inter-layer interactions. Curtis et al. [92]) introduced a real-time simulation of Tawaf crowds using finite state machines and the RVO algorithm [93] achieving realistic results but lacking group dynamics. Game engines like Unity and Unreal Engine have been pivotal in advancing simulations. Becker-Asano et al. [94] used Unity for 3D perception in airport terminals. Other platforms, such as the GAMA platform [95], support large-scale, multi-threaded simulations and integrate the Belief Desire Intention (BDI) paradigm for realistic human behavior modeling. These developments highlight the progression from basic models to scalable, behavior-rich systems for effective crowd management. One of the latest platforms, METIS [2, 96], is built on the Unity game engine and leverages reinforcement learning algorithms to simulate diverse human behaviors. It provides a dynamic, user-friendly interface, allowing users to create and customize their own environments for experimentation and exploration.

One outstanding issue in current simulators is the selfishness of the models that are employed: agents still act by themselves in order to fulfil their own goals. In the context of emergency situations considered in crowd management, the lack of cooperation between simulated agents precludes any form of first respondents' orchestration. When it comes to vehicles, this problem is more challenging and further compounded by their geometry and by that people or vehicles often make way for another to pass which goes against their own goals and thus leads to deadlocks. In this direction one very interesting research direction is to investigate multi-agent cooperative planning in the context of large multi-modal crowds.

Figure 11.7 illustrates a simulation approach within an internal environment where a crowd is distributed across various rooms of a large-scale building, emphasizing evacuation planning and strategy development. The individuals in the simulation are modelled as autonomous agents, categorized into different groups such as children, adults, and elderly individuals to reflect diverse behaviours and requirements. These agents utilize reinforcement learning [97] to exhibit customized and complex behaviours, enabling the simulation of realistic crowd dynamics under varying conditions. This advanced approach provides valuable insights for the development of efficient evacuation plans tailored to diverse scenarios. The simulation leverages the METIS system [2, 96], which empowers users to create custom environments through intuitive drag-and-drop functionalities. This makes it accessible for designing and executing complex experiments in both internal and external settings. Additionally, this simulation approach is versatile and can be extended to simulate crowd behaviour in open-air environments, further enhancing its applicability to urban planning, safety, and crisis management scenarios (Fig. 11.8).

Figure 11.7 illustrates a crowd simulation within an urban environment (external area), showcasing different levels of crowd density. The *a* image represents a scenario with a small number of individuals, the *b* image demonstrates a medium-sized crowd, and the *c* image depicts a densely populated area with a large crowd. These variations

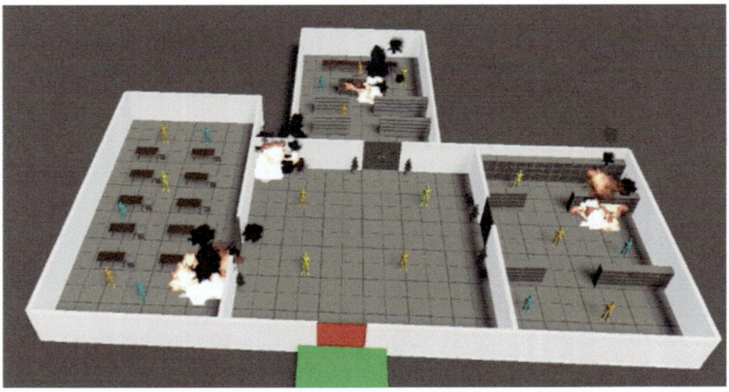

Fig. 11.7 Crowd simulation for internal environments

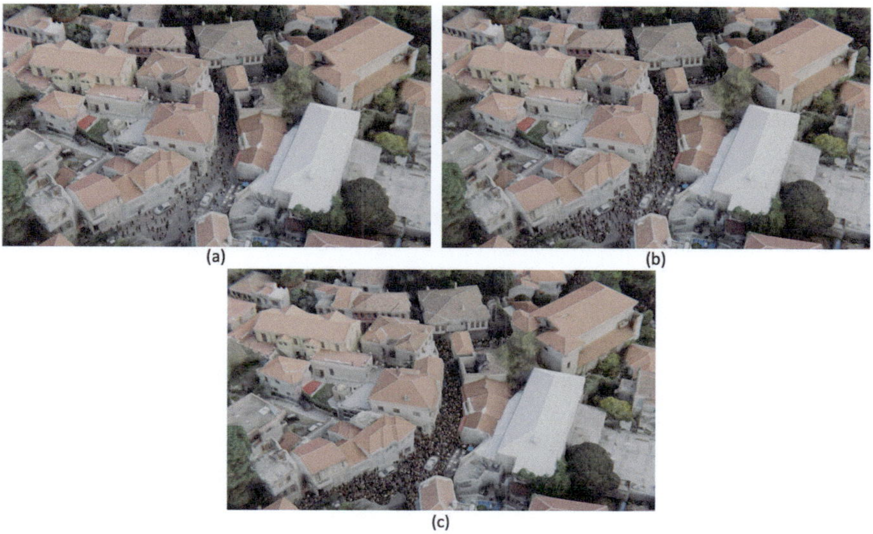

Fig. 11.8 Simulating varying crowd densities in urban environments

highlight the system's capability to model and adapt to diverse real-world crowd conditions effectively.

11.5 A General Framework

In this section, we propose a general framework for urban crowd management, leveraging advanced technologies such as RL, MAS, DT, and IIoT-enabled data integration. This framework aims to address key challenges in real-time adaptability, scalability, and trust in dynamic urban environments. The proposed framework is structured into four interconnected layers, each contributing to robust, adaptive, and scalable solutions for crowd management. The framework consists of the following components:

1. **Physical Environment Layer (PEL)**: This layer involves the collection of real-world data through a multimodal IoT infrastructure. Data sources include wearable IoT devices, unmanned aerial vehicles (UAVs), fixed and mobile sensors, and social media feeds. These sources can capture crowd density, pedestrian behaviours and characteristics, movement patterns, environmental conditions, and real-time sentiment, providing a comprehensive and diverse datasets for analysis.
2. **Simulation Layer (SL)**: Serving as the core of the framework, the digital twin creates a virtual representation of the urban environment and its features. It incorporates real-time data synchronization (from the PEL), integration of 3D

urban spaces, and simulation of pedestrian and crowd behaviors using RL and MAS. This layer enables scenario testing, predictive analysis, and validation of management strategies.

3. **Decision-Making Layer (DML)**: Advanced RL and MAS algorithms drive this layer, enabling predictive modeling, real-time optimization, and adaptive decision-making. MARL ensures efficient collaboration or competition among agents originating from the PEL and/or SL. While trust mechanisms enhance transparency and reliability in automated decisions.
4. **Human-in-the-Loop Interaction Layer (HLIL)**: To ensure trust and ethical oversight, this layer integrates real-time visualization tools, human intervention mechanisms, and community engagement strategies. It provides decision-makers with actionable insights and the capability to override automated decisions when necessary.
5. **Information Dissemination Layer (IDL)**: This layer focuses on the transmission of validated decisions from policymakers (from HLIL and DML) to first responders and citizens. Using IIoT systems, mobile devices, and public communication networks, the IDL ensures real-time dissemination of critical personalized information, such as evacuation instructions, emergency alerts, and updates. By bridging the gap between decision-making and on-the-ground implementation, this layer enhances coordination, situational awareness, and responsiveness.

The integration of these layers follows a structured workflow designed to ensure seamless interaction between the physical and virtual systems. First, *data collection* is performed using IoT devices, UAVs, and social platforms, which provide real-time data streams as inputs to the system. This data is then synchronized with the *digital twin*, enabling real-time updates and scenario simulations within the virtual environment. Subsequently, RL and MAS *optimization* occur, where virtual agents interact with the physical agents (IIoT nodes) through the digital twin to generate adaptive strategies for effective crowd management. Once optimized, *decisions* are validated and deployed in the physical world for implementation through various approaches such as coordination first responders or transferring the information through the IIoT and personal devices. Throughout this process, human oversight and feedback play a critical role, with real-time visualization tools ensuring transparency, ethical compliance, and the ability to make manual interventions if necessary.

This framework is designed for scalability, adaptable to various urban environments ranging from small-scale venues to city-wide implementations, encompassing both **internal areas**, such as large buildings, and **external areas**, including open urban spaces and outdoor event venues. By combining these technologies, the framework represents a significant step toward intelligent, trusted, and adaptive crowd management systems in smart cities.

11.5.1 *Future Trends and Research Directions*

This section explores future trends and research directions in Autonomous Crowd Management Systems (ACMS), focusing on advancements in RL, MAS, and decision-making technologies. It highlights key technological innovations, ethical considerations, and societal impacts, providing a roadmap for developing intelligent, adaptive, and trustworthy systems in dynamic environments. To address this, we introduce two main categories: one focused on technology-related advancements and another emphasizing strategies, societal impacts, and governance considerations.

11.5.1.1 Technological Related Advancment

Technological advancements play a pivotal role in shaping the future of ACMS. This category highlights critical areas of research and innovation necessary to enhance the effectiveness, reliability of these systems. The following points outline key technological research directions:

- Incorporating privacy-enhancing technologies like data minimization and anonymization into RL and MARL frameworks to ensure privacy when autonomous systems process real-world data, such as crowd monitoring and behavioral analysis.
- Developing RL and MARL systems with enhanced transparency to explain their decision-making processes in a manner understandable to non-experts. This is especially critical in MAS where coordination among agents must be interpretable and trustworthy.
- Addressing algorithmic biases in RL, MARL, and MAS by designing fairness-aware reinforcement learning algorithms and improving the diversity of training datasets. Ensuring these systems treat all populations equitably in high-stakes scenarios like surveillance and crowd management.
- Improving coordination mechanisms in MARL-based MAS to enable agents to effectively collaborate or compete in complex environments. This includes developing protocols for decentralized decision-making to ensure scalability and reliability in real-world applications.
- Develop more accurate human behavior models through advanced machine learning methods, enabling better predictions and decision-making in dynamic environments.
- Leveraging blockchain and secure communication protocols in MAS to ensure the integrity and trustworthiness of agent-to-agent and agent-to-human interactions. This will prevent data breaches and unauthorized control of decision-making systems.

11.5.1.2 Societal and Governance Considerations

While technological advancements are essential, societal and governance dimensions, including social, cultural, and regulatory aspects, are equally vital for the responsible development of ACMS. Significant ethical considerations arise from potential misuse, biases, and balancing security with individual liberties. Several studies [98–101] highlight the need to foster trust in autonomous systems by addressing these issues. Key research directions include:

- Establishing comprehensive legal and ethical frameworks to govern the use of autonomous systems, particularly in public and sensitive domains. These regulations should address informed consent, data ownership, and the limits of surveillance.
- Promoting awareness and education about ACMS capabilities, limitations, and ethical considerations. Engaging communities in discussions about surveillance and privacy to build trust and accountability.
- Encouraging international collaboration to establish uniform standards for ethical AI and ACMS. This includes sharing best practices for privacy, fairness, and security across nations.
- Conducting research to understand the long-term societal and cultural impacts of deploying AI and ACMS, ensuring alignment with diverse human values and beliefs.

11.6 Conclusion

In this chapter, we have presented a comprehensive framework for autonomous crowd management that integrates advanced technologies such as reinforcement learning, digital twins, multi-agent systems, and intelligent IoT nodes (devices). By leveraging these innovations, the framework addresses critical challenges in dynamic urban environments, including scalability, real-time adaptability, and trust among autonomous agents.

Reinforcement learning's adaptability and real-time decision-making capabilities serve as a foundation of the framework, showcasing its exceptional proficiency in simulating and effectively managing intricate and dynamic human behaviors. Combined with the predictive power of digital twins, these systems enable proactive and efficient crowd management strategies, ensuring public safety and operational efficiency. Multi-agent systems further enhance the framework by enabling collaborative and competitive behaviors among agents, effectively modeling the dynamics of human interactions during crowd movements. Furthermore, the integration of multimodal IoT data streams, including inputs from various devices and even those carried by pedestrians, into digital twin environments delivers exceptional situational awareness. This allows for precise monitoring, predictive analytics, and real-time responses to evolving crowd behaviors. The proposed framework also prioritizes human-in-the-loop interactions, ensuring ethical oversight, transparency, and trust in decision-making processes. While the presented framework demonstrates significant potential, it also highlights avenues for future research. Key areas include addressing algorithmic biases, enhancing privacy and security in IoT-enabled systems, and fostering global standards for ethical AI deployment in public domains.

To sumup, the fusion of reinforcement learning, digital twins, and multi-agent systems represents a transformative step toward intelligent and trusted crowd management systems. These technologies have the potential to revolutionize urban planning and public safety, creating smarter, safer, and more responsive cities that cater to the needs of their citizens.

Acknowledgements This work has been partially supported by the following projects: (a) MIS 5154714 of the National Recovery and Resilience Plan Greece 2.0 funded by the European Union under the NextGenerationEU Program and (b) ARTEMIS project funded by the European Union under Grant Agreement n.101188009. *Views and opinions expressed are however those of the author(s) only and do not necessarily reflect those of the European Union. Neither the European Union nor the granting authority can be held responsible for them.*

References

1. C. Celes, A. Boukerche, A.A. Loureiro, Crowd management: a new challenge for urban big data analytics. IEEE Commun. Mag. **57**(4), 20–25 (2019). https://doi.org/10.1109/MCOM.2019.1800640
2. G. Sidiropoulos, C. Kiourt, L. Moussiades, Crowd simulation for crisis management: the outcomes of the last decade. Mach. Learn. Appl. **2** (2020). https://doi.org/10.1016/j.mlwa.2020.100009
3. A.M. Al-Shaery, M.O. Khozium, Crowd management challenges: tackling approach for real time crowd monitoring. Int. J. Sci. Eng. Res. (2019)
4. A.M. Al-Shaery, S.S. Alshehri, N.S. Farooqi, M.O. Khozium, In-depth survey to detect, monitor and manage crowd. IEEE Access **8**, 209008–209019 (2020). https://doi.org/10.1109/ACCESS.2020.3038334
5. D. Fabian, W. Uwe, L. Mike, R. Michael, Y. Claudia, Urban digital twins for smart cities and citizens: the case study of Herrenberg, Germany. Sustainability **12**(6), 2307 (2020). https://doi.org/10.3390/su12062307
6. C. Martella, J. Li, C. Conrado, A. Vermeeren, On current crowd management practices and the need for increased situation awareness, prediction, and intervention. Saf. Sci. **91** (2017). https://doi.org/10.1016/j.ssci.2016.09.006
7. R. Narain, A. Golas, S. Curtis, M.C. Lin, Aggregate dynamics for dense crowd simulation. ACM Trans. Graph. **28**(5), 1–8 (2009). https://doi.org/10.1145/1618452.1618468
8. D. Darsena, G. Gelli, I. Iudice, F. Verde, Sensing technologies for crowd management, adaptation, and information dissemination in public transportation systems: a review. IEEE Sens. J. **23**(1), 68–87 (n.d.). https://doi.org/10.1109/JSEN.2022.3223297
9. F. J. Villanueva, C. Bolanos, A. Rubio, R. Cantarero, J. Fernandez-Bermejo, J. Dorado, Crowded event management in smart cities using a digital twin approach, in *IEEE International Smart Cities Conference* (Pafos, Cyprus, 2022), pp. 1–7. https://doi.org/10.1109/ISC255366.2022.9921923

10. P. Charalambous, J. Pettre, V. Vassiliades, Y. Chrysanthou, N. Pelechano, GREIL-crowds: crowd simulation with deep reinforcement learning and examples. ACM Trans. Graph. **42**(4), 1–15 (2023). https://doi.org/10.1145/3592459
11. R.S. Sutton, A.G. Barto, *Reinforcement Learning: An Introduction* (The MIT Press, 2018)
12. Lim, S. Y. (2015). *Crowd behavioural simulation via multi-agent reinforcement learning.* School of Computer Science and Applied Mathematics
13. A.M. Alasmari, N.S. Farooqi, Y.A. Alotaibi, Recent trends in crowd management using deep learning techniques: a systematic literature review. J. Umm Al-Qura Univ. Eng. Archit. (2024). https://doi.org/10.1007/s43995-024-00071-3
14. I. Ahmed, M. Ahmad, A. Ahmad, G. Jeon, IoT-based crowd monitoring system: using SSD with transfer learning. Comput. Electric. Eng. **93**(107226) (2021). https://doi.org/10.1016/j.compeleceng.2021.107226
15. D.A. Guastella, V. Campss, M.-P. Gleizes, A cooperative multi-agent system for crowd sensing based estimation in smart cities. IEEE Access **8**, 183051–183070 (2020). https://doi.org/10.1109/ACCESS.2020.3028967
16. Y. Wang, J. Ge, A. Comber, Simulation model of pedestrian flow based on multi-agent system and Bayesian Nash equilibrium. AGILE GIScience Ser. **2**(42) (2021). https://doi.org/10.5194/agile-giss-2-42-2021
17. Y Tang, X. Zhang, R. Wang, J. Xu, L. Hu, Y. Hao, Crowd intelligent grouping collaboration evacuation via multi-agent reinforcement learning, in *26th International Conference on Computer Supported Cooperative Work in Design* (2023), pp. 796–801. https://doi.org/10.1109/CSCWD57460.2023.10152607
18. R. Casadei, G. Delnevo, R. Girau, S. Mirri, Modelling groups of humans: towards crowd digital twins, in *IEEE Symposium on Computers and Communications* (Paris, France, 2024), pp. 1–6. https://doi.org/10.1109/ISCC61673.2024.10733562
19. S. Jaroslaw, Crowd management during mass events. Zeszyty Naukowe (2024)
20. M.A.-S. Ali, O.K. Mohamed, Crowd management challenges: tackling approach for real time crowd monitoring. Int. J. Sci. Eng. Res. (2019)
21. P. Sindhu, K.J. Prabuchandran, B. Shalabh, Reinforcement learning algorithm for non-stationary environments. Appl. Intell. **50**, 3590–3606 (2020). https://doi.org/10.1007/s10489-020-01758-5
22. S. Song, Ł Kidziński, X.B. Peng, C. Ong, J. Hicks, S. Levine, C.G. Atkeson, L.S. Delp, Deep reinforcement learning for modeling human locomotion control in neuromechanical simulation. J. Neuroeng. Rehabil. (2021). https://doi.org/10.1186/s12984-021-00919-y
23. A.J. Sathyamoorthy, J. Liang, U. Patel, T. Guan, R. Chandra, D. Manocha, DenseCAvoid: real-time navigation in dense crowds using anticipatory behaviors, in *IEEE International Conference on Robotics and Automation (ICRA)* (Paris, France, 2020). https://doi.org/10.1109/ICRA40945.2020.9197379
24. D. Zhang, W. Li, J. Gong, L. Huang, G. Zhang, S. Shen, J. Liu, H. Ma, DRLM3D: a deep reinforcement learning-based model with human-like perceptron and policy for crowd evacuation in 3D environments. ISPRS Int. J. Geo-Inf. **11**(4) (2022). https://doi.org/10.3390/ijgi11040255
25. C. Pang, L. Lyu, Q. Zhou, L. Zhou, Environment-sensitive crowd behavior modeling method based on reinforcement learning. Appl. Intell. **53**, 19356–19371 (2023). https://doi.org/10.1007/s10489-023-04509-4
26. Z. Liang, L. Li, L. Wang, Crowd-oriented behavior simulation:reinforcement learning framework embedded with emotion model, in *Artificial Intelligence. CICAI 2022. Lecture Notes in Computer Science* ed. by L. Fang, D. Povey, G. Zhai, T. Mei, R. Wang (n.d.). https://doi.org/10.1007/978-3-031-20503-3_16
27. G. Sidiropoulos, C. Kiourt, V. Sevetlidis, G. Pavlidis, Shaping the behavior of reinforcement learning agents, in *Proceedings of the 25th Pan-Hellenic Conference on Informatics* (2021), pp. 448–453. https://doi.org/10.1145/3503823.3503905
28. A. Wong, T. Bäck, A.V. Kononova, A. Plaat, Deep multiagent reinforcement learning: challenges and directions. Artif. Intell. Rev. **56**, 5023–5056 (2023). https://doi.org/10.1007/s10462-022-10299-x

29. J. Wang, Y. Hong, J. Wang, J. Xu, Y. Tang, Q.-L. Han, Cooperative and competitive multi-agent systems: from optimization to games. EEE/CAA J. Autom. Sin. **9**(5), 763–783 (2022). https://doi.org/10.1109/JAS.2022.105506
30. B. Ferguson, D. Paccagnan, B.M. Pradelski, Collaborative coalitions in multi-agent systems: quantifying the strong price of anarchy for resource allocation games, in *2nd IEEE Conference on Decision and Control* (Singapore, 2023), pp. 3238–3243. https://doi.org/10.1109/CDC49753.2023.10383819
31. C. Kiourt, *Collective Artificial Intelligence and Personalized Educational Interactions in Learning New Skills* (Hellenic Open University, School of Science and Technology, 2017)
32. F. Martinez-Gil, M. Lozano, F. Fernández, MARL-Ped: A multi-agent reinforcement learning based framework to simulate pedestrian groups. Simul. Model. Pract. Theory **47**, 259–275 (2014). https://doi.org/10.1016/j.simpat.2014.06.005
33. F. Martinez-Gil, M. Lozano, F. Fernández, Strategies for simulating pedestrian navigation with multiple reinforcement learning agents. Auton. Agent. Multi-Agent Syst. **29**(1), 98–130 (2014). https://doi.org/10.1007/s10458-014-9252-6
34. L. Jaedong, W. Jungdam, L. Jehee, Crowd simulation by deep reinforcement learning, in *MIG'18: Proceedings of the 11th ACM SIGGRAPH Conference on Motion, Interaction and Games* (2018), pp. 1–7. https://doi.org/10.1145/3274247.3274510
35. D. Hildreth, S. Guy, Coordinating multi-agent navigation by learning communication, in *Proceedings of the ACM on Computer Graphics and Interactive Techniques* (2019), pp. 1–17. https://doi.org/10.1145/3340261
36. Q. Wang, H. Liu, K. Gao, L. Zhang, Improved multi-agent reinforcement learning for path planning-based crowd simulation. IEEE Access **7**, 73841–73855 (2019). https://doi.org/10.1109/ACCESS.2019.2920913
37. Y. Chen, H. Wang, IntelligentCrowd: mobile crowdsensing via multi-agent reinforcement learning. IEEE Trans. Emerg. Top. Comput. Intell. **5**(5), 840–845 (2021). https://doi.org/10.1109/TETCI.2020.3042244
38. A. Cohen, E. Teng, V.-P. Berges, R.-P. Dong, H. Henry, M. Mattar, A. Zook, S. Ganguly, On the use and misuse of absorbing states in multi-agent reinforcement learning (2021). https://doi.org/10.48550/arXiv.2111.05992
39. J. Tweedale, P. Cutler, Trust in multi-agent systems, in *Knowledge-Based Intelligent Information and Engineering Systems* ed. by B. Gabrys, R. Howlett, L. Jain, vol. 4252 (2006). https://doi.org/10.1007/11893004_62
40. X. Chengxing, C. Canyu, J. Feiran, Y. Ziyu, L. Shiyang, S. Kai, G. Jindong, L. Guohao, Can large language model agents simulate human trust behavior? in *The Thirty-eighth Annual Conference on Neural Information Processing Systems* (2024). https://openreview.net/forum?id=CeOwahuQic
41. H. Gayathri, P. Aparna, A. Verma, A review of studies on understanding crowd dynamics in the context of crowd safety in mass religious gatherings. Int. J. Disaster Risk Reduct. 82–91 (2017). https://doi.org/10.1016/j.ijdrr.2017.07.017
42. X. Gong, M. Herty, B. Piccoli, G. Visconti, Crowd dynamics: modeling and control of multi-agent systems. Ann. Rev. Control Robot. Auton. Syst. **6**, 261–282 (2023). https://doi.org/10.1146/annurev-control-060822-123629
43. C. Burnett, T.J. Norman, K. Sycara, Trust decision-making in multi-agent systems, in *Proceedings of the Twenty-Second International Joint Conference on Artificial Intelligence* (2011). https://doi.org/10.5591/978-1-57735-516-8/IJCAI11-031
44. A. Al-Neaimi, S. Qatawneh, N. Al Saiyd, Conducting verification and validation of multi-agent systems. Int. J. Softw. Eng. Appl. (2012). https://doi.org/10.5121/ijsea.2012.3510
45. D. Jelenc, Toward unified trust and reputation messaging in ubiquitous systems. Ann. Telecommun. **76**, 119–130 (2020). https://doi.org/10.1007/s12243-020-00792-1
46. O. Samuel, N. Javaid, A. Khalid, M. Imrar, N. Nasser, A trust management system for multi-agent system in smart grids using blockchain technology, in *2020 IEEE Global Communications Conference* (2020), pp. 1–6. https://doi.org/10.1109/GLOBECOM42002.2020.9348231

47. C. Marche, M. Nitti, Trust-related attacks and their detection: a trust management model for the social IoT. IEEE Trans. Netw. Serv. Manage. **18**(3), 3297–3308 (2021). https://doi.org/10.1109/TNSM.2020.3046906
48. M. Amiri-Zarandi, R.A. Dara, Blockchain-based trust management in social internet of things, in *IEEE International Conference on Dependable, Autonomic and Secure Computing, International Conference on Pervasive Intelligence and Computing, International Conference on Cloud and Big Data Computing, International Conference on Cyber Science and Technology Congress (DASC/PiCom/CBDCom/CyberSciTech)* (2020), pp. 49–54. https://doi.org/10.1109/DASC-PICom-CBDCom-CyberSciTech49142.2020.00024).
49. Z. Tu, H. Zhou, K. Li, H. Song, Y. Yang, A blockchain-based trust and reputation model with dynamic evaluation mechanism for IoT. Comput. Netw. **218** (2022). https://doi.org/10.1016/j.comnet.2022.109404
50. R. Khalid, O. Samuel, N. Javaid, A. Aldegheishem, M. Shafiq, N. Alrajeh, A secure trust method for multi-agent system in smart grids using blockchain. IEEE Access **9** (2021). https://doi.org/10.1109/ACCESS.2021.3071431
51. E.M. Mianji, G.-M. Muntean, I. Tal, Trust and reputation management for data trading in vehicular edge computing: a DRL-based approach, in *IEEE International Symposium on Broadband Multimedia Systems and Broadcasting* (2024), pp. 1–7. https://doi.org/10.1109/BMSB62888.2024.10608229
52. S. Guleng, C. Wu, X. Chen, X. Wang, T. Yoshinag, Y. Ji, Decentralized trust evaluation in vehicular internet of things. IEEE Access **7**, 15980–15988 (2019). https://doi.org/10.1109/ACCESS.2019.2893262
53. W. Yuntao, S. Zhou, G. Shaolong, D. Minghui, H.L. Tom, L. Yiliang, A survey on digital twins: architecture, enabling technologies, security and privacy, and future prospects. IEEE Internet of Things J. **10**(17) (2023)
54. M.B.-S. Diego, M. Adriana-Simona, E.P.-G. Rodrigo, A.R.-M. Mauricio, A.R.-M. Ricardo, J.L.-S. Jorge de, Digital twin technology challenges and applications: a comprehensive review. Remote Sens. **14**(6) (2022). https://doi.org/10.3390/rs14061335
55. J.V. Felix, B. Cristina, R. Ana, C. Ruben, F.-B. Jesus, D. Javier, Crowded event management in smart cities using a digital twin approach, in *IEEE International Smart Cities Conference (ISC2)* (2022). https://doi.org/10.1109/ISC255366.2022.9921923
56. M. W. Grieves, Digital twins: past, present, and future, in *The Digital Twin*, ed. by N. Crespi, A.T. Drobot, R. Minerva (2023), pp. 97–121. https://doi.org/10.1007/978-3-031-21343-4_4
57. M. Grieves, *Digital Twin: Manufacturing Excellence through Virtual Factory Replication.* (White Paper; Dassault Systèmes: Vélizy-Villacoublay, 2015)
58. H. Banaee, F. Klügl, A. Loutfi, Challenges in using digital twins for modelling human behaviour in environments, neuro-cognitive modelling of humans and environments, in *Neuro-Cognitive Modeling of Humans and Environments, IJCAI 2020* (Yokohama, Japan, 2021)
59. B.R. Barricelli, E. Casiraghi, J. Gliozzo, A. Petrini, S. Valtolina, Human Digital twin for fitness management. IEEE Access **8**, 26637–26664 (2020). https://doi.org/10.1109/ACCESS.2020.2971576
60. S. Mihai, M. Yaqoob, V.H. Dang, D. William, T. Praveer, R. Mohsin, K. Mehmet, X.N. Huan, Digital twins: a survey on enabling technologies, challenges, trends and future prospects. IEEE Commun. Surv. Tutor. **23**(4), 2255–2291 (2022). https://doi.org/10.1109/COMST.2022.3208773
61. A. Salama, M. Patil, "YouWalk-UOS"–technology-enabled and user-centred assessment of urban open spaces. Open House Int. **49**(6), 1015–1029 (2024). https://doi.org/10.1108/OHI-01-2024-0021
62. J. Fernández-Alvarado, S. Fernández-Rodríguez, 3D environmental urban BIM using LiDAR data for visualisation on Google Earth. Autom. Constr. **138**, 104251 (2022). https://doi.org/10.1016/j.autcon.2022.104251
63. A. Koutsoudis, B. Vidmar, G.A. Ioannakis, F. Arnaoutoglou, G. Pavlidis, C. Chamzas, Multi-image 3D reconstruction data evaluation. J. Cult. Herit. **15**, 73–79 (2014). https://doi.org/10.1016/j.culher.2012.12.003

64. A. Tagarakis, E. Filippou, D. Kalaitzidis, L. Benos, P. Busato, D. Bochtis, Proposing UGV and UAV systems for 3D mapping of orchard environments. Sensors **22** (2022). https://doi.org/10.3390/s22041571
65. T. Wang, C. Tan, L. Huang, Y. Shi, T. Yue, Z. Huang, simplexity testbed: a model-based digital twin testbed. Comput. Ind. **145** (2023). https://doi.org/10.1016/j.compind.2022.103804
66. F. Remondino, L. Barazzetti, F. Nex, M. Scaioni, D. Sarazzi, UAV photogrammetry for mapping and 3D modeling-Current status and future perspectives, in *IISPRS-International Archives of the Photogrammetry, Remote Sensing and Spatial Information Sciences* (2011). https://doi.org/10.5194/isprsarchives-XXXVIII-1-C22-25-2011
67. G. Pavlidis, A. Koutsoudis, 3D digitization of tangible heritage, in *Handbook of Cultural Heritage Analysis*, ed. by S. D'Amico, V. Venuti. (Springer, Cham, 2022). https://doi.org/10.1007/978-3-030-60016-7_47
68. S. Batzdorfer, M. Bobbe, M. Becker, H. Harms, U. Bestmann, Multisensor equipped UAV/UGV for automated exploration, in *ISPRS-International Archives of the Photogrammetry, Remote Sensing and Spatial Information Sciences* (2017), pp. 33–40. https://doi.org/10.5194/isprs-archives-XLII-2-W6-33-2017
69. S. Javaid, N. Saeed, Z. Qadir, H. Fahim, B. He, H. Song, M. Bilal, Communication and control in collaborative UAVs: recent advances and future trends (2023). https://doi.org/10.48550/arXiv.2302.12175
70. T. Rosnell, E. Honkavaara, K. Nurminen, On geometric processing of multi-temporal image data collected by light UAV systems. Int. Arch. Photogramm. Remote. Sens. Spat. Inf. Sci. **28**, 63–68 (2012). https://doi.org/10.5194/isprsarchives-XXXVIII-1-C22-63-2011
71. A. Koutsoudis, G.A. Ioannakis, P. Pistofidis, F. Arnaoutoglou, N. Kazakis, G. Pavlidis, C. Chamzas, N. Tsirliganis, Multispectral aerial imagery-based 3D digitisation, segmentation and annotation of large scale urban areas of significant cultural value. J. Cult. Herit. **49**, 1–9 (2021). https://doi.org/10.1016/j.culher.2021.04.004
72. W. Yan, A. Shaker, N. El-Ashmawy, Urban land cover classification using airborne Lidar data: a review. Remote Sens. Environ. **158**, 295–310 (2015). https://doi.org/10.1016/j.rse.2014.11.001
73. G. Popescu, O. Balotă, D. Iordan, Increasing land classification accuracy using unmanned aerial vehicles (uavs) with multispectral lidar sensor, in *Scientific Papers. Series E. Land Reclamation, Earth Observation & Surveying, Environmental Engineering* (2017), pp. 181–188
74. B. Gibril, B. Kalantar, R. Al-Ruzouq, N. Ueda, V. Saeidi, A. Shanableh, S. Mansor, H. Shafri, Mapping heterogeneous urban landscapes from the fusion of digital surface model and unmanned aerial vehicle-based images using adaptive multiscale image segmentation and classification. Remote Sens. **1081** (2020). https://doi.org/10.3390/rs12071081
75. Y. Pan, X. Zhang, G. Cervone, L. Yang, Detection of asphalt pavement potholes and cracks based on the unmanned aerial vehicle multispectral imagery IEEE J. Select. Top. Appl. Earth Obs. Remote Sens. 1–12 (2018). https://doi.org/10.1109/JSTARS.2018.2865528
76. J. Jurado, J. Cárdenas-Donoso, C. Ogayar, L. Ortega, F. Feito, Semantic segmentation of natural materials on a point cloud using spatial and multispectral features. Sensors **20**, 2244 (2020). https://doi.org/10.3390/s20082244
77. M. Weinmann, M. Weinmann, Fusion of hyperspectral, multispectral, color and 3d point cloud information for the semantic interpretation of urban environments, in *ISPRS Geospatial Week 2019, 10–14 June 2019, Enschede, The Netherlands, The International Archives of the Photogrammetry, Remote Sensing and Spatial Information Sciences* (2019), pp. 1899–1906. https://doi.org/10.5194/isprs-archives-XLII-2-W13-1899-2019
78. H. Kerdegari, M. Razaak, V. Argyriou, P. Remagnino, Urban scene segmentation using semi-supervised gan, in *Proceedings Volume 11155, Image and Signal Processing for Remote Sensing*, vol. 26 (2019). https://doi.org/10.1117/12.2533055
79. R. Kemker, C. Salvaggio, C. Kanan, High-resolution multispectral dataset for semantic segmentation. CoRR (2017). https://doi.org/10.48550/arXiv.1703.01918.

80. A. Koutsoudis, G. Ioannakis, A.F. Petros Pistofidis, N. Kazakis, G. Pavlidis, C. Chamzas, N. Tsirliganis, Multispectral aerial imagery-based 3D digitisation, segmentation and annotation of large scale urban areas of significant cultural value. J. Cult. Herit. **49**, 1–9 (2021). https://doi.org/10.1016/j.culher.2021.04.004
81. R. Huang, D. Zhang, Y. Zhiwen, C. Nugen, W. Jia, K. Yang, Y. Zhang, Y. Sekimoto, B. Guo, Digital twins and intelligent IoT for a better world, in *International Conference on Artificial Intelligence of Things and Systems*, vol. 20 (2023). https://doi.org/10.1109/AIoTSys58602.2023.00013
82. V. Metuge, N. Valero, L. Zhao, V. Nino, D. Claudio, Preliminary data collection for collaborative emergency department crowd management using wearable device, in *IEEE International Conference on Digital Health* (Barcelona, Spain, 2022), pp. 34–36. https://doi.org/10.1109/ICDH55609.2022.00013
83. S. Ogunbunmi, Y. Chen, E. Blasch, G. Chen, A survey on reputation systems for UAV networks. Drones **8**(6), 253 (2024). https://doi.org/10.3390/drones8060253
84. A.M. Husman, W. Albattah, Z.Z. Abidin, Y.M. Mustafah, K. Kadir, S. Habib, M. Islam, A. Khan, Unmanned aerial vehicles for crowd monitoring and analysis. Electronics **10**(23), 2974 (2021). https://doi.org/10.3390/electronics10232974
85. V.X. Gong, W. Daamen, A. Bozzon, S.P. Hoogendoorn, Crowd characterization for crowd management using social media data in city events. Travel Behav. Soc. **20**, 192–212 (2020). https://doi.org/10.1016/j.tbs.2020.03.011
86. N. Mohammadi, A. Vimal, J. Taylor, Knowledge discovery in smart city digital twins, in *53rd Hawaii International Conference on System Sciences*, vol. I (2020). https://hdl.handle.net/10125/63943
87. J. Ondřej, J. Pettré, A.-H. Olivier, S. Donikian, A synthetic-vision based steering approach for crowd simulation, in *ACM SIGGRAPH 2010 Papers*, vol. 123 (2010). https://doi.org/10.1145/1833349.1778860
88. D. Wolinski, M.C. Lin, J. Pettré, Warpdriver: context-aware probabilistic motion prediction for crowd simulation. ACM Trans. Graph. (TOG), **35**(6), 1–11 (2016). https://doi.org/10.1145/2980179.2982442
89. D. Wolinski, C. Ming, Generalized warpdriver: unified collision avoidance for multi-robot systems in arbitrarily complex environments. Robot. Sci. Syst. (2018)
90. S. Van Den Hurk, I. Watson, A multi-layered flockinq system for crowd simulation, in *Computer Games, Multimedia and Allied Technology* (2010), pp. 184–191. https://doi.org/10.5176/978-981-08-5480-5_053
91. C. Reynolds, Flocks, herds and schools: a distributed behavioral model, in *Proceedings of the 14th Annual Conference on Computer Graphics and Interactive Techniques* (1987), pp. 25–34. https://doi.org/10.1145/37401.37406
92. S. Curtis, S. Guy, B. Zafar, D. Manocha, Virtual tawaf: a case study in simulating the behavior of dense, heterogeneous crowds, in *Proceedings of the IEEE International Conference on Computer Vision* (2011), pp. 128–135. https://doi.org/10.1109/ICCVW.2011.6130234
93. J. Van Den Berg, S. Guy, M. Lin, D. Manocha, Reciprocal n-body collision avoidance, in *Springer Tracts in Advanced Robotics*, pp. 3–19 (2011). https://doi.org/10.1007/978-3-642-19457-3_1
94. C. Becker-Asano, F. Ruzzoli, C. Hölscher, B. Nebel, A multi-agent system based on unity 4 for virtual perception and wayfinding. Transp. Res. Proc. 452–455 (2014) https://doi.org/10.1016/j.trpro.2014.09.059
95. P. Taillandier, B. Gaudou, A. Grignard, Q.-N. Huynh, N. Marilleau, P. Caillou, D. Philippon, A. Drogoul, Building, composing and experimenting complex spatial models with the GAMA platform. GeoInformatica **23**(2), 299–322 (2019). https://doi.org/10.1007/s10707-018-0033
96. G. Sidiropoulos, C. Kiourt, L. Moussiades, Metis: multi-agent based crisis simulation system, in *Games and AI, Technological, Cultural and Societal Aspects* (Athens, Greece: SETN 2020 Workshop, 2020)
97. C. Kiourt, D. Kalles, Synthetic learning agents in game-playing social environments. Adapt. Behav. **24**(16) (2016). https://doi.org/10.1177/1059712316679239

98. G. Eze, Ethical considerations in the deployment of AI surveillance during public health crises. J. Innov. Technol. (2023)
99. F. Catarina, H. Ellen, C.C. Caitlin, L. Christoph, AI-powered public surveillence systems: why we (might) need them and how we want them. Technol. Soc. (2022)
100. L.D. Lubna, M. Noorain, S.C. Bhawani, N. Thomas, Ethical dilemmas and privacy issues in emerging technologies: a review. Sensors **23**(3), 1151 (2023). https://doi.org/10.3390/s23031151
101. H.-P. Thokozani, L. Matheus, O. Arlene, B. Jennifer, B. Zahid, P.M. Plinio, Ethical considerations of public health surveillance in the age of the internet of things technologies: a perspective. 2024. Digital Health **10** (2024). https://doi.org/10.1177/2055207624129657

Chapter 12
A Reinforcement Learning System that Learns from Human-Smart City Interaction for Providing Better Quality of Services

Konstantinos Liagkouras and Konstantinos Metaxiotis

Abstract A smart city refers to an urban region that utilizes smart technologies and data to increase operational efficiency, share information with the public and improve both the quality of government services and citizen's welfare. Despite the ongoing research efforts, there are still many issues and challenges that face the citizens of cities, especially the inhabitants of the modern megacities. The size and density of modern megacities create unique challenges for citizens, such as housing issues, inequality, employment, water availability and environmental issues. Furthermore, modern megacities, due to their size make more difficult the citizens' participation in the decision-making process on issues that affect their area, such as waste management, housing affordability, energy efficiency, healthcare and suppression of crime. In this article, we propose a Reinforcement Learning System that learns from human-smart city interaction. This reinforcement learning system will allow citizens of megacities to participate and interact with the smart city and through this interaction, between humans and smart city, the system will learn optimal policies for a number of issues that affect people's life, leading to better quality of life and citizen satisfaction.

Keywords Reinforcement learning · Smart city · Waste management · Housing affordability · Energy efficiency · Healthcare · Suppression of crime

12.1 Introduction

Modern megacities face numerous challenges [33] due to their size, rapid growth, and complexity. Excessive population density strains housing, infrastructure, and services. The rapid population growth in megacities contributes to housing issues

K. Liagkouras (✉) · K. Metaxiotis
Department of Informatics, University of Piraeus, Piraeus, Greece
e-mail: kliagk@unipi.gr

G. A. Tsihrintzis et al. (eds.), *Human-Smart City Interactions and User-Citizen Experiences*, Learning and Analytics in Intelligent Systems 58,
https://doi.org/10.1007/978-3-032-06364-9_12

[32] due to the imbalance between the demand for housing and the availability of affordable and adequate housing options. Housing costs often outpace the income of lower and middle income groups, leading to homelessness or substandard living conditions. Modern megacities generate vast amounts of waste [21], often surpassing the capacity for collection and recycling. Big cities require massive amounts of energy, leading to higher carbon emissions and pressure on energy grids [53]. The crime rates are much higher in big cities [46], undermining the social fabric and affecting local communities. Finally, access to healthcare [57] in megacities can be challenging due to dense populations, leading to an overwhelming number of patients for healthcare facilities. The growth in healthcare facilities and resources [43] often lags behind population growth, resulting in overcrowded hospitals and long waiting times. Certain treatments or specialties may only be available at a few facilities, leading to bottlenecks.

In this article, we propose a reinforcement learning system that learns from human-smart city interaction [18] for providing better quality of services to the inhabitants of modern megacities. Reinforcement learning [34] is a subfield of machine learning that teaches an agent to take an action in an unknown environment that maximizes the reward overtime. The objective of a RL is to determine how an agent should take actions in an environment to maximize the cumulative reward. The Reinforcement learning (RL) has some essential components such as an agent, the environment, the states, the actions, the reward and the policy. The agent learns to make decisions by selecting actions and interacting with the environment, based on the rewards received [55]. The state is a representation of the current situation of the environment. An action allows the agent to transition between states in the environment. Rewards signal how good or bad an action was, guiding the agent to maximize cumulative rewards over time. Finally, the policy refers to the agent's behavior in the environment and which strategy is used to reach the goal, whether it is stochastic or deterministic policy. Table 12.1 illustrates the key components of a RL system.

Reinforcement learning (RL) systems are broadly categorized into two main types: model-free and model-based systems [35]. Model-free RL systems treat the environment as a black box and they learn optimal policies through trial-and-error. That means that the agent receives reward by performing the desired behaviour and

Table 12.1 Key components of a reinforcement learning system

Terms	Definitions
Agent	The unit that learns to make decisions within an environment
Environment	The external system with which the agent interacts
State	The current situation of the environment
Action	The mechanism by which the agent transitions between states
Reward	The immediate feedback from the environment that assesses the goodness or badness of an action
Policy	The strategy or rules guiding an agent's decision-making process within an environment

penalty for performing incorrectly. Q-learning [8] and Policy Gradient [19] methods are the two most common model-free approaches. In Q-learning the agent learns how to behave into an unknown environment to maximize rewards. On the other hand, Policy Gradient method focus directly on optimizing the policy that governs actions. Model-free systems [54] are primarily used for performing tasks in complex or dynamic environments.

Model-based RL systems [36] create a simulation of the environment to predict the outcomes of actions. In more detail, model-based RL systems simulate future states and rewards, for better decision-making. According to a number of studies [13, 36] model-based approaches can be more efficient than model-free methods, as they incorporate the concept of planning and forecasting. However, modelling the environment can be a complex and challenging task. Monte Carlo Tree Search [13] is one of the most popular model-based RL techniques and it is having several prevalent real-world applications such as the autonomous driving.

In this article we examine, how a reinforcement learning system can learn and benefit from the human—smart city interaction for providing better services to all members of the community. A RL system can actively involve humans during its learning process for guidance or feedback [12]. The feedback from humans can help the agent understand the desired behavior. Also, human feedback can assist the agent in ambiguous situations. Finally, the humans can evaluate the overall performance of the RL system and provide feedback to refine its learning [48]. Below, we present how a reinforcement learning system interacts with humans for providing better quality of services to the inhabitants of smart cities.

This work is organized as follows. In Sect. 12.2, we discuss how a reinforcement learning (RL) system that learns from human-smart city interaction can be used for providing better quality of services to the citizens of modern megacities. In particular, we consider the implementation of a reinforcement learning system to critical aspects of a city's life, such as waste management, housing affordability, energy efficiency, healthcare and the suppression of crime. Finally, in Sect. 12.3, we present the conclusions and we discuss future research.

12.2 A Reinforcement Learning System that Learns from Human-Smart City Interaction for Providing Better Quality of Services

Modern metropolitan areas face numerous challenges [1] due to their size and rapid growth. The lack of affordable housing is one of the major issues of big cities. Due to the rapid population growth the inhabitants of modern megacities face also limited access to healthcare services. Social inequalities [22], including economic disparities often lead to higher rates of crime. Finally, inefficient waste management is increasingly becoming one of the greatest challenges, particularly for highly urbanized metropolitan areas. A Reinforcement Learning (RL) system that learns from human-smart city interaction [42] can significantly improve the quality of services to inhabitants of a city by optimizing decision-making in dynamic and complex urban environments. Below, we present how such a Reinforcement Learning (RL) system can be applied to a variety of aspects of citizen's daily life such as, crime prevention, energy efficiency, improving healthcare services, optimizing the housing market and providing more efficient waste management for all inhabitants of modern megacities.

12.2.1 Application of Reinforcement Learning in Crime Prediction

The inhabitants of modern megacities can benefit from a Reinforcement Learning (RL) system in crime prevention. In a Reinforcement Learning (RL) system for crime prevention the agent learns by interacting with its environment and receiving feedback through rewards or penalties [51]. The agent aims to maximize cumulative rewards by executing the desired behaviour. That means that in the case of crime prediction, RL could be used in a variety of ways, such as optimizing policing strategies, resource allocation, and improving crime prevention efforts [17].

A RL system for predictive policing by analysing data can help to forecast when and where crime will happen [20]. The main benefit from such a RL system is that it can help to the allocation of police resources with the most efficient way for preventing crimes or responding to crimes. The RL system for crime prevention can integrate human feedback into the learning cycle. Figure 12.1 illustrates the proposed RL system for crime prevention. The environment could be represented by a city grid. Respectively, the states represent the levels of crime in various neighborhoods. Finally, the agent learns to allocate resources effectively by considering how the deployment affects crime levels [46]. The system can also utilize data from social media [5]. The RL system could learn to identify potential crime hotspots and predict areas where authorities could prevent crime incidents.

As shown in Fig. 12.2, the RL system for predictive policing should be connected with an additional subsystem for collecting and analysing data. The proposed subsystem could collect data not only through surveillance devices, but also through human feedback. The subsystem should be able to classify the threats into categories, based on the nature of the crime. Finally, it should be connected and share its intelligence with the law enforcement authorities in real time.

A RL system can be a powerful tool in the hands of law enforcing authorities. However, despite the undeniable benefits, these systems can also pose some threats. Algorithmic bias is one of the most important concerns. If the system is fed with biased historical data, such as racial profiling or biased arrest data, it can replicate existing biases in the criminal justice system [45]. This could lead to unfair targeting of certain groups of people or areas, based on stereotypes. Additionally, the implementation of surveillance systems could raise concerns about the privacy of our data. The prediction of potential crimes based on profiling, could infringe on civil

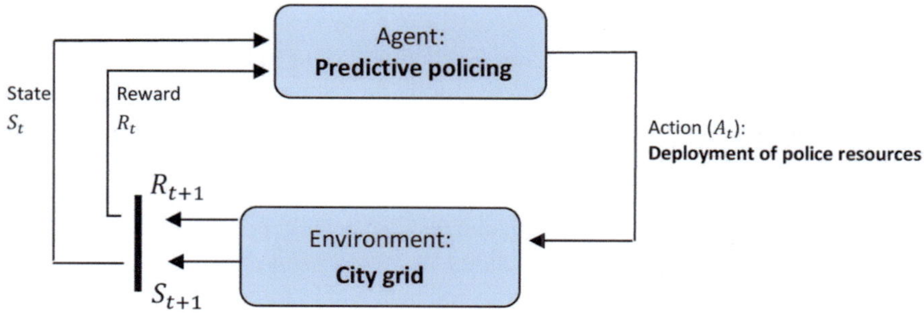

States: Represent the levels of crime in various regions
Reward: How the deployment affects crime levels

Fig. 12.1 A reinforcement learning system for crime prevention

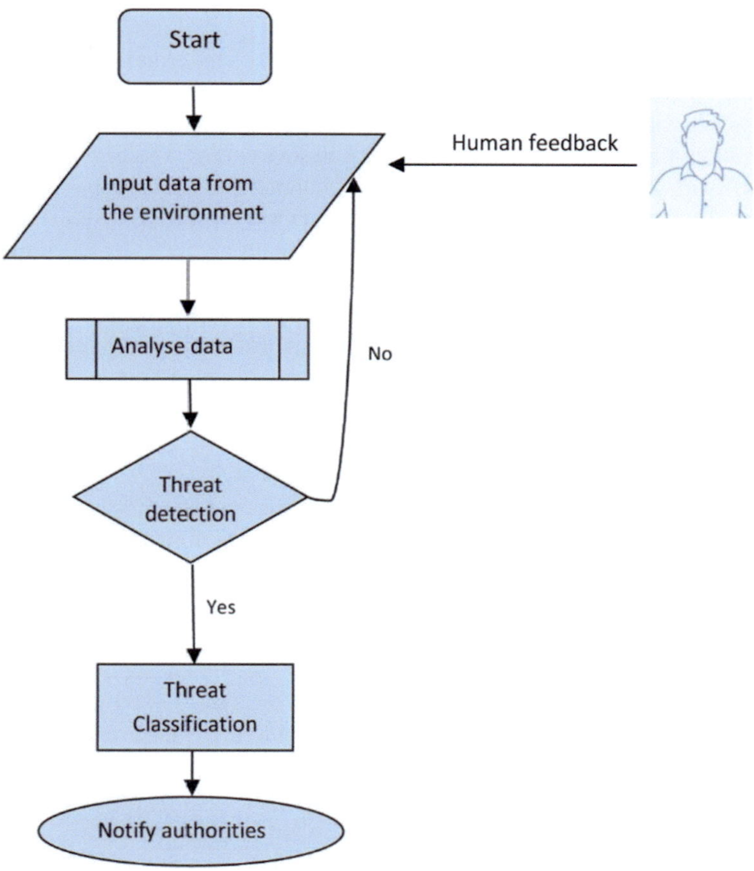

Fig. 12.2 Flowchart of the proposed subsystem for collecting and analysing data for crime prevention

rights. Another area of concern is the quality of the available data, as inaccurate or incomplete data could lead to poor predictions, misallocation of resources, or ineffective interventions. A RL predictive policing system should be able to adapt to a continuously changing environment. This would require continuous learning and updates, as new data comes in, which can be challenging, especially in dynamic urban environments. A crime prediction system would require to store to data centers huge amount of data. Finally, crime-related data are usually in different formats such as text, images, graph, audio, unstructured data, and semi-structured data, so, the process of transforming these data to the understandable format is also a challenge. [45].

12.2.2 Application of Reinforcement Learning for Energy Efficiency

The inhabitants of modern megacities can benefit from a Reinforcement Learning (RL) system for energy efficiency [44]. Reinforcement Learning (RL) can be a powerful tool for enhancing energy efficiency in smart cities. The first step in developing a smart grid management system is to collect data on energy consumption patterns, weather conditions, and other relevant factors [7]. The RL system for energy efficiency can integrate human feedback into the learning cycle. Energy efficiency can significantly benefit from human feedback by integrating insights from real-world user behavior and preferences into energy systems. Feedback allows energy solutions to be tailored to actual usage patterns, improving their effectiveness [41]. As shown in Fig. 12.3, this data can be obtained from Internet of Things (IoT) devices, smart meters, weather sensors, human feedback, and other monitoring devices. Then, data preprocessing techniques such as data cleaning, data transformation and data integration can be used to prepare the data for further analysis.

A RL system for energy efficiency can help to lower costs, make optimal use of the available financial resources [23, 25, 28, 31] and minimize the environmental impact. For example, a RL system can learn how to control energy flows in real-time to balance supply and demand and reduce losses [30]. Figure 12.4 illustrates the proposed RL system for energy efficiency. The environment could be represented by a smart grid, where states represent the levels of energy consumption during the daytime and nighttime, and the agent learns how to control energy flows in real-time to balance supply and demand [2]. In addition to that, since renewable energy such as wind power, solar power and bioenergy is variable, a RL system could help balancing renewable energy sources, ensuring that the electricity grid remains stable and energy is used efficiently.

Energy efficiency systems can be utilized by both local communities and households alike. For example, RL systems can learn how to adjust heating or cooling within buildings based on the current occupancy and weather forecast [14]. Also, smart lighting systems can be controlled by RL to reduce energy consumption. The

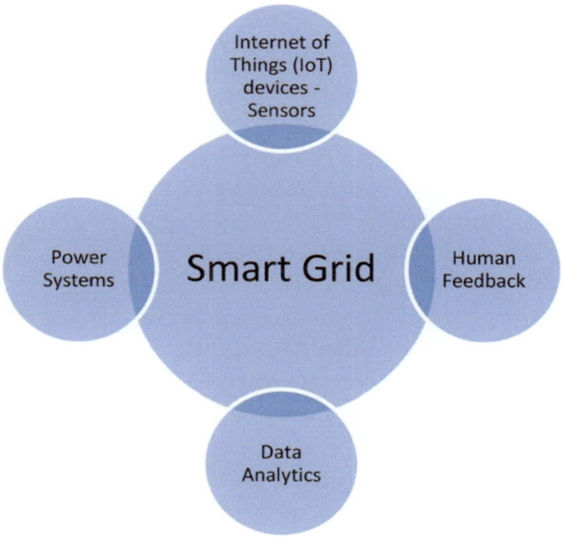

Fig. 12.3 A smart grid for energy efficiency

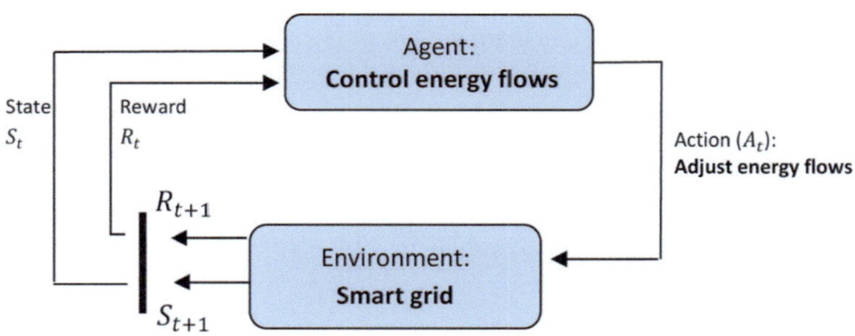

States: Represent the levels of energy consumption
Reward: How to balance energy supply and demand, reduce losses, and minimize costs

Fig. 12.4 A reinforcement learning system for energy efficiency

implementation of RL systems for energy efficiency requires large amounts of high-quality data for training the system. These data should be collected from various sources such as sensors, meters and weather stations [14]. Finally, a RL system for energy efficiency should be integrated with the existing city infrastructure.

12.2.3 Application of Reinforcement Learning in Healthcare

Hospitals in major cities are under enormous pressure. The inhabitants of smart cities can benefit from a Reinforcement Learning (RL) system in healthcare. Below, we present some real-world applications of a RL system in healthcare, such as personalized treatment plans and more efficient allocation of the available medical resources [50]. In particular, a RL system for healthcare can be used to personalize treatment plans for individual patients [56]. The system will take into consideration patient's medical data and recommend the most effective treatment protocols. Patient's feedback is a valuable input into the learning cycle as will allow to adjust treatments accordingly [15]. Additionally, feedback from healthcare professionals can identify operational bottlenecks, leading to improvements in healthcare services and making optimal use of the available scarce financial resources [24, 26, 27]. Figure 12.5 illustrates the proposed RL system for optimizing patients' treatment.

People with chronic diseases such as diabetes, hypertension, or asthma can benefit from a RL for chronic disease management. This system can learn to adjust medication dosages and monitor protocols to ensure optimal health outcomes [3]. The agent can learn how different actions, such as changes in diet, exercise, or medication, impact the patient's condition and adapt accordingly. Providing personalized medical services, such as finding the optimal dosage of medication for a patient is a challenging task. A RL agent can learn which dosages yield the best outcomes while minimizing side effects or adverse reactions [11]. The same apply with patients that must take a combination of different drugs. RL can assist in learning how to combine drugs most effectively based on a patient's unique needs, optimizing treatment regimens for conditions like cancer or autoimmune diseases [16]. Despite the undeniable benefits of RL systems in healthcare, these systems in order to be able to work correctly require vast amounts of high-quality data, to learn and make decisions,

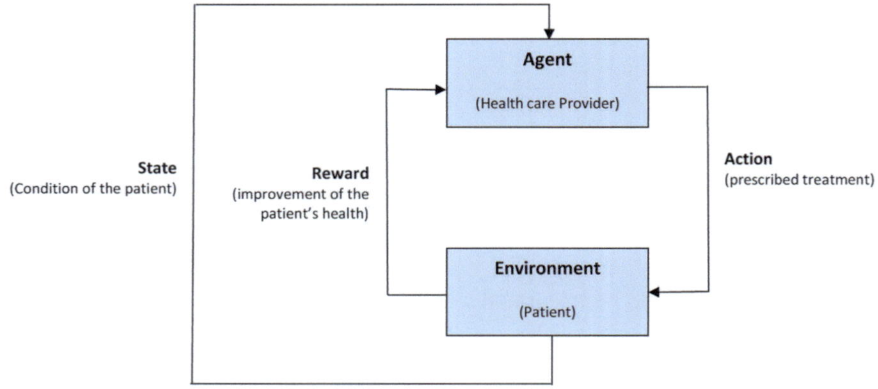

Fig. 12.5 A reinforcement learning system for optimizing patients' treatment

while respecting patients' privacy. Finally, RL-based treatment decisions in healthcare must be transparent and explainable, as mistakes can have serious consequences in patients' health and well-being [49].

12.2.4 Application of Reinforcement Learning for Optimizing the Housing Market

Insufficient housing supply, affect prices and make housing largely inaccessible and unaffordable for many inhabitants of modern megacities [4]. The adoption of RL systems by local authorities can help smart cities to optimize housing development, allocation, and to improve overall living conditions and manage housing demand [52]. A RL system for housing market optimization can greatly benefit from human feedback by aligning its offerings with the evolving needs and preferences of buyers, renters, and investors [37]. Feedback from prospective buyers and tenants can help developers and landlords better understand desirable features, such as energy-efficient designs, proximity to amenities, or modern layouts. Real estate agents and policymakers can also use this input to identify gaps in affordability, availability, or housing quality. Also, machine learning models can help optimizing housing supply based on the current demand and future trends. As shown in Fig. 12.6, a machine learning system could be used to analyse the available data from the housing market and human feedback and recommend where and when new housing projects should be initiated to address areas with high housing demand, as reflected by the forecasted home prices.

A RL system for optimizing the housing market could be connected with a machine learning subsystem (see Fig. 12.6) that can be used for analysing data from the housing market in real-time based demand, housing prices, availability, and socio-economic factors. The RL system can learn the most effective pricing strategies to ensure housing affordability, while maximizing the utilization of available housing stock [39]. For instance, it can help optimize rent prices for different neighbourhoods, ensuring that they remain affordable for lower-income residents, while encouraging new developments in high-demand areas to prevent housing shortages [47]. Despite the undeniable benefits of RL systems for optimizing the housing market, these systems in order to be able to work correctly require large and high-quality datasets to be effective. In particular, finding accurate data on land use, demographics, housing prices, and construction costs can be challenging task. Additionally, the successful implementation of RL for housing requires coordination among various stakeholders, including urban planners, policymakers, developers, and residents [9]. Finally, the system must be transparent and explainable to gain trust from these stakeholders.

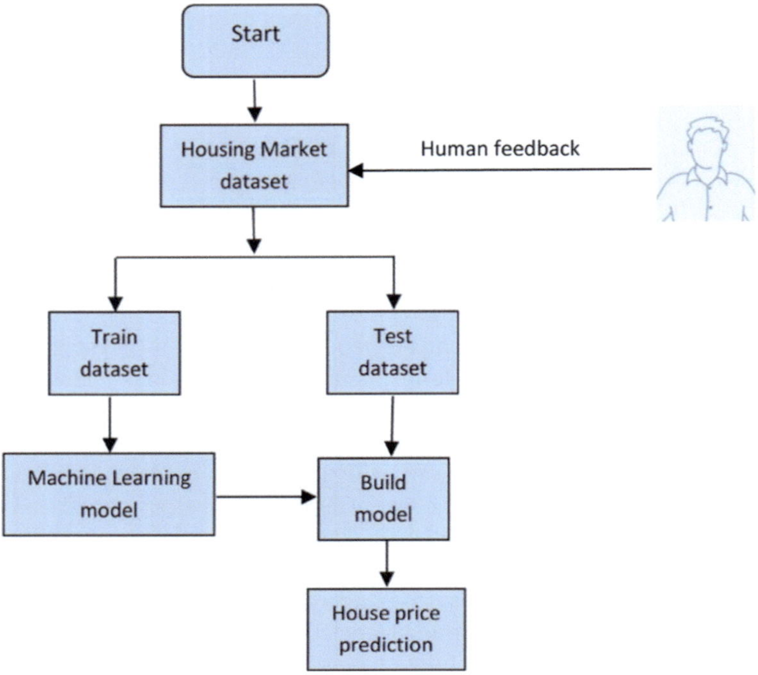

Fig. 12.6 A machine learning system for forecasting home prices

12.2.5 Application of Reinforcement Learning in Waste Management

Modern megacities are increasingly facing waste management challenges due to rapid urbanization. A RL system can optimize many aspects of a waste management system such as waste collection, sorting, recycling, and resource allocation [10]. By applying RL to waste management, smart cities can reduce costs, increase operational efficiency, and improve sustainability [58]. Furthermore, waste management can significantly benefit from human feedback by improving the efficiency and effectiveness of waste reduction, collection, and recycling processes [6]. Feedback from residents and local businesses can highlight challenges such as inconvenient collection schedules, insufficient recycling options, or lack of clarity about waste segregation rules. The feedback from the community allows waste management systems to address specific needs, such as optimizing routes [29], and developing targeted educational campaigns. Different types of waste management techniques must be applied depending on the kind of waste (liquid, biological, solid, hazardous or organic waste). Traditionally, the waste segregation method was relied on manual hand-picking [59], however nowadays manual selection is not considered a viable option as it is heavily labor intensive and prone to errors. For that reason,

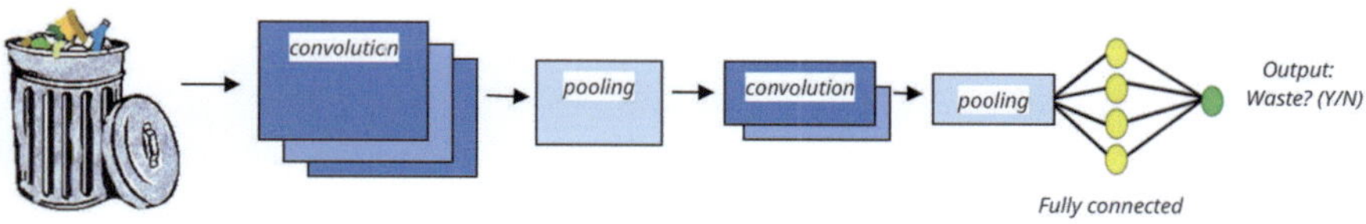

Fig. 12.7 A convolutional neural network for waste detection and classification

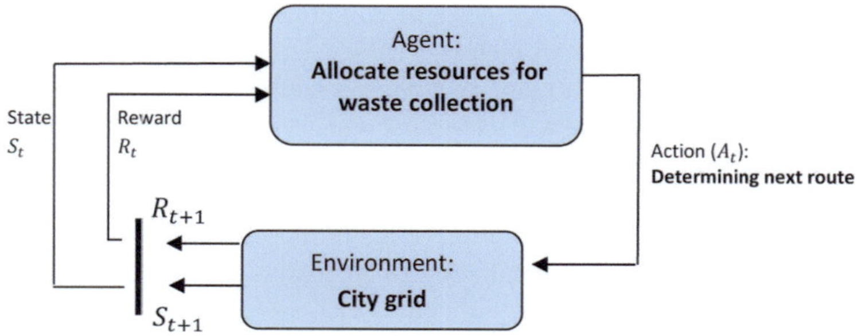

States: Represent the locations of the waste trucks

Reward: How to reduce fuel consumption, minimize operational costs, and provide timely waste pickup

Fig. 12.8 A reinforcement learning system for optimizing routes for waste trucks

as shown in Fig. 12.7, we are proposing a Convolutional Neural Network (CNN) for waste detection and classification.

The Convolutional Neural Network (CNN) for waste detection and classification could be connected with a RL system that will be able to find the optimal routes for waste trucks in real-time based on waste container fullness and the current traffic conditions. Figure 12.8 illustrates the proposed RL system for optimizing routes for waste trucks.

As shown in Fig. 12.8 the RL system will allocate resources for optimizing waste collection. A reward can be given, based on factors such as reduced fuel consumption, timely waste pickup, minimized operational costs, and improving overall efficiency. Waste management requires proper scheduling of waste collection, maintenance of equipment, and allocation of human resources [40]. Traditional systems follow fixed routes to pick up trash and lack the ability to adapt to sudden spikes in waste production, such as after holidays or large events. On the contrary, a RL system can be used to dynamically allocate resources, such as vehicles, workers, and equipment to different areas based on real-time data, such as waste volume, weather conditions, human feedback and operational constraints [38]. This can reduce the chances of over or under staffing and optimize the use of resources.

12.3 Conclusions

Modern megacities face numerous challenges due to their size, rapid growth, and complexity. The size of modern megacities makes more difficult the citizens' participation in the decision-making process on issues that affect their area, such as waste management, housing affordability, energy efficiency, healthcare and suppression of crime. In this article, we propose a Reinforcement Learning System that learns from human-smart city interaction. A smart city collects data not only from Internet of Things (IoT) devices, smart meters, weather sensors and cameras, but also through human feedback. In this article, we propose a reinforcement learning system that will allow citizens of megacities to participate and interact with the smart city and through this interaction, between humans and smart city, the system will learn optimal policies for a number of issues that affect people's life, leading to better quality of life and citizen satisfaction. The innovation of the proposed reinforcement learning system is that it integrates human feedback into the learning cycle allowing citizens to vote on or prioritize city initiatives, influencing RL system's objectives.

Despite the undeniable benefits, RL systems can also pose some threats. Algorithmic bias is one of the most important concerns. Biased historical data, can perpetuate existing biases. This could lead to unfair targeting of certain groups of people or areas, based on stereotypes. Additionally, the implementation of surveillance systems could raise concerns about the privacy of our data and the protection of our civil rights. Another area of concern is the quality of the available data, as inaccurate or incomplete data could lead to poor predictions, misallocation of resources, or ineffective interventions. A RL system should be able to adapt to a continuously changing environment. This would require continuous learning and updates, as new data comes in, which can be challenging, especially in dynamic urban environments.

References

1. A. An, Growing liveable cities: an indicators study of Melbourne's fast-growing metropolitan area. Geo J. **89**, 224 (2024)
2. N. Anand, P. Parwekar, V. Bali, Enhancing energy efficiency in smart cities through robust deep learning frameworks, in *Recent Advancements in Artificial Intelligence. ICRAAI 2023. Innovations in Sustainable Technologies and Computing* ed. by R. Nayak, N. Mittal, M. Kumar, Z. Polkowski, A. Khunteta (Springer, Singapore, 2024). https://doi.org/10.1007/978-981-97-1111-6_10
3. E. Baccour, A. Erbad, A. Mohamed, M. Hamdi, M. Guizani, Reinforcement learning-based dynamic pruning for distributed inference via explainable AI in healthcare IoT systems. Futur. Gener. Comput. Syst. **155**(2024), 1–17 (2024)
4. S.L. Barrera, Understanding informal housing in the Mississippi delta: lessons from Latin American informal settlements, in *Informality and the City* ed. by G. Marinic, P. Meninato. (Springer, Cham, 2022). https://doi.org/10.1007/978-3-030-99926-1_25
5. M.A. Bhat, Comparative review of different techniques for predictive analytics in crime data over online social media, in *Advances in Data and Information Sciences. ICDIS 2023*, ed. by

S. Tiwari, M.C. Trivedi, M.L. Kolhe, B.K. Singh. Lecture Notes in Networks and Systems, vol. 796. (Springer, Singapore, 2024). https://doi.org/10.1007/978-981-99-6906-7_12

6. Z. Cai, H. Fang, J. Yang, L. Fan, T. Ji, Y. Hu, X. Wang, Application of hyperspectral band selection method based on deep reinforcement learning to low-value recyclable waste classification. Process Saf. Environ. Prot. **192**(2024), 1138–1150 (2024)
7. M.M. Dhanvijay, S.C. Patil, Energy efficient deep reinforcement learning approach to control the traffic flow in iot networks for smart city. J. Ambient. Intell. Hum. Comput. **15**, 3945–3961 (2024)
8. L. Ding, W. Du, J. Zhang et al., Better value estimation in Q-learning-based multi-agent reinforcement learning. Soft. Comput. **28**, 5625–5638 (2024)
9. K.A. Farnish, E.A. Schoenfeld, Implications of the COVID-19 pandemic for youth housing and homelessness services. Child Adolesc. Soc. Work J. **41**, 75–94 (2024)
10. J. Gao, A. Wahlen, C. Ju et al., Reinforcement learning-based control for waste biorefining processes under uncertainty. Commun. Eng. **3**, 38 (2024)
11. R. Gayathri, S.K.B. Sangeetha, M. Sandeep Kumar, R. Hariharan, M.B. Benjula Anbu Malar, M. Saurav, Q. Hong, Enhancing heart disease prediction with reinforcement learning and data augmentation. Syst. Soft Comput. **6** 200129
12. Z. Guo, T.J. Norman, E.H. Gerding, Multi-trainer binary feedback interactive reinforcement learning. Ann. Math. Artif. Intell. (2024). https://doi.org/10.1007/s10472-024-09956-4
13. C.F. Hayes, M. Reymond, D.M. Roijers et al., Monte Carlo tree search algorithms for risk-aware and multi-objective reinforcement learning. Auton. Agent. Multi-Agent Syst. **37**, 26 (2023)
14. M. Huotari, A. Malhi, K. Främling, Machine learning applications for smart building energy utilization: a survey. Arch. Computat Methods Eng. **31**, 2537–2556 (2024)
15. A. Ismail, M. Naeem, U.B. Khalid et al., Improving adherence to medication in an intelligent environment using reinforcement learning. J. Reliable. Intell. Environ. **11**, 3 (2025)
16. R. Jain, P. Singh, A. Kaur, An ensemble reinforcement learning-assisted deep learning framework for enhanced lung cancer diagnosis. Swarm Evolut. Comput. **91**, 101767 (2024)
17. K. Jenga, C. Catal, G. Kar, Machine learning in crime prediction. J. Ambient. Intell. Hum. Comput. **14**, 2887–2913 (2023)
18. H.S. Jeong, H. Chung, Bridging smart technologies and healthy cities: a scoping review using WHO's 6P framework. Sustain. Cities Soc. **116**, 105888 (2024)
19. C. Jia, F. Zhang, T. Xu et al., Model gradient: unified model and policy learning in model-based reinforcement learning. Front. Comput. Sci. **18**, 184339 (2024)
20. S.P. Kale, S. Raghunathan, DReLNet: abnormal human activity classification in theft surveillance video using reinforcement learning based fusion. SIViP **19**, 208 (2025). https://doi.org/10.1007/s11760-024-03677-1
21. K. Kocsis, J. Kövendi, B. Bokor, Waste collection route optimisation for the second waste-to-energy plant in Budapest. Sustain. Cities Soc. **117**, 105953 (2024)
22. R. Kumari, R. Raman, R.K. Patel, Regional disparities in social, environmental, and economic indicators among the Indian States. Geo J. **88**, 4351–4371 (2023)
23. K. Liagkouras, A new three-dimensional encoding multiobjective evolutionary algorithm with application to the portfolio optimization problem. Knowl. Based Syst. **163**(2019), 186–203 (2019)
24. K. Liagkouras, K. Metaxiotis, Improving multi-objective algorithms performance by emulating behaviors from the human social analogue in candidate solutions. Eur. J. Oper. Res. **292**(3), 1019–1036 (2021)
25. K. Liagkouras, K. Metaxiotis, G. Tsihrintzis, Incorporating environmental and social considerations into the portfolio optimization process. Ann. Oper. Res. **316**, 1493–1518 (2022)
26. K. Liagkouras, K. Metaxiotis, Multi-period mean–variance fuzzy portfolio optimization model with transaction costs. Eng. Appl. Artif. Intell. **67**(2018), 260–269 (2018)
27. K. Liagkouras, K. Metaxiotis, Improving the performance of evolutionary algorithms: a new approach utilizing information from the evolutionary process and its application to the fuzzy portfolio optimization problem. Ann. Oper. Res. **272**, 119–137 (2019)

28. K. Liagkouras, K. Metaxiotis, Stock market forecasting by using support vector machines, in *Machine Learning Paradigms. Learning and Analytics in Intelligent Systems*, ed. by G. Tsihrintzis, L. Jain, vol. 18. (Springer, Cham, 2020). https://doi.org/10.1007/978-3-030-49724-8_11
29. K. Liagkouras, K. Metaxiotis, Re-examining the optimal routing problem from the perspective of mobility impaired individuals, in *Advances in Assistive Technologies. Learning and Analytics in Intelligent Systems* ed. by G.A. Tsihrintzis, M. Virvou, A. Esposito, L.C. Jain, vol. 28. (Springer, Cham, 2022). https://doi.org/10.1007/978-3-030-87132-1_9
30. K. Liagkouras, K. Metaxiotis, Sources of uncertainty and risk quantification methods in supply chain management: a literature study. Oper. Res. Forum **4**, 90 (2023)
31. K. Liagkouras, K., Metaxiotis, Extracting sentiment from business news announcements for more efficient decision making, in *Advances in Artificial Intelligence-Empowered Decision Support Systems. Learning and Analytics in Intelligent Systems* ed. by G.A. Tsihrintzis, M. Virvou, H. Doukas, L.C. Jain, vol. 39. (Springer, Cham, 2024). https://doi.org/10.1007/978-3-031-62316-5_11
32. F. Liao, A.D. Palma, Q. Tian, Special issue on housing and transportation: affordability, accessibility, and equity. Transp. Policy **125**(2022), 56–58 (2022)
33. Y. Lin, X. Wu, Enhancing knowledge-based professional management of auxiliary police in mega-cities: challenges and solutions. J. Knowl. Econ. **15**, 12216–12252 (2024)
34. U. Lorenz, Reinforcement learning as a subfield of machine learning, in *Reinforcement Learning from Scratch.* (Springer, Cham, 2022). https://doi.org/10.1007/978-3-031-09030-1_1
35. F. Louati, F.B. Ktata, I. Amous, Enhancing intrusion detection systems with reinforcement learning: a comprehensive survey of RL-based approaches and techniques. SN Comput. Sci. **5**, 665 (2024)
36. F.M. Luo, T. Xu, H. Lai et al., A survey on model-based reinforcement learning. Sci. China Inf. Sci. **67**, 121101 (2024). https://doi.org/10.1007/s11432-022-3696-5
37. S.N.M. Moghaddam, H. Cao, Housing, affordability, and real estate market analysis, in *Artificial Intelligence-Driven Geographies. City Development: Issues and Best Practices* (Springer, Singapore, 2024). https://doi.org/10.1007/978-981-97-5116-7_12
38. M.T. Munir, B. Li, M. Naqvi, Revolutionizing municipal solid waste management (MSWM) with machine learning as a clean resource: opportunities, challenges and solutions. Fuel **348**, 128548 (2023)
39. J. Park, J. Baek, Y. Song, Optimizing smart city planning: a deep reinforcement learning framework. ICT Express (2024). https://doi.org/10.1016/j.icte.2024.11.005. ISSN 2405-9595
40. W. Peng, O.K. Sadaghiani, A systematic review on smart waste biomass production using machine learning and deep learning. J. Mater. Cycles Waste Manag. **25**, 3175–3191 (2023)
41. B. Priya, J. Malhotra, Intelligent multi-connectivity based energy-efficient framework for smart city. J. Netw. Syst. Manag. **31**, 48 (2023)
42. T. Pushkar, D. Serogina, N. Matvieieva, H. Sobolieva, H. Zhovtyak, Human-centered approach to shaping smart city development strategies, in *Smart Technologies in Urban Engineering. STUE 2023*, ed. by O. Arsenyeva, T. Romanova, M. Sukhonos, I. Biletskyi, Y. Tsegelnyk. Lecture Notes in Networks and Systems, vol. 808. (Springer, Cham, 2023). https://doi.org/10.1007/978-3-031-46877-3_3
43. M.F. Samsudin, Y.C. Lim, T.N. Rochmah et al., Assessing the performance of non-specialised private hospitals in Malaysia-an upper-middle-income medical tourism destination country using the Pabón-Lasso model. BMC Health Serv. Res. **24**, 1414 (2024)
44. K. Seerangan, M. Nandagopal, T. Govindaraju et al., A novel energy-efficiency framework for UAV-assisted networks using adaptive deep reinforcement learning. Sci. Rep. **14**, 22188 (2024)
45. N. Shah, N. Bhagat, M. Shah, Crime forecasting: a machine learning and computer vision approach to crime prediction and prevention. Vis. Comput. Ind. Biomed. Art **4**, 9 (2021)
46. H.K. Sharma, T. Choudhury, A. Kandwal, Machine learning based analytical approach for geographical analysis and prediction of Boston City crime using geospatial dataset. GeoJ. **88**(Suppl 1), 15–27 (2023)

47. S.Ö. Subaşı, T. Baycan, Impacts of the COVID-19 pandemic on private rental housing prices in Turkey. Asia-Pac J Reg Sci **6**, 1177–1193 (2022)
48. Q. Sun, Y.W. Si, Supervised actor-critic reinforcement learning with action feedback for algorithmic trading. Appl. Intell. **53**, 16875–16892 (2023)
49. P. Tam, I. Song, S. Kang, S. Kim, Privacy-aware intelligent healthcare services with federated learning architecture and reinforcement learning agent, in *Advances in Computer Science and Ubiquitous Computing. CUTECSA 2022*, ed. by J.S. Park, L.T. Yang, Y. Pan, J.H. Park. Lecture Notes in Electrical Engineering, vol. 1028. (Springer, Singapore, 2023). https://doi.org/10.1007/978-981-99-1252-0_78
50. P. Tamrakar, G.R. Pathak, M. Lal, A. Goel, M. Bhende, Personalized healthcare recommendations with Q-learning reinforcement learning, in *Machine Learning Algorithms. ICMLA 2024. Communications in Computer and Information Science*, M. Khurana, A. Thakur, P. Kantha, C.S. Shieh, R.K. Shukla, vol. 2238 (Springer, Cham, 2025). https://doi.org/10.1007/978-3-031-75861-4_16
51. D.J. Vimala, K.S. Kavitha, Adaptive deep Q learning network with reinforcement learning for crime prediction. Evol. Intel. **16**, 685–696 (2023)
52. J. Wang, M. Das, S. Tappert, Applying reinforcement learning to estimating apartment reference rents, in *Artificial Intelligence and Machine Learning in the Travel Industry*, ed. by B. Vinod. (Palgrave Macmillan, Cham, 2023). https://doi.org/10.1007/978-3-031-25456-7_10
53. Q. Wang, Y. Li, R. Li, Ecological footprints, carbon emissions, and energy transitions: the impact of artificial intelligence (AI). Humanit. Soc. Sci. Commun. **11**, 1043 (2024)
54. W. Wang, Reinforcement learning paradigm, in *Principles of Machine Learning*. (Springer, Singapore, 2025). https://doi.org/10.1007/978-981-97-5333-8_10
55. X. Wang, F. Meng, X. Liu et al., Causal explanation for reinforcement learning: quantifying state and temporal importance. Appl. Intell. **53**, 22546–22564 (2023)
56. J. Yang, R. El-Bouri, O. O'Donoghue et al., Deep reinforcement learning for multi-class imbalanced training: applications in healthcare. Mach. Learn. **113**, 2655–2674 (2024)
57. P. Zhao, J. Li, M. Zhang, Unequal roles of cities in the intercity healthcare system. Nat. Cities (2025). https://doi.org/10.1038/s44284-024-00185-8
58. W. Zhouwen, X. Chen, Y. Lin, J. Wen, Y. Chen, A smart home energy management system based on human activity recognition and deep reinforcement learning. Energy Build. **325**, 114951 (2024)
59. M. Zuccotto, A. Castellini, D.L. Torre et al., Reinforcement learning applications in environmental sustainability: a review. Artif. Intell. Rev. **57**, 88 (2024)

Chapter 13
A Diagnostic Clinical Decision Support System Which Leverages the Medical Information of Patients Suffering from Pneumonia

Evgenia Psarra and Dimitris Apostolou

Abstract In today's era, there has been a surge in the risk of pneumonia attributed to modern lifestyles emerging as a prevalent concern. Various factors contribute to the development of this disease, including bacteria, viruses, fungi, inhalation into the lungs of food, liquids, or other substances, environmental factors such as the exposure to certain environmental factors such as smoke, pollutants, or chemicals, immune system weakness, age, malnutrition, smoking chronic health conditions including conditions such as chronic obstructive pulmonary disease, asthma, diabetes, heart disease, or kidney disease. Artificial intelligence now plays a crucial role in assisting medical professionals in diagnosing such severe conditions which can result in chronic effects such as long-term lung damage or worsen existing chronic lung conditions. This study aims to analyze patients' health metrics to determine if they are affected by pneumonia, utilizing machine learning techniques. Specifically, we have developed an algorithm that leverages Convolutional Neural Networks, while processing medical data to predict potential health issues associated with pneumonia.

CCS Concepts Information system · Machine learning · Artificial intelligence

Additional Keywords and Phrases Pneumonia · Convolutional neural networks · Machine learning · Decision support system

E. Psarra (✉) · D. Apostolou
Department of Informatics, University of Piraeus, Piraeus, Greece
e-mail: jennypsarra@unipi.gr

D. Apostolou
e-mail: dapost@unipi.gr

D. Apostolou
Institute of Communications and Computer Systems, Athens, Greece

G. A. Tsihrintzis et al. (eds.), *Human-Smart City Interactions and User-Citizen Experiences*, Learning and Analytics in Intelligent Systems 58,
https://doi.org/10.1007/978-3-032-06364-9_13

13.1 Introduction

In the 21st century, aspiration pneumonia is prevalent among older patients and has a high mortality rate. Aspiration pneumonia is diagnosed by confirming inflammatory findings in the lungs along with evident aspiration or the presence of dysphagia. It is particularly common in hospitalized cases of community-acquired pneumonia, nursing and healthcare-associated pneumonia, and hospital-acquired pneumonia. The incidence of aspiration pneumonia continues to rise annually [1].

Pneumonia is categorized into three subgroups based on the clinical setting in which the patient develops infection symptoms: community-acquired pneumonia, hospital-acquired pneumonia, and ventilator-associated pneumonia [2]. The study by Koivula et al. [3] provides new insights into the long-term outcomes for patients with community-acquired pneumonia, revealing that excess mortality rates persist many years after the initial infection. Torres et al. [4] evaluated the outcomes of elderly patients with community-acquired pneumonia treated at an acute-care hospital, focusing on the severity of pneumonia, functional status, comorbidities, and frailty.

Increasing age was independently linked to risk-adjusted short-term and long-term mortality in critically ill patients with pneumonia [5]. To be noted that in low-risk patients with community-acquired pneumonia, there was a specific correlation between the severity of the illness and its cause [6].

This work aims to analyze patient's chest x-ray to predict whether an individual suffers from pneumonia disease using machine learning techniques. We developed an algorithm that leverages deep neural network technology to examine data features and determine the patient's health condition. In our research, we also explored the optimal design of the neural network's architecture, achieving high prediction accuracy. We compared the performance of our architecture with others architectures (VGG16, VGG19, ResNet) in predicting pneumonia disease based on the medical data.

We utilized the "Chest X-ray images (pneumonia)" dataset, which includes x-rays of patients who suffer or not from pneumonia [7]. The patient's medical X-ray serves as input for the prediction algorithm, and the medical diagnosis is the output generated by our Convolutional Neural Network.

We experimented on the parameters of the CNN architecture and we achieved satisfactory results by taking into consideration the following metrics: (i) accuracy, (ii) precision, (iii) recall, (iv) f1 score (a combination of precision and recall scores), (v) specificity, (vi) roc-auc score (Compute Area Under the Receiver Operating Characteristic Curve) and the cohen's kappa score. All these metrics showed satisfactory results about the efficiency of our prediction CNN model.

Additionally, we relied on our previous research which employs contextual information [8] to establish attribute-based access control policies [9], where the machine learning component is integrated into the attribute-based access control system for patient medical records, as explored in our previous works [10–17]. This integration aims to enhance the healthcare system by prioritizing patient well-being and data security while providing doctors with comprehensive insights into the patient's medical status.

13.2 Background and Related Work

The diagnosis of pneumonia is both straightforward and complex. Recent studies are challenging traditional views of pneumonia and the radiological gold standards that have been the basis of research for decades. Specifically, the reliability of chest radiographs in diagnosing pneumonia is now highly questionable compared to computed tomography (CT) scans. Pneumonia can be defined in various contexts, such as clinical, pathological, radiological, or microbiological, depending on the question at hand, and often it is a combination of these definitions [18]. In summary, despite pneumonia being a very common and serious disease, its diagnosis remains difficult, with numerous studies indicating that clinicians' accuracy is not particularly high when based on traditional radiological standards. Additionally, traditional chest radiography gold standards are now being questioned by CT studies that reveal up to 40% of pneumonia cases may be misdiagnosed, either falsely identified or missed entirely [18].The key takeaway for clinicians, amidst all the uncertainty about the most accurate way to diagnose pneumonia, is that uncertainty exists.

Kermany et al. [19] present the broad applicability of their AI system, by using Deep Neural Networks in diagnosing pneumonia using chest X-ray images [7]. When comparing chest X-rays showing pneumonia to those that are normal, their system achieved an accuracy rate of 92.8%. Yee et al. [20] present the development of a pneumonia diagnosis system utilizing convolutional neural network-based feature extraction, by using the dataset [7]. Three models were employed: K-Nearest Neighbor, Neural Network, and Support Vector Machine. The findings reveal that the Neural Network model attained the highest sensitivity at 84.1%, followed by Support Vector Machine at 83.5% and the K-Nearest Neighbor Algorithm at 83.3%, by using the dataset of [7]. In the work of Sharma et al. [21], a deep learning (DL) model utilizing VGG16 was employed to detect and classify pneumonia using two chest X-ray image datasets. The VGG16 with Neural Networks (NN) achieved an accuracy of 92.15%, for the first dataset [7] and 95.4% for a second dataset.

In the study of Sorić et al. [22], a Convolutional Neural Network machine learning model was developed using a dataset comprising both pneumonia and non-pneumonia images that the model needed to classify accurately, by using dataset [7]. Ultimately, the model achieved impressive results with a high accuracy of 90.38%. Acharya et al. [23] presented a novel approach for identifying and classifying different types of pneumonia. Utilizing the Kaggle dataset [7], they focused on specific features relevant to the disease. They trained their model using both the right and left parts of chest X-rays, enhancing the robustness and improving the accuracy of the classification process. The accuracy rate of their approach isn't recorded. Abubeker et al. [24] proposed Bek-Bas network (B2-Net) model which can differentiate between normal, bacterial, and viral pneumonia in chest x-ray images Their proposed model demonstrates an accuracy of 97.69%, effectively distinguishing between normal, bacterial, and viral pneumonia in chest X-ray images by utilizing the dataset [7]. Kundu et al. [25] implemented deep transfer learning to address the limited availability of data and developed an ensemble comprising three convolutional neural network models: GoogLeNet, ResNet-18, and DenseNet-121. This approach was evaluated on two publicly available pneumonia X-ray datasets, provided by Kermany et al. [7] and the Radiological Society of North America (RSNA), respectively, using a five-fold cross-validation scheme. The proposed method achieved accuracy rates of 98.81% and 86.85% respectively.

13.3 Methods

13.3.1 Convolutional Neural Network Model

Convolutional neural networks are a type of multilayer perceptron designed to mimic the visual context's behavior. They address the challenges of the Multi-layered Perceptron architecture by leveraging the strong spatially local correlations in natural images. Unlike Multi-layered Perceptron, convolutional neural networks have distinct features:

3D Volumes of Neurons: The layers in a convolutional neural network have neurons organized in three dimensions: width, height, and depth. Each neuron in a convolutional layer connects to a small region of the previous layer, known as a receptive field. Various layer types, both locally and fully connected, are stacked to build the CNN architecture.

Local Connectivity: Using the concept of receptive fields, convolutional neural networks maintain spatial locality by enforcing local connections between neurons in adjacent layers. This design ensures that the learned "filters" respond strongly to local input patterns. Stacking multiple layers results in nonlinear filters that progressively become more global, creating representations from small input parts to larger areas.

Shared Weights: Filters in convolutional neural networks are replicated across the visual field, with these replicated units sharing the same parameters (weight vector and bias) to form a feature map. Consequently, all neurons in a convolutional layer respond to the same feature within their receptive field. This replication grants translational equivariance, ensuring that the layer's activation map remains consistent under shifts in the input feature locations.

Pooling: In pooling layers, feature maps are divided into rectangular sub-regions, and features within each rectangle are down-sampled to a single value, typically the average or maximum value. Pooling reduces the feature map sizes and provides a degree of local translational invariance, enhancing the convolutional neural networks robustness to variations in feature positions.

These features enable convolutional neural networks to generalize better on vision tasks. Weight sharing significantly reduces the number of parameters, decreasing memory requirements and allowing for the training of larger, more powerful networks.

In the beginning of our prediction procedure and after loading the dataset we implement an image normalization procedure where we apply a grayscale normalization by converting the range from 0–255 to a new range of decimal numbers from 0 to 1. By this way the Convolutional Neural Network (CNN) will converge faster to the optimum solution. Afterwards, the architecture of our CNN model is defined and all its parameters are appropriately set. Then, after the compilation phase where we experimented on several parameters of the compiler, we proceeded to the fitting phase of the process and finally we evaluated this model by implementing k-fold cross validation on the dataset.

13.4 Implementation

13.4.1 The Proposed Neural Network Architecture

After experimentation with a number of CNN parameters, which are presented in detail in the following paragraph 4.2, we arrived at the final and most effective design of the CNN architecture, which consists of the layers as shown in Table 13.1.

In this model the input layer consists of a convolutional two-dimensional layer with 32 parallel filters, where the images of the dataset are loaded. The dimensions of the images, when loaded, are adjusted to (240, 240) pixels so as to fit the standards of all the architectures we study. Additionally, all the originally gray images, where loaded as 3 colors channel images so as to be appropriately used in all the models we study. It should be noted that all these images are loaded as three-dimensional arrays (rows, columns, channels), where the parameter of channels equals to number 3 (red, blue, green).

The two-dimensional convolutional layers of our neural network comprise several parallel filters ranging from 32 to 512, as shown in Table 13.1. The dimension of all these filter patches are (3, 3) pixels. In these convolutional layers the movement of the filters between applications, which is referred as stride, is set to (1, 1) pixels. Additionally, the non-linear activation function ReLU is applied to the feature maps output of every convolutional layer.

Furthermore, the architecture of our model, as shown in Table 13.1, comprises pooling layers (MaxPool2D) so as to handle the problem of downsampling. The pool size of these Max Pooling layers is set to (2, 2). Additionally, our model comprises several dropout layers which serve to prevent overfitting by randomly disabling a proportion of neurons during training. Finally, the last part of our model consists of a series of dense layers culminating in the final dense layer that includes a single neuron and a sigmoid activation function that serves for the final binary classification decision.

Table 13.1 Architecture design of our CNN model

Layers	Layer's details
Conv2D, 32 filters (3*3)	input_shape = (240,240,3) strides = 1 activation = 'relu' BatchNormalization
MaxPool2D 2*2	strides = 2
Conv2D, 64 filters (3*3)	strides = 1 activation = 'relu' BatchNormalization
Dropout	(0.15)
MaxPool2D 2*2	strides = 2
Conv2D, 128 filters (3*3)	strides = 1 activation = 'relu' BatchNormalization
Dropout	(0.15)
MaxPool2D 2*2	strides = 2
Conv2D, 256 filters (3*3)	strides = 1 activation = 'relu' BatchNormalization
Dropout (0.2)	
MaxPool2D 2*2	strides = 2
Conv2D, 512 filters (3*3)	strides = 1 activation = 'relu' BatchNormalization
Dropout	(0.2)
MaxPool2D 2*2	strides = 2
Flatten	
Dense (units = 256)	activation = 'relu'
Dropout	(0.25)
Dense (units = 128)	activation = 'relu'
Dropout	(0.25)
Dense (units = 164)	activation = 'relu'
Dense (units = 32)	activation = 'relu'
Dense (units = 1)	activation = 'sigmoid'

13.4.2 The CNN Architecture Study

The various stages of the study on the configuration of the parameters of the architecture of our CNN are presented in detail in the following Table 13.2.

In the design Number 1 the CNN architecture consists of sequential convolutional layers comprising 64, 64, 128, 128, 256 and 256 filters (3*3 pixels) respectively. Every block of identical layers is followed by a MaxPool2d (2*2) layer. The next design (number 2) turned out to be more effective than the previous one. In this structure, layers with corresponding filter numbers of 32 and 512 were attached while in this design Batch Normalization and sparse Dropout of neurons were considered. The design number 3, where the double layers are respectively reduced to single ones and furthermore an additional convolutional layer of 32 filters is considered, shows better accuracy than both the previous ones An additional layer of 512 filters, as shown in number 4, proved to provide better performance than all the previous designs, while the inclusion of batch normalization along with sparse dropout layers

Table 13.2 CNN model architecture study

No	CNN model architecture	Accuracy (%)	Precision (%)	Recall (%)	F1 score (%)	Specificity (%)	Roc-auc (%)	Kappa (%)
1	64–64–128–128–256–256 without Dropout and Batch Normalization	95.27	96.25	97.33	96.78	89.71	93.52	87.88
2	32–64–64–128–128–256–256–512–512 with Dropout and Batch Normalization	95.31	95.19	98.64	96.86	86.29	92.46	87.57
3	32–64–128–256 without Dropout and Batch Normalization	96.20	97.69	97.09	97.39	93.78	95.44	90.40
4	32–64–128–256–512 without Dropout and Batch Normalization	96.35	97.21	97.82	97.51	92.38	95.10	90.69
5	***32–64–128–256–512 with Dropout and Batch Normalization***	***96.76***	***97.56***	***98.03***	***97.79***	***93.33***	***95.68***	***91.75***

results in the best accuracy in our study on the specific consideration of this dataset, as shown in design number 5.

13.4.3 The Dataset

We trained this model on the dataset [7] which comprises a total of 5,840 X-Ray images in jpeg format, across two categories: Pneumonia and Normal. This dataset consists of 5,840 patients from which 1575 (26.97%) were diagnosed as healthy (they were classified in the normal category) and 4265 (73.03%) as infected from pneumonia (they were classified in the pneumonia category).

The chest X-ray images were sourced from retrospective cohorts of pediatric patients aged one to five years old at Guangzhou Women and Children's Medical Center, Guangzhou. These images were obtained as part of the patients' routine clinical care. All chest radiographs underwent an initial quality control screening to eliminate low-quality or unreadable scans. Subsequently, two expert physicians graded the diagnoses for the images before they were used to train the AI system. To mitigate potential grading errors, a third expert also evaluated the evaluation set. As it is illustrated below in Fig. 13.1, the left x-ray illustrates the lungs of a healthy person, where the x-ray in the middle, and the x-ray to the left characterize a person who suffers from pneumonia. To be noted that that x-rays as demonstrated in the following figure have been taken from the dataset [7].

We created a Convolutional Neural Network (CNN) to forecast the likelihood of pneumonia disease and conducted experiments to refine its architecture. We employed the x-ray per patient of the dataset as the input, and the probability assessment of patient's lungs to be infected or not, as the output of the network.

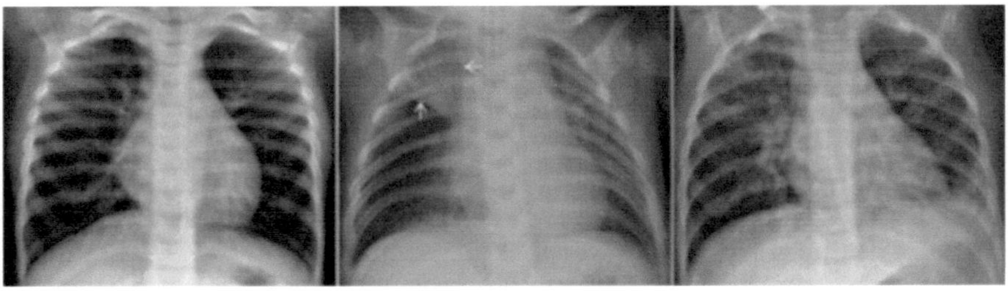

Fig. 13.1 Examples of samples from the dataset [7]

Table 13.3 Dataset splits for 5-fold cross validation

k-folds =5	Normal	Pneumonia
Train = 80%	1260	3412
Test = 20%	315	853
Total Records	1575	4265

13.4.4 Five-Fold Cross Validation

To evaluate the performance of our model we employed a 5-fold cross-validation statistical method and thus having lower bias than other methods. The structure of these splits is shown in Table 13.3.

It should be noted that we implemented the StratifiedKFold which is a special variation of the kfold cross validation and which takes class information into account to avoid building folds with imbalanced class distributions and which is the most suitable for our binary classification problem ("normal" or "pneumonia"). Thus, the folds are made by preserving the percentage of samples for each class.

13.4.5 Criteria For Performance Evaluation

In this work we firstly evaluated the performance of the models we developed with the performance index of accuracy (Eq. 13.2)

$$accuracy = \frac{TP + TN}{TP + TN + FP + FN} \tag{13.2}$$

where TP and TN denote the true positives and true negatives, while FP and FN denote the false positive and false negative respectively.

We additionally implemented the following performance evaluation indices:

$$precision = \frac{TP}{TP + FP} \tag{13.3}$$

$$recall = \frac{TP}{TP + FN} \tag{13.4}$$

$$f1 = \frac{2 * precision * recall}{precision + recall} = \frac{2 * TP}{2 * TP + FP + FN} \tag{13.5}$$

$$specificity = \frac{TN}{TN + FP} \tag{13.6}$$

$$cohen's\ kappa = \frac{2 * (\mathrm{TP} * \mathrm{TN} - \mathrm{FN} * \mathrm{FP})}{(\mathrm{TP} + \mathrm{FP}) * (\mathrm{FP} + \mathrm{TN}) + (\mathrm{TP} + \mathrm{FN}) * (\mathrm{FN} + \mathrm{TN})} \tag{13.7}$$

Furthermore, we implemented the measure roc-auc (receiver operating characteristic curve). The area under the ROC curve (AUC) represents the probability that the model, given a randomly chosen positive and negative example, will rank the positive higher than the negative.

13.5 Results

13.5.1 Performance of Our Model

In Fig. 13.2 the model performance metrics of our model are demonstrated. All these measures (accuracy, precision, recall, f1, specificity, roc-auc, cohen's kappa) are illustrated according to each k-fold cross validation. In the last two rows of the accompanying table the mean of each performance measure along with the corresponding standard deviation are shown. It should also be noted that in the compiling phase of the model, we leveraged the Adam compiler with the "learning rate" parameter equal to 1e-4. According to [25] the Adam optimizer is the most suitable to train the k-fold consideration of this dataset. Additionally, during model training, we implement callbacks to locate and store the best model structure during the iteration process.

We experimented on several values of the "learning rate" parameter of the Adam compiler and got the results shown in Table 13.4. Finally, we chose the "learning rate" = 0.0001 it produces the best performance results.

13.5.2 Performance of VGG-19 Model

Furthermore, we implemented the VGG19 algorithm (with color channels per image = 3) on the same dataset and with the same dimensions (240,240) during loading, we additionally applied exactly the same compiler (Adam, learning rate = 1e-4) and finally we got the following evaluation results, as shown in Fig. 13.3.

13.5.3 Performance of VGG-16 Model

Additionally, we implemented the VGG19 algorithm on the same dataset and with the same dimensions (240,240) during loading, and with the same compilation parameters, as shown in Fig. 13.4.

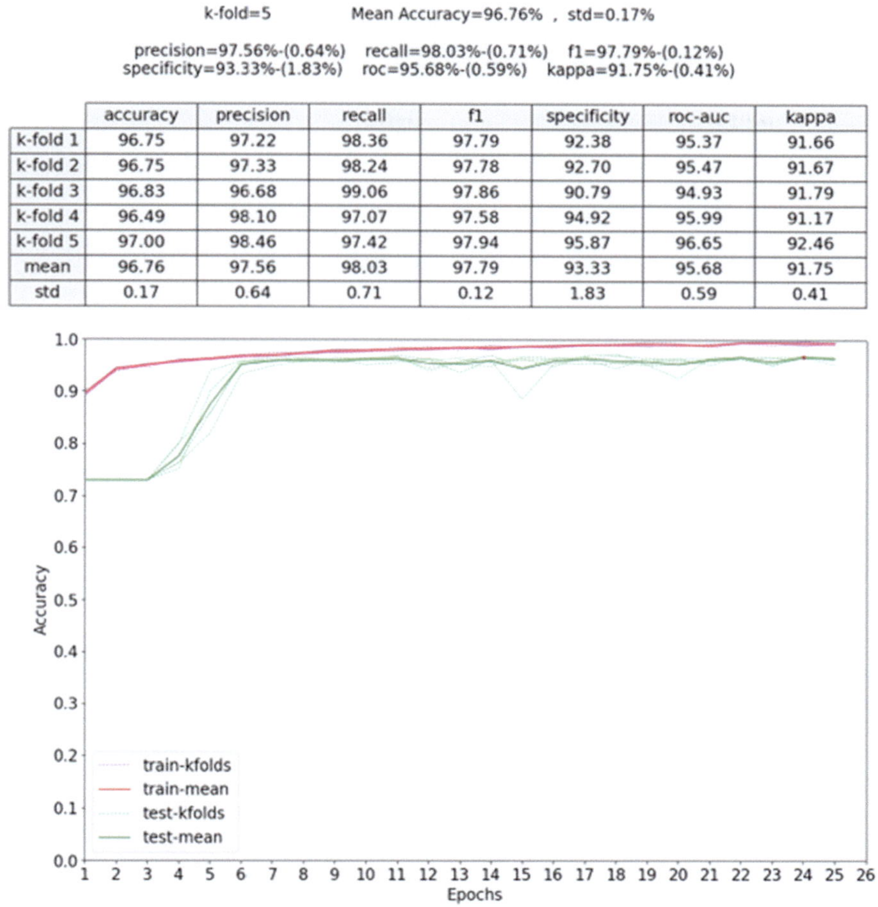

	accuracy	precision	recall	f1	specificity	roc-auc	kappa
k-fold 1	96.75	97.22	98.36	97.79	92.38	95.37	91.66
k-fold 2	96.75	97.33	98.24	97.78	92.70	95.47	91.67
k-fold 3	96.83	96.68	99.06	97.86	90.79	94.93	91.79
k-fold 4	96.49	98.10	97.07	97.58	94.92	95.99	91.17
k-fold 5	97.00	98.46	97.42	97.94	95.87	96.65	92.46
mean	96.76	97.56	98.03	97.79	93.33	95.68	91.75
std	0.17	0.64	0.71	0.12	1.83	0.59	0.41

Fig. 13.2 Performance of our model with Adam optimizer (learning rate = 1e-4)

Table 13.4 Model performance metrics for several learning rates of Adam optimizer (at 25 epochs)

Learning rates	Learning rates	Accuracy (%)	Precision (%)	Recall (%)	Fi score (%)	Specificity (%)	Roc-auc (%)	Kappa (%)
0.01	1e-2	89.97	91.26	86.98	93.69	70.98	83.98	68.44
0.001	1e-3	93.82	94.24	97.96	95.95	82.60	90.28	82.83
0.0001	1e-4	*96.76*	*97.56*	98.03	*97.79*	*93.33*	*95.68*	*91.75*
0.00001	1e-5	95.63	95.41	*98.78*	97.06	87.11	92.95	88.56

k-fold=5 Mean Accuracy=96.13% , std=1.00%

precision=97.07%-(0.73%) recall=97.66%-(1.32%) f1=97.36%-(0.70%)
specificity=92.00%-(2.12%) roc=94.83%-(1.11%) kappa=90.14%-(2.45%)

	accuracy	precision	recall	f1	specificity	roc-auc	kappa
k-fold 1	96.83	97.44	98.24	97.84	93.02	95.63	91.90
k-fold 2	94.26	97.01	95.08	96.03	92.06	93.57	85.68
k-fold 3	95.89	95.69	98.83	97.23	87.94	93.38	89.27
k-fold 4	96.83	97.44	98.24	97.84	93.02	95.63	91.90
k-fold 5	96.83	97.78	97.89	97.83	93.97	95.93	91.95
mean	96.13	97.07	97.66	97.36	92.00	94.83	90.14
std	1.00	0.73	1.32	0.70	2.12	1.11	2.45

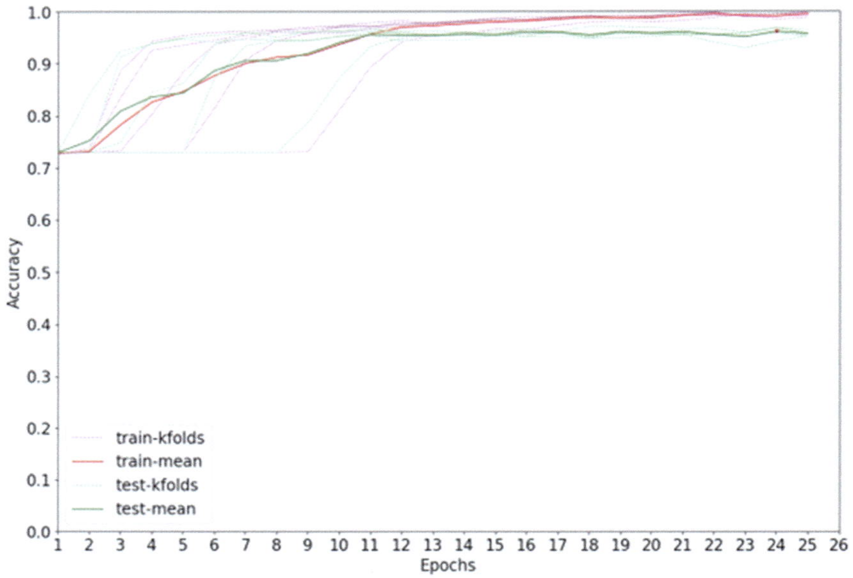

Fig. 13.3 Performance of the VGG19 model compiled with Adam optimizer (learning rate = 1e-4)

13.5.4 Performance of ResNet-50 Model

Additionally, we implemented the ResNet-50 algorithm on the same dataset and with the same dimensions (240,240) during loading, and with the same compilation parameters, as illustrated in Fig. 13.5.

k-fold=5 Mean Accuracy=96.25% , std=0.46%

precision=97.03%-(0.37%) recall=97.87%-(0.55%) f1=97.44%-(0.32%)
specificity=91.87%-(1.05%) roc=94.87%-(0.56%) kappa=90.41%-(1.16%)

	accuracy	precision	recall	f1	specificity	roc-auc	kappa
k-fold 1	96.92	97.56	98.24	97.90	93.33	95.79	92.13
k-fold 2	96.58	97.21	98.12	97.67	92.38	95.25	91.24
k-fold 3	96.23	96.44	98.48	97.45	90.16	94.32	90.26
k-fold 4	95.63	97.07	96.95	97.01	92.06	94.51	88.93
k-fold 5	95.89	96.86	97.54	97.20	91.43	94.48	89.50
mean	96.25	97.03	97.87	97.44	91.87	94.87	90.41
std	0.46	0.37	0.55	0.32	1.05	0.56	1.16

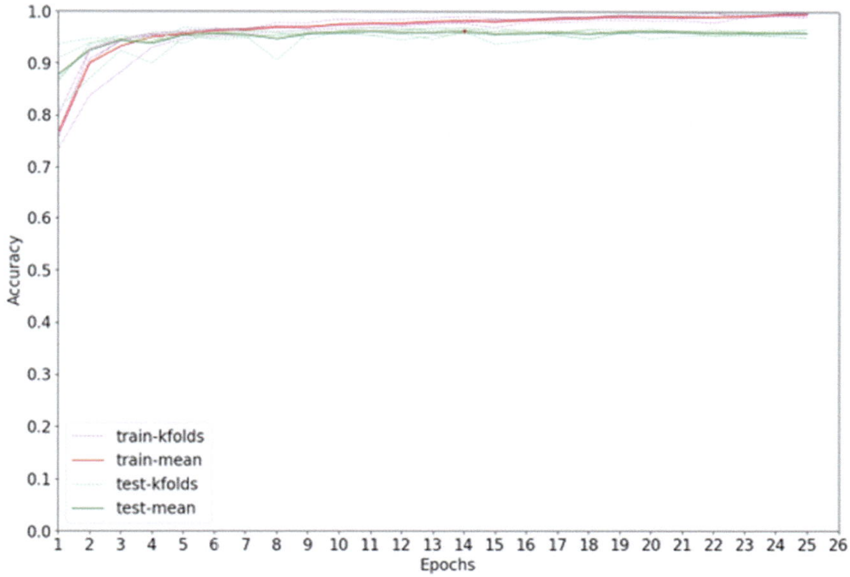

Fig. 13.4 Performance of the VGG16 model compiled with Adam optimizer (learning rate = 1e-4)

13.6 Discussion

The following Table 13.5 summarizes all the performance metrics (accuracy, precision, recall, f1 score, specificity, roc-auc score and cohen's kappa). Most of these metrics showed satisfactory results regarding the effectiveness of our prediction CNN model compared to the models we examined.

We deduce that our model outperforms the VGG16, VGG19 and ResNet50 architectures architecture on the specific dataset. Furthermore, our literature research revealed that our accuracy of this model (96.76%), is better regarding the classic convolutional neural networks' architecture compared with the works of Kermany et al. [19], and Sharma et al. [21], with the equivalent accuracy percentages of 92.8% and 92.15% respectively. On the contrary, regarding hybrid methods we observe that they perform better, as seen from the work of Abubeker et al. [24] who created

k-fold=5 Mean Accuracy=95.92% , std=0.78%

precision=97.75%-(0.25%) recall=96.65%-(0.96%) f1=97.19%-(0.55%)
specificity=93.97%-(0.67%) roc=95.31%-(0.68%) kappa=89.77%-(1.89%)

	accuracy	precision	recall	f1	specificity	roc-auc	kappa
k-fold 1	96.49	97.99	97.19	97.59	94.60	95.89	91.15
k-fold 2	94.52	97.59	94.84	96.20	93.65	94.25	86.42
k-fold 3	96.32	97.54	97.42	97.48	93.33	95.38	90.66
k-fold 4	95.63	97.51	96.48	96.99	93.33	94.91	89.01
k-fold 5	96.66	98.11	97.30	97.70	94.92	96.11	91.58
mean	95.92	97.75	96.65	97.19	93.97	95.31	89.77
std	0.78	0.25	0.96	0.55	0.67	0.68	1.89

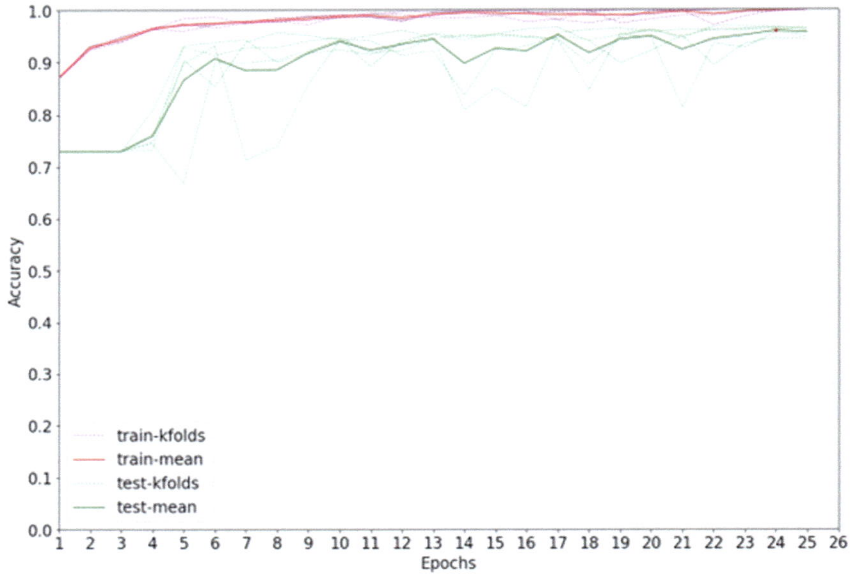

Fig. 13.5 Performance of the ResNet-50 model compiled with Adam optimizer (learning rate = 1e-4)

Table 13.5 Classifiers comparison on predicting pneumonia

Models	Accuracy	Precision	Recall	Fi score	Specificity	Roc-auc	Kappa
VGG-19	96.13	97.07	97.66	97.36	92.00	94.83	90.14
VGG-16	96.25	97.03	97.87	97.44	91.87	94.87	90.41
ResNet-50	95.92	***97.75***	96.65	97.19	***93.97***	95.31	89.77
This work	***96.76***	97.56	***98.03***	***97.79***	93.33	***95.68***	***91.75***

an Bek-Bas network (B2-Net) model with accuracy 97.67%, and Kundu et al. [25] who developed a network comprising three convolutional neural network models: GoogLeNet, ResNet-18, and DenseNet-121, with accuracy 98,81%.

13.7 Conclusions

Predicting pneumonia using decision support mechanisms offers valuable assistance to medical staff, enabling them to accurately assess the disease and treat patients more promptly and effectively. We deduce that our model outperforms the VGG16, VGG19, and ResNet50 architectures on dataset [7]. In future work, we intend to explore the prediction mechanisms for other serious diseases.

Acknowledgments This work has been partly supported by the University of Piraeus Research Center.

References

1. S. Teramoto, The current definition, epidemiology, animal models and a novel therapeutic strategy for aspiration pneumonia. Respir. Investig. **60**(1), 45–55 (2022). https://doi.org/10.1016/j.resinv.2021.09.012
2. N.P. Dueck, S. Epstein, T. Franquet, C.C. Moore, J. Bueno, Atypical pneumonia: definition, causes, and imaging features. RadioGraphics **41**(3), 720–741 (2021). https://doi.org/10.1148/rg.2021200131
3. I. Koivula, M. Stén, P.H. Mäkelä, Prognosis after community-acquired pneumonia in the elderly. Arch. Intern. Med. **159**(14), 1550. https://doi.org/10.1001/archinte.159.14.1550
4. O.H. Torres et al., Outcome predictors of pneumonia in elderly patients: Importance of functional assessment. J. Am. Geriatr. Soc. **52**(10), 1603–1609 (2004). https://doi.org/10.1111/j.1532-5415.2004.52492.x
5. W.I. Sligl, D.T. Eurich, T.J. Marrie, S.R. Majumdar, Age still matters: prognosticating short- and long-term mortality for critically ill patients with pneumonia. Crit. Care Med. **38**(11), 2126–2132 (2010). https://doi.org/10.1097/ccm.0b013e3181eedaeb
6. M. Falguera et al., Nonsevere community-acquired pneumonia. Arch. Intern. Med. **161**(15), 1866 (2001). https://doi.org/10.1001/archinte.161.15.1866
7. P. Mooney, Chest X-ray images (pneumonia). (Mar 2018). https://www.kaggle.com/datasets/paultimothymooney/chest-xray-pneumonia. Accessed 20 May 2024
8. E. Psarra, Y. Verginadis, I. Patiniotakis, D. Apostolou, G. Mentzas, A context-aware security model for a combination of attribute-based access control and attribute-based encryption in the healthcare domain. Adv. Intell. Syst. Comput. **2020**, 1133–1142 (2020). https://doi.org/10.1007/978-3-030-44038-1_104
9. E. Psarra, I. Patiniotakis, Y. Verginadis, D. Apostolou, G. Mentzas, Securing access to healthcare data with context-aware policies, in *2020 11th International Conference on Information, Intelligence, Systems and Applications (IISA)* (July 2020). https://doi.org/10.1109/iisa50023.2020.9284393
10. E. Psarra, Y. Verginadis, I. Patiniotakis, D. Apostolou, G. Mentzas, Accessing electronic health records in critical incidents using context-aware attribute-based access control. Intell. Decis. Technol. **15**(4), 667–679 (2022). https://doi.org/10.3233/idt-210214

11. E. Psarra, D. Apostolou, Y. Verginadis, I. Patiniotakis, G. Mentzas, Context-based, predictive access control to Electronic Health Records. Electronics **11**(19), 3040 (2022). https://doi.org/10.3390/electronics11193040
12. E. Psarra, Clinical decision support systems–Diagnostic and prognostic attribute-based access control in acute care. Dissertation (2023)
13. E. Psarra, D. Apostolou, Y. Verginadis, I. Patiniotakis, G. Mentzas, Permissioned blockchain network for proactive access control to electronic health records. BMC Med. Inf. Decis. Mak. **24**, 1 (2024). https://doi.org/10.1186/s12911-024-02708-8
14. M.T. de Oliveira, Y. Verginadis, L.H.A. Reis, E. Psarra, I. Patiniotakis, S.D. Olabarriaga, AC-ABAC: attribute-based access control for electronic medical records during Acute Care. Exp. Syst. Appl. **213**, 119271 (2023). https://doi.org/10.1016/j.eswa.2022.119271
15. E. Psarra, D. Apostolou, Dynamic and personalized access control to electronic health records, in *Learning and Analytics in Intelligent Systems* (2024), pp. 129–153. https://doi.org/10.1007/978-3-031-62316-5_5.
16. E. Psarra, D. Apostolou, A neural network-based context handling mechanism for accessing the medical information of patients suffering from coronary heart disease, in *2024 15th International Conference on Information, Intelligence, Systems & Applications (IISA)* (July 2024), pp. 1–6. https://doi.org/10.1109/iisa62523.2024.10786708
17. E. Psarra, D. Apostolou, A multi-layer Perceptron neural network-based access control mechanism to breast cancer patients' Health Data, in *2024 15th International Conference on Information, Intelligence, Systems & Applications (IISA)* (July 2024), pp. 1–4. https://doi.org/10.1109/iisa62523.2024.10786721
18. G. Waterer, What is pneumonia? Breathe **17**(3), 210087 (2021). https://doi.org/10.1183/20734735.0087-2021
19. D.S. Kermany et al., Identifying medical diagnoses and treatable diseases by image-based Deep Learning. Cell **172**(5) (Feb 2018). https://doi.org/10.1016/j.cell.2018.02.010
20. S. Lee Yee, W. Jee Raymond, Pneumonia diagnosis using chest X-ray images and machine learning, in *Proceedings of the 2020 10th International Conference on Biomedical Engineering and Technology* (Sept 2020). https://doi.org/10.1145/3397391.3397412
21. S. Sharma, K. Guleria, A deep learning based model for the detection of pneumonia from chest X-ray images using VGG-16 and neural networks. Proc. Comput. Sci. **218**(2023), 357–366 (2023). https://doi.org/10.1016/j.procs.2023.01.018
22. M. Soric, D. Pongrac, I. Inza, Using convolutional neural network for chest X-ray image classification, in *2020 43rd International Convention on Information, Communication and Electronic Technology (MIPRO)* (Sept 2020). https://doi.org/10.23919/mipro48935.2020.9245376
23. A. Kumar Acharya, R. Satapathy. A deep learning based approach towards the automatic diagnosis of pneumonia from chest radio-graphs. Biomed. Pharmacol. J. **13**(1), 449–455 (Mar 2020). https://doi.org/10.13005/bpj/1905
24. K.M. Abubeker, S. Baskar, B2-net: An artificial intelligence powered machine learning framework for the classification of pneumonia in chest X-ray images. Mach. Learn. Sci. Technol. **4**(1), 015036 (Mar 2023). https://doi.org/10.1088/2632-2153/acc30f
25. R. Kundu, R. Das, Z. Woo Geem, G.-T. Han, R. Sarkar, Pneumonia detection in chest X-ray images using an ensemble of deep learning models. PLOS One **16**(9) (2021). https://doi.org/10.1371/journal.pone.0256630

MIX
Papier aus verantwortungsvollen Quellen
Paper from responsible sources
FSC® C105338

If you have any concerns about our products,
you can contact us on
ProductSafety@springernature.com

In case Publisher is established outside the EU,
the EU authorized representative is:
Springer Nature Customer Service Center GmbH
Europaplatz 3, 69115 Heidelberg, Germany

Printed by Libri Plureos GmbH
in Hamburg, Germany